Creative Tourism

Activating Cultural Resources and Engaging Creative Travellers

Creative Tourism

Activating Cultural Resources and Engaging Creative Travellers

Edited by

Nancy Duxbury
Centre for Social Studies, University of Coimbra

Sara Albino
Interdisciplinary Centre for History, Culture and Societies, University of Évora

Cláudia Pato de Carvalho
Centre for Social Studies, University of Coimbra

CABI

CABI is a trading name of CAB International

CABI
Nosworthy Way
Wallingford
Oxfordshire OX10 8DE
UK

Tel: +44 (0)1491 832111
Fax: +44 (0)1491 833508
E-mail: info@cabi.org
Website: www.cabi.org

CABI
WeWork
One Lincoln St
24th Floor
Boston, MA 02111
USA

Tel: +1 (617)682-9015
E-mail: cabi-nao@cabi.org

A catalogue record for this book is available from the British Library, London, UK.

Library of Congress Cataloging-in-Publication Data

Names: Duxbury, Nancy, editor. | Albino, Sara, editor. | Carvalho, Cláudia Pato, editor.
Title: Creative tourism : activating cultural resources and engaging creative travellers / Nancy Duxbury, Centre for Social Studies, University of Coimbra, Sara Albino, Interdisciplinary Centre for History, Culture and Societies, University of Évora, Cláudia Pato Carvalho, Centre for Social Studies, University of Coimbra, Editors.
Description: Boston : CAB International, 2021. | Includes bibliographical references and index. | Summary: "This book provides a synthesis of current research and international best practice, including practitioner perspectives and a range of case study examples. It covers types of creative tourist, trends, designing and implementing creative tourism products, embedding activities in a community and place, and addressing sustainability challenges"-- Provided by publisher.
Identifiers: LCCN 2021018823 (print) | LCCN 2021018824 (ebook) | ISBN 9781789243536 (hardback) | ISBN 9781789243543 (paperback) | ISBN 9781789243550 (ebook) | ISBN 9781789243567 (epub)
Subjects: LCSH: Tourism--Social aspects. | Tourism--Social aspects--Case studies. | Cultural industries. | Cultural industries--Case studies.
Classification: LCC G155.A1 C74 2021 (print) | LCC G155.A1 (ebook) | DDC 338.4/791--dc23

LC record available at https://lccn.loc.gov/2021018823
LC ebook record available at https://lccn.loc.gov/2021018824

References to Internet websites (URLs) were accurate at the time of writing.

ISBN-13: 9781789243536 (hardback)
9781789243543 (paperback)
9781789243550 (ePDF)
9781789243567 (ePub)

DOI: 10.1079/9781789243536.0000

Commissioning Editor: Claire Parfitt
Editorial Assistant: Lauren Davies
Production Editor: Marta Patiño

Typeset by SPi, Pondicherry, India
Printed and bound in the USA by Integrated Books International, Dulles, Virginia

Contents

List of tables

List of figures

Acknowledgements

Making a book is a creative journey, one that unfolds and reveals itself *en route*. On one hand, it is a learning process filled with continuously appearing details and challenges. On the other hand, it is a collection of delightful gifts offering different perspectives, experiences, insights, reflections, and aspirations. We wish to formally thank all the authors who contributed to this book – we are so grateful to have been entrusted with your chapters. The value embedded in each of the chapters has been inspirational as we worked together toward the overall compilation of the volume.

We wish to acknowledge the chapters contributed by members of the CREATOUR® research-and-application project, both researchers and practitioners. Thank you to the CREATOUR® pilot project entrepreneurs for your generosity and openness to share your stories – and lessons learned – in this book (see Chapters 6, 16, 17, 21, 26). Thank you also to the CREATOUR® researchers who were asked to revisit and reconsider their research for a non-academic audience – and to add key 'take-aways' for practice that derive from their research (see Chapters 4, 9, 13, 28, and 29). We also included our policy recommendations on creative tourism development in small cities and rural areas which are based on our experiences with the CREATOUR® project in Portugal (Chapter 31).

We thank the funders of the CREATOUR® project, who made the national creative tourism experiment possible across Portugal. This book was developed within the CREATOUR® project (no. 16437), which is funded by the Portuguese Foundation for Science and Technology (FCT/MEC) through national funds and cofunded by FEDER through the Joint Activities Programme of COMPETE 2020 and the Regional Operational Programmes of Lisbon and Algarve.

Finally, we thank the members of the editorial and production teams at CABI who first suggested the idea for this book, and then accompanied us in various ways on the journey.

Cover image: 'Basketry for Birds' (Oficina de Cestaria) was a construction workshop organized by MARCA ADL, a local development agency located in Montemor-a-Novo in the Alentejo region of Portugal. During the workshop, participants learn traditional basketry techniques applied to the creation of bird nests and feeders. The workshop, held in 2019, was offered in partnership with Ciranda Associação de Artesãos de Montemor-o-Novo. MARCA was a participating organization in the CREATOUR® project, developing a series of creative tourism pilot activities that linked creative tourism with environmental action. The cover image is from the documentary 'creatour.pt – creative tourism in portugal' (Available at: https://www.youtube.com/watch?v=kWf__o6RfX0, accessed 12 August 2021).

Contributors

Geetika Agrawal has 11+ years of experience leading teams and designing digital products and services. She has worked for global clients such as Lincoln Center for the Performing Arts, MasterCard, MTV, Disney, Samsung, Nokia, Microsoft, AARP, Barnes & Noble, and Walmart. She is part of the R/GA Accelerator (powered by Techstars) Services Team that helped launch more than a dozen connected devices and IoT start-ups in 2014 and 2015. She is on the faculty of the undergraduate Graphic Design and Advertising Department of the School of Visual Arts in New York City. She holds a master's degree in Industrial Design from the Art Center College of Design, California, and a bachelor's degreee in Interior Architecture from CEPT, Ahmedabad, India. She is the founder and CEO of the organization Vacation with an Artist (VAWAA), an experiences marketplace for art-hungry travellers to book apprenticeships with artists, designers, and creatives around the world. E-mail: geetika@vawaa.com

Sara Albino is an integrated researcher at CIDEHUS, University of Évora, Portugal, and invited adjunct professor at the Polytechnic Institute of Beja, Portugal. She is also an associate researcher at CIEBA – Center for Research and Fine Arts Studies, University of Lisbon. She develops research in the areas of intangible heritage, cultural mapping, and educational tourism. She is the co-founder of Buinho Creative Hub, the rural FabLab of Baixo Alentejo, promoting an international artistic residency programme and a citizen science programme for primary and secondary schools. Sara has a bachelor's degree in European Studies, a master's degree in European Policies, and holds a PhD in Tourism Planning. E-mail: albino.sara@gmail.com

Carlos Alcobia is the president of the direction of Buinho and is responsible for its artistic and educative activities. He is an integrated researcher at CIDEHUS, University of Évora, Portugal. With a special interest in the areas of innovation economics and creative industries, his current research explores the intersection between creativity, business, and innovation. Carlos has experience in the creation and management of FabLabs with extensive experience in technology for digital education, including the areas of physical computing, robotics, programming, and digital manufacturing. Carlos has a bachelor's degree in Economics, a master's degree in Curatorial Studies, and holds a PhD in Product Design. E-mail: info@buinho.pt

Fernando Almeida completed his bachelor's degree in the Faculty of Fine Arts of the University of Porto, Portugal. He is currently the artistic director of Encontrarte Amares, a festival that focuses on the dialogues between artistic creation and communities, which takes place in Amares, Portugal. E-mail: fernandolouroalmeida@gmail.com

Larissa Almeida completed her bachelor's degree in Hotel Management and her masters' degree in Business Management both at Universidade Federal de Pernambuco, Brazil. She has been working on tourism and local development for the past eight years. As an entrepreneur she launched Loa Experiences, a tourism agency specialized in offering authentic cultural experiences. As a consultant she has worked developing creative tourism business in several communities in north-eastern Brazil. She is a teacher and researcher on creative tourism and social innovation. She is also the co-founder and member of the management board of RECRIA – Brazilian Experiences and Creative Tourism Network. E-mail: larissal.almeida@gmail.com

Juliana Alves is a researcher at Lab2PT (Landscape, Heritage and Territory Laboratory), University of Minho, Portugal. She holds a PhD in Human Geography from the University of Minho. She also has a post-doctorate from the University of Minho in the field of creative tourism. Her main subjects of research are in the areas of sustainable management and tourism, creative tourism, and local and regional development. E-mail: jalves.geografia@gmail.com

Fiona Eva Bakas is a critical tourism researcher with international teaching experience. She holds a PhD in Tourism from (Otago University, New Zealand, (2014), and has 20 years of varied work experience (corporate and academic), and is currently a lecturer at Lusofona University, Lisbon. Before this she was a postdoctoral researcher in a nation-wide project on creative tourism in rural areas and small cities (CREATOUR®), at the Centre for Social Studies, University of Coimbra, Portugal. Fiona is a member of the research groups: GOVCOPP (Aveiro), and ETEM (University of the Aegean) and is an associate member of the NGO Equality in Tourism. Her research interests are: sustainable tourism, creative and cultural tourism, connection to place, events in tourism, gender in tourism labour, feminism, qualitative methodologies, cultural mapping, handicrafts, entrepreneurship, rural tourism, local development, community engagement and ecotourism. E-mail: fiona.bakas@ulusofona.pt

Sónia Moreira Cabeça is a sociologist, PhD. She is a researcher at the Research Centre for Tourism, Sustainability and Well-being (CinTurs), University of Algarve, and at the UNESCO chair in Intangible Heritage and Traditional Know-How, University of Évora, Portugal. She is specialist in ethnosociology and intangible heritage. She has devoted several years to the study of Cante, a traditional chant from the Alentejo, and its bearers, having written a doctoral thesis (published) about the subject. She took part in national and transnational research projects on intangible heritage, smart specialization, and tourism. She was a post-doctorate researcher in the CREATOUR® project, working on creative tourism, cultural mapping, heritage, and sustainability. She has edited a journal and two books on creative tourism, and authored a book about creative innovation. E-mail: soniacabeca@hotmail.com

Cláudia Pato de Carvalho is a researcher at the Centre for Social Studies (CES) of the University of Coimbra, Portugal, where she is co-coordinator of the 'Cities, Cultures and Architecture' research group (CCArq). She co-coordinates the Ciência Viva CES programme, an extension programme for science education in schools. She participates in national action-research projects in the field of community arts and regional development (REDE ARTÉRIA, 2018–2021) and creative tourism (CREATOUR®: Creative Tourism Destination Development in Small Cities and Rural Areas, 2016–2020), and a European project on the societal value of culture (UNCHARTED: Understanding, Capturing and Fostering the Societal Value of Culture, 2020–2024). E-mail: claudiacarvalho@ces.uc.pt

Pedro Costa is an associate professor at the Department of Political Economy at ISCTE – Instituto Universitário de Lisboa and director of DINAMIA'CET–ISCTE (Research Center on Socioeconomic Change and Territory), where he coordinates the 'Cities and Territories' research group. He is an economist and holds a PhD in Urban and Regional Planning. He works primarily in the areas of territorial development, planning, and cultural economics. He is the author of many publications and has been involved in multiple research projects in these areas, at national and international levels. E-mail: pedro.costa@iscte-iul.pt

Caroline Couret is the Founder and Director of the Creative Tourism Network® and the Barcelona Creative Tourism programme. Caroline is an expert in creative tourism. Born in France, she holds a post-graduate in the Management of Culture, Cultural Policies and Museology. Caroline regularly speaks at conferences and also lectures at universities and business schools worldwide. She has published articles about creative tourism in different languages, is an external expert for the European Union, and regularly collaborates with the UNWTO. Since 2001, she has lived and worked in different countries and managed a wide array of international projects, including festivals commissioned by the Barcelona City Council, the European-funded projects La Casa de Velázquez (Madrid) and Festival de Cannes (France), and various projects in Louisiana, Mexico, and Morocco. E-mail: carolinecouretv@gmail.com

Ana Rita Cruz is a researcher at DINÂMIA'CET–ISCTE where she has been working on the study of territorial cultural and creative dynamics and their interrelations with tourism. Ana Rita is a sociologist and holds a PhD in Tourism Studies and a master's degree in Tourism Destination Management and Development from the Faculty of Economics of the University of Algarve, Portugal, where she also now collaborates as an invited assistant professor. Additionally, Ana Rita Cruz has been collaborating on several projects as an external expert in the areas of culture, creativity, and tourism. E-mail: ana.rita.cruz@iscte-iul.pt

Marie-Andrée Delisle is a tourism development strategist with over 30 years of experience. Her travels as consultant, workshop facilitator, lecturer, and travel trade writer to 65 countries have shaped her knowledge and comprehension of the tourism and travel industry. Her firm, Marie-Andrée Delisle

Tourisme-Conseil, specializes in tourism product enhancement, development and marketing (www.madelis.com). She is a member of the CREATOUR® advisory council. She holds a master's degree in tourism planning and management and is currently working on her PhD on the topic of creative tourism at Université du Québec à Montréal, Canada. She co-authored a book on alternative tourism published in 2007 entitled *Un autre tourisme est-il possible?* (*Is another tourism possible?*) and has written a chapter, 'Montréal: a creative tourism destination?', in *A Research Agenda for Creative Tourism* published by Edward Elgar Publishing in 2019. E-mail: info@madelis.com

Nancy Duxbury, PhD, is a senior researcher, lead coordinator of the transdisciplinary thematic line 'Urban Cultures, Sociabilities and Participation', and a former co-coordinator of the 'Cities, Cultures and Architecture' research group (CCArq) at the Centre for Social Studies, University of Coimbra, Portugal. She is the principal investigator of 'CREATOUR®: Creative Tourism Destination Development in Small Cities and Rural Areas', a national research-and-application project (2016–2020) that catalysed and developed creative tourism in small cities and rural areas across four regions of Portugal. She also leads the CES research team within the European research project 'UNCHARTED: Understanding, Capturing and Fostering the Societal Value of Culture' (2020–2024). Other research interests include cultural mapping, culture in local sustainable development, and culture-based development models in smaller communities. She is a member of the European Expert Network on Culture. E-mail: duxbury@ces.uc.pt

Margaret Feeney is an artist currently working on research project *Fosca: It Started With a Kiss*, supported by Creative New Zealand, the Blumhardt Foundation, and The Arts Foundation. She also works on their collective invention project *Water Systems,* with an established exhibition record, in which she looks at information space and the expanding influence of form over content in the human built world. She also works at Māpura Studios, an outsider art studio for artists living with diversity, adversity, and disability, and runs the Māpura Art Programmes at Auckland Prison, Paremoremo, New Zealand. She has a bachelor's degree in Art History from Victoria University, Wellington, New Zealand, and a master's degree in Fine Arts from RMIT, Melbourne, Australia. E-mail: mfeeneynz@gmail.com

Maria Assunção Gato is a researcher at DINÂMIA'CET–ISCTE and holds a PhD in Cultural and Social Anthropology. Her main research interests have been focused on the fields of urban anthropology and sociology, exploring topics such as spatial consumption, values and meanings, social recompositions, lifestyles, and residential dynamics in urban territories. In parallel, she has been developing a second line of research focused on cultural, participative, and creative dynamics, taking into account socio-economic change and its impacts on a territory and local communities. She is currently the editor of the scientific journal *Cities, Communities and Territories*. E-mail: maria.gato@iscte-iul.pt

Alexandra R. Gonçalves was formerly the Regional Director of Culture of the Algarve region, Portugal, and councillor of the Municipality of Faro with responsibilities for culture, social action, health, advertising, economy, and tourism. She holds a PhD in Tourism from the University of Évora and a master's degree in Management and Administration of Cultural Heritage from the Faculty of Human and Social Sciences of University of Algarve and University of Paris-8. She is currently the head director of the School of Management, Hospitality and Tourism at the University of Algarve. She has several articles and research published in tourism, cultural heritage, and events. She collaborates in several research projects, highlighting the CREATOUR® project, where she is the lead researcher for the regional team of the University of Algarve. E-mail: marodrig@ualg.pt

Diana Guerra is a Colombian business administrator from the Pontifical Xavierian University in Bogotá, Colombia, with 25 years of experience as an entrepreneurship consultant through the Inter-American Development Bank and the Foundation for Sustainable Development in Latin America. She taught for five years at the Colombian Tourism and Hotel Management School. She is a co-founder of 5Bogota, a creative tourism alternative in Colombia, where she focuses on budget management and financial analysis, thus creating value and protecting corporate assets. She has led the group of hosts at 5Bogota for seven years, overseeing tours for over 3,000 travellers. She is also a winning participant in social entrepreneurship contests through 5Bogota. Moreover, she is a co-author of the book *Bogotá through the 5 Senses*. E-mail: dpazguerra@gmail.com

Maria Huhmarniemi works as an installation artist, dealing with issues relating to the North as well as environmental issues such as human–nature connectedness and environmental responsibility, policy, and ethics. Her artworks are space-related installations in exhibition halls and in nature. She often collaborates with researchers from different disciplines and develops methods for art-based research. As a researcher, she is interested in eco-activism and socio-politically engaged contemporary and environmental education on how art can change our relationships with nature. She focuses on thinking about our future and the nature and culture in the North and the Arctic. She has exhibited in various exhibitions, curated international art events, and published research articles. E-mail: maria.huhmarniemi@ulapland.fi

Jan Ifversen is associate professor in European Studies at the Department of Global Studies, Aarhus University, Denmark. He has written extensively on European identity, history, and heritage politics. He was PI of the cross-disciplinary research project 'Rethinking Tourism in a Coastal City – Design for New Engagements' (2016–2019) funded by Innovation Fund Denmark. Currently, he is co-PI of 'ECHOES: European Colonial Heritage Modalities in Entangled Cities' (2018–2021), funded by Horizon 2020, and director of the cross-disciplinary research programme 'Uses of the Past', funded by the Faculty of Arts, Aarhus University. E-mail: jif@cas.au.dk

Britta Timm Knudsen is professor at the Department of Nordic Studies and Experience Economy, Culture and Communication, Aarhus University, Denmark. She has authored and edited numerous publications on difficult heritage and tourism. She has been part of the cross-disciplinary research project 'Rethinking Tourism in a Coastal City – Design for New Engagements' (2016–2019) funded by Innovation Fund Denmark. Currently, she is part of the H2020 project 'ECHOES: European Colonial Heritage Modalities in Entangled Cities' (2018–2021). E-mail: norbtk@cc.au.dk

Outi Kugapi is a doctoral candidate in tourism research at the Multidimensional Tourism Institute, University of Lapland, Finland. She knits and purls together her PhD, warm mittens, and two projects concerning cultural sensitivity (ARCTISEN, NPA) and handicraft tourism (Handmade in Lapland, ESF). Her research interests are tourism experiences in craft tourism and she has co-authored several articles about the topic, such as 'Affective entanglements with travelling mittens' (*Tourism Geographies*, 2020), 'Context mapping in creative tourism' (*Annals of Tourism Research*, 2020), as well as a chapter about crafts as employment and cultural experience service in *Tourism Employment in Nordic Countries: Trends, Practices and Opportunities* (Palgrave Macmillan, 2020). Moreover, she has co-authored chapters on Indigenous tourism in several books. E-mail: outi.kugapi@ulapland.fi

Laura Laivamaa is a PhD student at University of Lapland, Finland, interior and textile designer (MA), industrial designer (MA), and fourth-generation crafter. During the last few years she has been working in multiple service design projects at the University of Lapland. Laivamaa is also an active textile designer and she has designed several Arctic patterns for Finnish jacquard-weaving mill Lapuan Kankurit. E-mail: laura.laivamaa@ulapland.fi

Marília Lúcio holds a master's degree in Environmental Engineering – Profile Management of Environmental Systems and a Post-Graduation in Spatial Planning and Environmental Planning, both from the Faculty of Sciences and Technology, New University of Lisbon, Portugal. She was Head of the Environment and Sustainable Development Division and later Coordinator of the Sustainability, Local Economy, Tourism and Employment Project Team, which is part of the Support Office for Economic Activity and Entrepreneurship, of the Municipality of Loulé. She is currently Head of the Local Economy, Commerce and Tourism Division and of the Loulé Criativo project at the Municipality of Loulé. E-mail: loulecriativo@cm-loule.pt

João Filipe Marques holds a PhD in Sociology from the École des Hautes Études en Sciences Sociales of Paris and a master's degree in Cultural Anthropology from the New University of Lisbon. He is assistant professor at the Faculty of Economics of the University of Algarve and integrated researcher at the Research Center for Tourism Sustainability and Well-being (CinTurs – UAlg), University of Algarve, Portugal. Currently, he is the director of the master's programme in Sociology. He has been teaching social sciences and sociological theories since the 1990s and his main field of research has been the sociology of racism and ethnicity. Recent research interests include the sociology of tourism, leisure, and travel. He teaches sociology of tourism in the Sociology programmes of the University of Algarve. E-mail: jfmarq@ualg.pt

Lénia Marques is assistant professor of Cultural Organizations and Management at the Erasmus University Rotterdam, The Netherlands. Previously she worked as a researcher and lecturer in different countries (Portugal, United Kingdom, and the Netherlands). Lénia Marques is also a member of the board of directors of the World Leisure Organization, since 2016. She has published extensively in creativity and innovation in tourism, leisure, and events. She is co-editor of 'Leisure and Innovation', a special issue of the *World Leisure Journal* (2019); *Event Design: Social Perspectives and Practices* (2015); *Intercultural Crossings: Conflict, Memory and Identity* (2012); and 'Exploring Creative Tourism', a special issue of the *Journal of Tourism Consumption and Practice* (2012). She has also been advising different countries, regions, and cities on culture and tourism policies. E-mail: marques@eshcc.eur.nl

Noémi Marujo holds a PhD in Tourism from the University of Évora, Portugal. She concluded her master's degree in Sociology at the University of Évora and a bachelor's degree in Social Communication at the Beira Interior University, Portugal. She is assistant professor of Tourism Studies in the School of Social Sciences at the University of Évora. She is an integrated researcher of the Interdisciplinary Centre for History, Culture and Societies (CIDEHUS) at University of Évora and a member of the management and scientific team of Alentejo Sustainable Tourism Observatory, integrated in the UNWTO-INSTO. She was the lead of the University of Évora research team in the CREATOUR® project. She is a member of the UNESCO Chair team in 'Intangible Heritage and Traditional Know-How: Linking Heritage', University of Évora, Portugal. E-mail: noemi@uevora.pt

Satu Miettinen is a professor of service design at the University of Lapland, Finland. She works as a dean of the Faculty of Art and Design. For several years she has been working with service design research and authored number of books and research publications in this area. She is working with the SINCO service prototyping and simulation research environment. Her research interests are in the areas of service design including the areas of social and public service development, citizen engagement, and digital service development. She is a visiting lecturer for service design for PUC in Chile and Hokkaido University in Japan. She is PI and co-ordinator in several national and international service design research projects funded by TEKES (Finnish Fund for Innovation and Technology), European Commision Horizon 2020, MSCA, RISE, etc. Satu has worked as a visiting professor at Stanford University in the United States, Tongji University in China, and the University of Trento in Italy. Satu also has a strong design research interest for the complex, extreme, and marginal contexts. She co-ordinates the Arctic design lab that is part of DESIS network. E-mail: satu.miettinen@ulapland.fi

Ana Osredkar is a service designer, consultant, and trainer. She helps organizations deliver better experiences for their users and establish processes and relationships that support those experiences. She holds a master's degree in Marketing Management from the Vlerick Business School in Ghent, Belgium (2005) and a bachelor's degree in Economics from the Faculty of Economics in Ljubljana, Slovenia (2004). After a decade of working in the corporate realm she decided to follow her passion for service design and in 2014 co-founded Servis 8, an institute for service design innovation, education, and workshops. She's been working with public and private organizations across sectors on developing policies, digital services, and cultural experiences. E-mail: ana.osredkar@gmail.com

Elena Paschinger holds a master's degree in international tourism management and speaks seven languages fluently. A world traveller and author, she chronicles her adventures and professional work on www.creativelena.com, the only bilingual blog with a special focus on creative tourism. She has published *The Creative Traveler's Handbook* and currently teaches at the University of Applied Sciences FH Joanneum in Styria, Austria. She also works as a tourism and social media consultant, translator, speaker, and networking partner for several EU projects. E-mail: reachme@creativelena.com

Margarida Perestrelo is an assistant professor at the Department of Social Research Methods at ISCTE – Instituto Universitário de Lisboa and a researcher at DINÂMIA'CET–ISCTE. Margarida holds a PhD in Sociology, specializing in theories and methods of sociology. Her main areas of research are data analysis applied to social sciences and prospective methodologies applied to the territory, structural analysis, analysis of subsystems, actors' strategy analysis, and scenarios. Author of many publications, she has been involved in multiple research projects in the area of territorial prospective. E-mail: margarida.perestrelo@iscte-iul.pt

Sabrina V. Pratt is the owner of SVPratt Creative Strategies, a consulting firm working at the intersection of the arts, tourism, and economic development. In 2019, her firm's arts and cultural district plan for the city of Las Cruces, New Mexico, was recognized with the international Creative Tourism Award. She served as the director of the City of Santa Fe Arts Commission for 22 years, was interim co-director and team leader overseeing economic development and affordable housing and other programmes for two years, and worked at the Santa Fe Convention and Visitors Bureau for five years. She obtained a bachelor's degree from Vassar College, Poughkeepsie, New York, and a Master of Public Administration from the University of New Mexico, Albuquerque, New Mexico, United States. E-mail: sabrina@svprattcs.com

Paula Remoaldo is a full professor, with a PhD in Human Geography. She has been the head of Lab2PT (Landscape, Heritage and Territory Laboratory) at the University of Minho, Portugal, since 2017, and the leader of the tourism and development group in this Laboratory. In the last 30 years, she has been working on and/or coordinating several national and international projects. Her main scientific research fields are cultural and creative tourism and regional and local development. E-mail: premoaldo@geografia.uminho.pt

Guida Rolo graduated in Cultural Heritage Management from Oporto's Polytechnic Institute. After her studies in cultural production, she collaborated with a diverse set of cultural organizations in Oporto and acted as production coordinator. Between 2007 and 2012, she turned to communication and marketing, working as campaign coordinator for cultural organizations and multinational companies. Throughout her career, she worked as a producer and cultural agent, and has been responsible for several projects related to the artistic field. From 2012 to 2017, she was the production coordinator responsible for the organization of the Fujifilm International Photography Festival, in Viseu, Portugal. Guida has a collaborative participation in the project New Hand Lab as member of its board of directors and she is executive director of the project Criss Cross, which she created in 2019, working in the areas of cultural production and project management, with several artistic structures and projects at the national and international levels. E-mail: guidarolo@gmail.com

Maria do Rosário Borges holds a degree in Tourism Management and Planning (1995), a master's degree in Information Management (2002), and a PhD in Tourism (2016), all from the University of Aveiro. Maria do Rosário has been a lecturer in higher education since 1998, and currently lectures at the University of Évora. She is a researcher member of the CIDEHUS Research Unit and co-founder of ASTO – Alentejo Sustainable Tourism Observatory, part of UNWTO-INSTO since 2018, and currently a member of the management and scientific team. Her research interests focus on issues about international tourism, sustainable development of tourism, public policies and governance in the tourism field, GIS for tourism planning, and pedagogical issues in higher tourism education. E-mail: mrborges@uevora.pt

Greg Richards is professor of Placemaking and Events at Breda University of Applied Sciences and professor of Leisure Studies at Tilburg University in the Netherlands. E-mail: G.W.Richards@tilburguniversity.edu

Rita Salvado has been director of the Wool Museum of the University of Beira Interior (UBI), Portugal, since December 2018. She is assistant professor at the Textile Science and Technology Department of UBI and coordinator of the PhD programme on Fashion Design. She received her PhD in Textile Engineering in 2002 with the European Label (Portugal, France, and Sweden). Her research and teaching activities are focused on the synergy between textile engineering and design, applied to wearable technology and cultural valorization. She is committed to the collective valorization of the wool heritage and its impact on local development. E-mail: rita.salvado@ubi.pt

Melanie Sarantou is adjunct professor in Social Design at the University of Lapland, Finland, investigating the role of arts, narrative practices, and service design in marginalized communities. She currently coordinates a research project that investigates the role of art in marginalized communities across European countries. E-mail: melanie.sarantou@ulapland.fi

Kathleen Scherf is professor of Communication at Thompson Rivers University in British Columbia, Canada, where she teaches communication theory, travel media, and deep mapping. She holds a bachelor's degree from the University of Toronto, and a master's degree and PhD from the University of British Columbia, Canada. Her edited collection, *Creative Tourism in Smaller Communities: Place, Culture, and Local Representation*, was published by the University of Calgary Press (2021). E-mail: kath@tru.ca

Jaime Serra holds a PhD in Tourism from the Faculty of Economics, University of Algarve, Portugal. He is assistant professor of the School of Social Sciences at the University of Évora in the field of tourism studies. He is also a researcher (G1: Societal Changes) and assistant director at the Interdisciplinary Centre for History, Culture and Societies (CIDEHUS), at the University of Évora, Portugal. His research interests are in the fields of tourism demand trends, tourist behaviour, tourism and demography, and applied quantitative methods. He is coordinator of the Tourism Creative Lab at the CIDEHUS research centre and a member of the management and scientific team of Alentejo Sustainable Tourism Observatory (ASTO), integrated in the UNWTO-INSTO. He is a researcher in the CREATOUR® project. E-mail: jserra@uevora.pt

Catharina Sligting holds a bachelor's degree in English language and literature from the Free University of Amsterdam, and an MBA in international politics from Erasmus University in Rotterdam, the Netherlands. She was director of the national historic museum Slot Loevestein, a medieval castle, where she developed an extensive cultural and educational programme. She also initiated large-scale theatrical productions on location in nature, abandoned factories, and shipyards. Since 2007, she has been based on her own biodynamic farm in the Serra da Estrela nature park in Portugal, where she is president of the Association Domínio Vale do Mondego, organizing creative tourism activities such as art workshops, nature field trips, philosophical convivia, residencies for artists, and the annual small-scale art festival Angel Festival/ESTIVAL with international artists, resident participants, and the local community. E-mail: info@dominiovaledomondego.com

Mirian Tavares is associate professor at the University of Algarve, Portugal. With academic studies in communication sciences, semiotics, and cultural studies (PhD in Communication and Contemporary Culture from Federal University of Bahia, Brazil), her research is related to cinema, literature, and other arts, and to artistic and aesthetic film studies. As a lecturer at the University of Algarve, she has participated in the development of the Visual Arts degree, the master's programmes in Communication, Culture and Arts and Cultural Management, and the PhD programmes in Communication, Culture and Arts and Digital Media and Arts. She is the coordinator of CIAC – Arts and Communication Research Centre, and the director of the Digital Media and Arts PhD. She was assistant to the State Secretary for Culture of the State of Sergipe, in Brazil. E-mail: mtavares@ualg.pt

Elisabete Tomaz is a researcher at DINÂMIA'CET–ISCTE, collaborating in projects in the field of culture and creativity planning and development policies. She holds a PhD in Sociology from the School of Sociology and Public Policy at ISCTE–IUL. She is a member of the international network COST Action 'The Geography of New Working Spaces and the Impact on the Periphery' and has participated in the COST Action 'Investigating Cultural Sustainability' (2011–2015). She collaborated with INTELI (a Lisbon innovation centre) on several European and municipal council projects in territorial development, urban planning, and innovation policies. E-mail: elisabete_tomaz@iscte-iul.pt

Tiago Vinagre de Castro is a cultural manager, urbanist, and researcher. He holds a bachelor's degree in Tourism Management and Planning (2006) and a master's degree in Urban and Regional Planning (2012) from the University of Aveiro, and a specialization in Show Production and Management from the University Lusófona of Porto (2007). Over the past 15 years, he has been working mainly in the cultural sector as a producer, manager, and creator within national and international projects. In 2012, he cofounded the NGO 4iS –Platform for Social Innovation where he develops urban transformation projects through culture and participation as well as projects that combine heritage and arts with community development. He is a researcher within the project CREATOUR®: Creative Tourism Destination Development in Small Cities and Rural Areas, and a PhD student in Sociology – Cities and Urban Cultures with the research 'The role of cultural intermediaries in the dynamics of urban regeneration in small cities'. E-mail: tiagocastro@ces.uc.pt

Jutamas (Jan) Wisansing is a founder and managing director, leading an innovative team, at Perfect Link Consulting Group 'A Consortium of Experts' in Thailand. She specializes in community innovation and transformative empowerment programmes, creative tourism, and sustainable culinary supply chain management. Jan has had a leading role in establishing the Creative Tourism Brain Bank Network in Thailand and ASEAN Gastronomy Tourism Network and representing ASEAN voices as a board member of IGCAT (International Institute of Gastronomy, Culture, Arts and Tourism). E-mail: jutamas@perfectlink.co.th

Mélanie Wolfram was born in Munich, Germany, with a French nationality and arrived in Lisbon, Portugal, at the age of two. She holds a bachelor's degree in History of Art and Archeology, and a master's degree in History and Archeology. In 2003, she started her PhD in Paris, in co-supervision between the University of Lisbon and Sorbonne – Paris IV dedicated to the study of the Christianization of the Alentejo in Roman and Visigothic times. She returned to Portugal in 2007 and in the last year of her thesis decided to live in Alentejo, where she still lives. In 2012, she started a new professional stage: she is the mentor of the Vagar Walkingtours project, a cultural touring company in the Alentejo. In 2018, she created a game, 'Play Évora', within the CREATOUR® project. She is a member of the board of the Regional Tourism Promotion Agency in Alentejo, since 2018, and, since 2020, has been collaborating with Eventiz Media Group (France), at the first International Forum – A World For Travel, to be held in Évora in 2021. E-mail: melanie@vagarwalkingtours.com

Diana Zuluaga is a Colombian advertising professional with a degree from the Palermo University in Buenos Aires, Argentina, and a specialization in Creative Entrepreneurship Management from Cordoba University, in Argentina. With seven years of experience in the tourism industry, she is a co-founder of 5Bogota, a creative tourism alternative in Colombia. She is experienced in public relations and in the creation of entrepreneurial networks through the conception and development of innovative tourism products. She is also a winning participant in social entrepreneurship contests through 5Bogota. She is co-author of the book *Bogotá through the 5 Senses*. E-mail: dianazuluagag@gmail.com

1 An introduction to creative tourism development: Articulating local culture and travel

NANCY DUXBURY,[a*] CLÁUDIA PATO CARVALHO,[a] AND SARA ALBINO[b]

[a]*Centre for Social Studies, University of Coimbra, Portugal;* [b]*CIDEHUS, University of Évora, Portugal*

Creative tourism is a dynamic tourism niche that has emerged both as a development of cultural tourism and in opposition to the emergence of 'mass cultural tourism'. On the one hand, creative tourism demand is driven by travellers seeking more active and participative cultural experiences in which they can use and develop their own creativity. On the other hand, creative tourism provides avenues for communities' desire to accentuate their distinctive elements and develop new value-added initiatives for local benefit.

Responding to these demands and striving to provide alternative approaches to tourism development, interest in creative tourism has been rising in many places, both urban and rural. Through the CREATOUR® creative tourism research-and-application project in Portugal (see Box 1.1), which was the impetus for the current book, we observed that creative tourism is an inspiring trajectory for agencies, organizations, and entrepreneurs involved in advancing local culture-based development and cultural tourism.

The original definition of creative tourism put forward by Greg Richards and Crispin Raymond in 2000 continues to resonate in the field: 'Tourism which offers visitors the opportunity to develop their creative potential through active participation in learning experiences which are characteristic of the holiday destination where they are undertaken' (p. 18). While commonly associated with traditional crafts, creative tourism encompasses a wide array of creative experiences. In this vein, it can be seen as an approach rather than set types of activities. With a diverse range of activities encompassed within creative tourism, one can view its 'transversality' as a strength, which can complement and extend other types of tourism (Gonçalves *et al.*, 2020, Chapter 31).[1]

Based on an extensive review of definitions of creative tourism, the CREATOUR® project defined creative tourism as active creative activity encouraging personal self-expression and interaction between visitors and local residents, inspired by local endogenous resources (place and people), and designed and implemented by local residents. This approach to creative tourism highlighted four dimensions: active participation, learning, creative self-expression, and community engagement, all underlined by an immersive connection to place (Bakas *et al.*, 2020; see also Scherf, 2021).

Today, creative tourism is situated among an array of locally focused alternatives-to-mass-tourism that have been emerging in recent years, such as relational tourism, conscious travel, situated tourism, and community-based tourism. It is viewed as a type of sustainable and responsible form of travel that functions on a small scale and focusses on local actors. It provides an avenue for rethinking and reconfiguring visitor relations with an eye to local benefits. It provides a platform for developing new initiatives based on local cultural resources and specificities, both tangible and intangible.

*Corresponding author: duxbury@ces.uc.pt

 Creative Tourism: Activating Cultural Resources and Engaging Creative Travellers (eds. N. Duxbury, S. Albino, and C. Pato de Carvalho)
DOI: 10.1079/9781789243536.0001

Box 1.1. About CREATOUR®

CREATOUR® is a large, multidisciplinary research-and-application project that has been encouraging experimental practices and pilot trials to more fully understand the issues and potential of creative tourism in small cities and rural contexts in Portugal. Interlinking culture, tourism, and local development perspectives and methodologies, the project catalysed and researched creative tourism in small cities and rural areas. The project enabled and empowered an array of bottom-up development approaches and initiatives across the Algarve, Alentejo, Centro, and Norte regions of Portugal.

The initial phase of CREATOUR® operated from November 2016 to June 2020. It was funded by the Portuguese Foundation for Science and Technology (FCT/MEC) through national funds and was co-funded by FEDER through the Joint Activities Programme of COMPETE 2020 and the Regional Operational Programmes of Lisbon and Algarve. It was an ambitious experiment, involving five research centres and 30 researchers – from the Centre for Social Studies (CES), University of Coimbra (coordinator); Laboratory of Landscapes, Heritage and Territory (Lab2PT), University of Minho; Interdisciplinary Centre for History, Culture and Societies (CIDEHUS), University of Évora; Centre for Spatial and Organizational Dynamics (CIEO), University of Algarve; and Centre for Socioeconomic and Territorial Studies (DINÂMIA'CET-ISCTE), ISCTE – Instituto Universitário de Lisboa – working with 40 participating 'pilot' organizations, including art, culture, environmental, local, and regional development associations; science and education organizations; municipalities; and entrepreneurs.

CREATOUR® promoted an integrated approach – crossing regions, disciplines, sectors, and organizational types, with the imperative to bridge research and practice made central to the project. Within this framework, the diversity of perspectives among CREATOUR® participants fostered knowledge exchanges, provided important learning moments, and sparked new ways of understanding and interpreting developments. Through this collaborative and experimental approach, we found that insights from front-line experiences are invaluable to advance knowledge about development strategies for creative tourism in smaller places, to better understand the dynamics of creative tourism as a field of practice, and to foster culturally sensitive and place-based initiatives that benefit both local communities and visitors.

Among the project's findings, we learned that creative tourism development outside large cities is feasible. We also found that creative tourism has significant potential for inspiring new ideas for revitalizing local culture and heritage resources, reimagining community self-representation for tourism, and providing social, cultural, and economic added value for smaller places. However, challenges of distributed geography and the diversity of small-scale initiatives require significant attention to capacity building, training, and mentorship; dedicated resources for collective marketing; and time to deepen knowledge networking and the development of partnerships.

For further information, visit: www.creatour.pt (accessed 26 April 2021).

(Adapted from Duxbury and Silva, 2020.)

And, beyond this, creative tourism can be used to pursue broader societal and sustainable development goals.

Yet creative tourism is a young field and there are few books available about it, with most directed towards an academic audience. Recently, a few publications directed towards practitioners have emerged: *Creating Creative Tourism Toolkit* (Richards *et al.*, 2019), *Creative Tourism: The CREATOUR Recipe Book* (Gonçalves *et al.*, 2020), and *Creative Tourism: Guide for Practitioners* (Vinagre de Castro *et al.*, 2020). However, our experiences in the CREATOUR® project highlighted that there is a general lack of accessible knowledge available to practitioners about types of creative tourists, trends in creative tourism, how to design and implement a creative tourism product, embedding creative tourism activities in a community and place, and addressing challenges of innovation and sustainability. These key topics inspired and informed this book, which was developed with the input of the participating pilot organizations involved in the CREATOUR® project and further elaborated by the editors in collaboration with colleagues in Colombia, Brazil, and elsewhere.

This book brings together current research and leading practices internationally in creative tourism and travelling, offering perspectives and reflections from both practitioners and researchers. It includes 32 chapters, featuring 46 authors, writing about creative tourism initiatives in 15 countries: Austria, Brazil, Canada, Colombia, Denmark, Finland,

Kenya, Namibia, Portugal, The Netherlands, New Zealand, Slovenia, Spain, Thailand, and the United States. The book is intended for entrepreneurs and public agencies interested in developing creative tourism activities and programmes, with a complementary interest expected from students and researchers in creative tourism, cultural tourism, and community-based tourism fields.

The book aims to offer theoretical approaches as well as to inform practical implementation, presenting a wide range of examples, experience-based insights, and advice. It offers guidance for practitioners in planning, operationalizing, and iteratively improving their creative tourism projects and adapting them to changing local situations. The book also aims to situate creative tourism within local development, and to show how it can contribute to local economic benefit, community engagement, social inclusion, empowerment, cultural vitality and sustainability, cross-cultural exchange, and responsible travel.

To contextualize the collection, a few words about the book's approach to creative tourism entrepreneurship and its subtitle elements – activating cultural resources and engaging creative travellers – are in order. In this introductory chapter, we present insights relating to these themes, woven together from the contributions to this volume. We then provide an overview of the six parts of the book (summaries of the individual chapters are included in an introduction in each section). In closing, we offer some thoughts on how community-engaged creative tourism can foster broader local impacts.

Creative tourism entrepreneurship

Creative tourism presents 'an encouraging opportunity for entrepreneurs with low entry hurdles' (see Zuluaga and Guerra, Chapter 2, p. 105), but successfully developing and managing a creative tourism initiative requires attention to many planning and operational aspects, dynamic opportunities, and contextual challenges. To inform creative tourism development in a wide variety of situations, this book provides strategies and processes that have been used to design and implement creative tourism initiatives in Thailand (Chapter 12), Slovenia (Chapter 13), Colombia (Chapter 14), and Brazil (Chapter 24), as well as case study accounts of creative tourism projects in Portugal (Chapters 16, 17, 21, 26, and 28), Denmark (Chapter 18), the United States (Chapter 22), Kenya (Chapter 23), Namibia (Chapter 27), and Lapland, Finland (Chapter 30).[2]

The approaches and pathways innovated in each situation were developed out of particular contexts and circumstances. This book uniquely brings together these multiple perspectives and experiences to guide entrepreneurs and agencies in developing their own situationally relevant and locally resonant approaches to creative tourism. We note that the elaboration of these practice-based accounts also valuably informs creative tourism research, addressing a lack of information on the concrete processes undertaken in conceiving, planning, and operationalizing creative tourism initiatives.

The book also acknowledges the important roles that public agencies and municipalities can take as 'public entrepreneurs'. Public entities such as regional development agencies and municipalities can play key roles in facilitating local and regional networks of independent actors, supporting new ideas and initiatives, and (re)positioning tourism to pursue diverse public benefits (see Gato *et al.*, Chapter 29). Two examples are provided in this book in some detail: the Santa Fe Creative Tourism Initiative in New Mexico, USA (Chapter 22 by Sabrina V. Pratt) and Loulé Criativo in Portugal (Chapter 21 by Marília Lúcio). These public initiatives both created the conditions and propelled the development of multifaceted creative tourism programmes. Related to this, in Chapter 31 Gonçalves *et al.* outline public policy recommendations to nurture and support creative tourism development in small cities and rural areas, based on findings from the CREATOUR® project.

Understanding how to implement creative tourism initiatives in specific contexts requires deep understanding of local scenes and dynamics (see Marques, Chapter 23). Local contextual and 'how-to' knowledge is essential for effective and meaningful implementation and development (Isar, 2013). The degree of social embeddedness, generally defined as 'the nature, depth, and extent of an individual's ties into a community', also plays an important role (see Bakas, Chapter 28, p. 218).

Four key dimensions of creative tourism – *active participation, learning, creative self-expression,* and *community engagement* – served as a useful framework for developing creative tourism activities in the CREATOUR® project. Jutamas (Jan) Wisansing's experiences with the DASTA creative tourism project in Thailand (see Chapter 12) placed accents on the 3Ss – story and storytelling, senses, and sophistication – and drew attention to the design of the *spaces* in which the activities are realized. Altogether, these aspects collectively contribute to meaningfully engaging travellers through *in situ* creative experiences immersed in the place where they are realized. Some insights on these dimensions, pulled from chapters in this book, are presented in Box 1.2.

The importance of nurturing relationships and networks underlies many of the messages in this book. As Caroline Couret of the Creative Tourism Network® remarks, creative tourism is valued for 'its capacity to create a value chain at the local and regional scale' (see Chapter 8, p. 59). In developing and promoting creative tourism, 'local people' are becoming 'more and more a part of the marketing mix, since creative tourism experiences and workshops are about local people, their artists and artisans, their culture, their place, and their role as ambassadors' (see Delisle, Chapter 19, p. 146). Diana Zuluaga and Diana Guerra concur, stressing the importance of 'fostering a network of hosts with the ability to showcase the innermost and most representative aspects of a community, which is accomplished through exploring the destination and scouting not only the places, but also the people who could belong to the network' (see Chapter 2, p. 15). Geetika Agrawal (see Chapter 10) and Margaret Feeney (see Chapter 5) highlight how both local artists and travelling artists can play key roles in creative tourism.

Collaborative networks require an effective operating platform. For Loulé Criativo, the local network of creative tourism partners is equipped with 'a simple, agile, and fast operating model, not only to respond to requests that arise from tourists and others interested in activities, but also to follow up on other initiatives related to strengthening the network and promotion, among other aspects' (see Lúcio, Chapter 21, p. 156). The core of initiatives like 5Bogota (Colombia), Loulé Criativo (Portugal), Santa Fe's Creative Tourism Initiative (United States), and Recria (Brazil) is the establishment of local partnership networks and clusters in which collaboration, knowledge sharing, and capacity building are central.

This highlights the necessity to develop cross-sectoral bridges and manage partnerships, and the need to build understanding and skills that enable these collaborations. For example, beyond craft-making and teaching skills, craftspeople and artisans may 'need an understanding of how to be involved and work in tourism. This might need new education and training to learn new skills' (see Huhmarniemi *et al.*, Chapter 30, p. 245). Intermediators that connect artisans with tourism through organizing and promoting creative tourism initiatives can also facilitate these connections (see Bakas, Chapter 28).

Processes of trust building take time, ongoing communication, the development of mutual understanding, and respect. As an example, Rita Salvado and Guida Rolo's creative tourism development partnership, COOLWOOL, provides a reflective learning experience about the challenges and discoveries involved in forging an innovative working relationship between two organizations, and the importance of knowing the strengths and capacities as well as the weaknesses of each partner to find 'a solution of real complementarity' and 'a common ground to support a coherent programme' (see Chapter 17, p. 128). From this base of 'sharing, knowledge, trust, and generosity between the two institutions', they note, it is possible 'to iteratively involve other local partners who share this spirit in this creative project' (p. 128).

Finally, multiple chapters stress how the development journey should be accompanied by an ongoing process of reflection towards a process of iterative and continuous improvement, adaptation, and evolution (e.g. see Sligting, Chapter 16, and Salvado and Rolo, Chapter 17). An integrated and systematic self-assessment process, such as that elaborated by Maria Assunção Gato *et al.* in Chapter 29, can be usefully applied to better understand the links between motivations for action, networking synergies, and desired results. It can also help identify the impacts generated by creative tourism activities for the organizers, stakeholders, and the wider community, and thus 'support decision making for their sustainability' (Gato *et al.*, Chapter 29, p. 225).

We now turn our attention to the two cross-cutting aspects of creative tourism development in the book's subtitle. We believe these two aspects – activating cultural resources and engaging creative travellers – together form an interconnected framework for conceiving, operationalizing, and sustaining creative tourism initiatives in diverse contexts.

Activating cultural resources

The multifaceted dimensions of a place, its cultures, its communities, and its identity – both tangible and intangible aspects – provide points of inspiration and meaningful settings for creative tourism activities. In this context, culture is understood to

Box 1.2. Key dimensions of creative tourism

Active participation. Not just watching someone else create or buying something handmade by someone else, creative tourism provides opportunities for active involvement in creative activities, for DIY experiences, and for bringing home self-made souvenirs and keepsakes. The depth of immersion in creative experiences can vary widely, 'from time-limited "snack culture" activities to immersive mini-apprenticeships and programmes' (see Richards and Duxbury, Chapter 7, p. 53). Regardless of the duration, however, focusing on production processes moves the participants 'from consumers of goods to gathers of experiences and wisdom' (see Huhmarniemi *et al.*, Chapter 30, p. 240). Creative tourism that focuses on dialogue and 'maker cultures such as crafting' also offers a site for 'cultural encounter, dialogue, and knowledge exchange' (Huhmarniemi *et al.*, p. 240).

Learning. At its core, creative tourism is associated with learning about a local culture as well as the techniques involved in making an artefact or other type of creative expression. This can include direct instruction and mentorship or indirect learning approaches more akin to facilitating processes. As artist Margaret Feeney writes, artist-led collaborative creative tourism activities can nurture a 'creative space' that is 'open to interaction. It is anarchic and tangential. Discoveries and experiences are made rather than directed' (see Chapter 5, p. 44). Yet, as Larissa Almeida points out, this is only the beginning: creative tourism 'can be used as a strategy to teach many abilities to people at any age, developing creative capacities as well as notions of history, language, biology, and even teamwork abilities, empathy, and other social skills' (see Chapter 24, p. 179). Furthermore, through personal encounters, 'cross- and intercultural experiences may come about, including learning about one another, such as historical and social contexts' (see Sarantou and Miettinen, Chapter 27, p. 207; see also Cabeça *et al.*, Chapter 9).

Creative self-expression. Creative tourism activities provide opportunities for visitors to engage in experiences of creative self-expression. The learning that takes place during a creative tourism experience can 'develop the personal skills of the traveller, enabling them to enhance their own creative capabilities and potential' (see Richards and Duxbury, Chapter 7, p. 53). Beyond copying a pattern or technique, it is essential that participants can 'develop their own narratives and engage their imagination in the process' (see Huhmarniemi *et al.*, Chapter 30, p. 246, referencing Richards and Wilson, 2006). These opportunities to develop one's creative potential can, in turn, lead to the emergence of new ideas and innovation (see Wisansing, Chapter 12). The cultivation of personally meaningful ideas can be transformative: 'tourists must be able to develop their own narratives, relations, and kinship into the hosts and places of encounter. This is a way for the traveller-learner to change their old ways of experiencing and find new meanings for life' (see Huhmarniemi *et al.*, Chapter 30, p. 250).

Community engagement. Creative tourism has its deepest and most meaningful impacts if it is 'integrated into the social fabric' of the place in which it occurs (see Wisansing, Chapter 12, p. 86). In many contexts, creative tourism development requires community commitment and direct involvement to sustain it (see Marques, Chapter 23). Kathleen Scherf notes that resident-based creative tourism initiatives must be embedded, led, and driven locally – if tourism is not embedded in the community of local residents, 'creative tourism activities will not be sustainable, will not be genuinely place-based, and simply will not work' (see Chapter 25, p. 193).

Spaces. Three types of 'spaces for creativity' are prevalent among travellers: 'spaces to conduct their creative work "on the road", creative residencies, and more personal "escape" spaces where one can retreat from the busyness and pressures of urban life and re-connect personally' (see Richards and Duxbury, Chapter 7, p. 54). Creative cultural spaces serve as a connector between local creators, the broader community, and visitors (see Wisansing, Chapter 12). In both urban and rural areas, formal or informal creative spaces provide opportunities for consumers and producers to gather, to exchange ideas, and to co-create experiences. Over the last two decades, 'new typologies of creative spaces and support services' have emerged (see Albino and Alcobia, Chapter 11, p. 74). Creative spaces are also constructed temporally through the intentional development of small-scale participatory art festivals (see F. Almeida, Chapter 26; Sligting, Chapter 16). All these different types of spaces play roles in hosting and enabling creative tourism.

include behaviours, beliefs, stories, traditions, and rituals that support social connections, expression, and communication (de Munck and Bennardo, 2019). These are closely intertwined with other features of a place. As Maria Huhmarniemi *et al.* point out in Chapter 30, 'the distinction between cultural and natural heritage is blurry: human and non-human nature are in continuous interaction' (p. 242).

Successful and sustainable creative tourism products are those that are 'genuinely embedded in the community and its people, are willingly shared with visitors by locals, and are extensions of place-based skills and knowledge that exist regardless of the tourism product' (see Scherf, Chapter 15, p. 111; see also Chapter 21 on Loulé Criativo). At the same time, it is important to connect internal understandings with external perceptions of a place. As Diana Zuluaga and Diana Guerra note, a destination's externally known 'public' attributes are often the motivation for travellers to choose a place, and determining these can provide important insights for creating experiences that 'incorporate in some way the generally recognized attributes' (see Chapter 14, p. 106).

Identifying local cultural assets in the context of creative tourism can be facilitated through cultural and deep mapping projects (see Scherf, Chapter 15) and emotional mapping exercises (see Bakas *et al.*, Chapter 13). Mapping can incorporate a wide variety of activities, including reviewing secondary information, personal observations, inquiries and conversations with local residents, learning the stories attached to places, identifying intangible assets, and talking with passionate residents (see L. Almeida, Chapter 24). Embedding mapping projects within a community-led development framework can foster an inclusive process that can 'identify a community's unique characteristics: people, traditions, and organizations; networks and clusters of interest and expertise; skills, knowledge, and resources; and special locations, histories, and stories' (see Scherf, Chapter 15, p. 112). The resulting directories of local skills and unique local content comprise an important initial step in identifying potential attractions (see L. Almeida, Chapter 24; Wisansing, Chapter 12).

More generally, heightened attentiveness to a place and its specificities informs the process of creative tourism development. Britta Timm Knudsen and Jan Ifversen advise to 'look for overseen, unnoticed, infra-ordinary resources – also non-human resources' to provide inspiration (see Chapter 18, p. 134). They speak of the creative tourism experience depending on 'a heightened awareness of the created and co-created poetry of that which is simply there' (p. 133). Diana Zuluaga and Diana Guerra note that oftentimes 'the intangible heritage essence' that is so enticing to today's travellers 'is synonymous with the ordinary' (see Chapter 14, p. 107).

Stories and identity are important. The identity of a place is constructed through the narratives and meanings that people give to it, based on its physical and social aspects and its history (Eräranta *et al.*, 2016). Emotional mapping exercises that relate emotions, experiences, and perceptions with specific spaces can inform and inspire creative tourism activities that embody 'the significant cultural information, traditions, stories, values, and hopes that locate people in their place(s)' (see Bakas *et al.*, Chapter 13, p. 94). An ethics of care for culture, place, and people should permeate these activities (Duxbury, 2021). As Melanie Sarantou and Satu Miettinen note, 'the cultural value of place ... is deeply rooted in cultural meanings and traditions, which are often rendered invisible or silent due to hegemonic forces at play' (see Chapter 27, p. 209).

Creative tourism initiatives should then integrate meaningful elements that connect visitors to the place where the activities occur, from the perspective(s) of the people guiding them. Previous research within CREATOUR® identified four key ways in which creative tourism projects can connect to place: the specific site and endogenous materials used in the activity; the nature of the activity itself and how it connects to that place; storytelling about the history, narratives, and local characters; and moments for socializing within the creation activities and in social moments such as a picnic or walk, which can 'play an important role in allowing visitors to slow-down, relax, and absorb the "feeling of place"' (Bakas *et al.*, 2020, p. 119).

Arts-based methods used within creative tourism initiatives can serve to 'express and share the richness of culture, knowledge systems, heritage, and nature connectedness' (see Huhmarniemi *et al.*, Chapter 30, p. 240; see also Knudsen and Ifversen, Chapter 18). Embracing storytelling through imaginative and interpretation processes to tell 'the story of the locality' (see Wisansing, Chapter 12, p. 87) can also serve as inspiring platforms for articulating individual narratives of place-attachment and personal expression.

In this book, the chapters specifically outlining creative tourism development processes (see Chapters 12, 13, 14, and 24) all include advice regarding identifying and using local resources in creative tourism initiatives. More widely, the creative tourism initiatives presented across the chapters in this book provide a range of examples of place-specific creative tourism development building on local resources. These examples demonstrate different approaches to activating cultural and place-specific resources in ways that aim to nourish, articulate, and contribute to the vibrancy and sustainability of their place, its specificities, and its array of resources.

Engaging creative travellers

It is widely acknowledged that a 'new generation' of travellers has emerged in recent years. They want to be active participants in the 'everyday' experiences of the places they visit as well as in exceptional and personally meaningful experiences. In their travels, they are focusing on self-development and searching for 'transformative' experiences (World Tourism Organization, 2016).

Within this context, the creative traveller is seen as 'a person who approaches the act of travelling as a transformational experience in their life ... searching for exceptional experiences that will bring about change and allow them to widen their perspective on their own values and beliefs' (see Zuluaga and Guerra, Chapter 2, p. 15). Creative travellers seek meaningful, personally significant, and 'unique' experiences – experiences that 'make them become a better person, that help them grow' (see Agrawal, Chapter 10, p. 68). They pursue 'genuine engagement' with activities that centralize 'the interaction of the [creative] practitioner and the travelling mind' (see Feeney, Chapter 5, p. 44). As Fernando Almeida (see Chapter 26, this volume) puts it,

> We are collecting beings. We collect memories, experiences, knowledge gathered between relationships established in a globalized world. The activities in which we choose to participate contribute effectively to the construction of our individual and collective identity. (p. 199)

Yet creative tourists comprise a very diverse profile within the new generation of travellers and are also found within 'existing segments, like MICE, singles, seniors, linguistic, and so on' (see Couret, Chapter 8, p. 60). Participants in creative tourism may be international or near-by visitors; people new to your activity, experiencing 'a "first contact" with the local cultural identity' (see Salvado and Rolo, Chapter 17, p. 127), repeat visitors to your locale, or part-time residents; they may be newcomers to an activity or professional artists looking for new sources of inspiration and to extend their skills, among other characteristics.

Finding most creative tourism participants were accompanied, Jaime Serra *et al.* suggest that designers of creative tourism activities should also keep in mind 'how to attract and facilitate the participation of small groups, and – if it can be accommodated – the involvement of children' (see Chapter 4, p. 32). Tackling the challenge, 'How can you discover a city with kids without boring them?', creative tourism entrepreneur Melanie Wolfram developed an interactive and creative game, Play Évora (see Chapter 6).

Creative tourism entrepreneurs must develop a 'special understanding of the activity you are offering and its attractiveness to different types of travellers' (see Zuluaga and Guerra, Chapter 2, p. 18). Just as one size does not fit all, not all travellers will be attracted to a particular experience. Some travellers will fall into 'a "creative spectator" category, appreciating staged experiences, while others, passionate creative travellers, are longing to actively curate their artistic experiences with the artists' (see Wisansing, Chapter 12, p. 90). And, of course, we must also acknowledge that 'not all tourists are open to creative tourism' (see Zuluaga and Guerra, Chapter 14, p. 110).

As Marie-Andrée Delisle details in Chapter 19, a market segmentation exercise can help identify specific target markets for a creative tourism experience. Then, on the basis of insights gained through analysing prospective participants, aspects of the creative tourism experience can be personalized for different target markets. Social media interest groups have emerged as important means of market segmentation. Creative tourism activities are linked to personal passions and pursuits, with travel being an extension of these interests.

Many of the book's authors highlight the importance of social media. Diana Zuluaga and Diana Guerra (see Chapter 14) point to the importance of internet presence and communications; specialized platforms, publications, and influencers in the tourism industry; international networks; and reviews by previous clients. Kathleen Scherf suggests that social media can also be used 'as a shared meeting site for visitors and locals' (see Chapter 25, p. 195). Elena Paschinger remarks that creative tourism may be particularly well suited to social media due to 'its interactive nature and lively cultural exchange between locals and travellers' (see Chapter 20, p. 153).

But who are the creative travellers visiting *your* locale? How do you identify and learn more about them? More specifically, how do you identify who is a prospective participant in a particular creative tourism activity? These are essential questions to move from a general understanding of the 'creative traveller' to identifying who is a likely participant in your specific creative tourism activity.

To help readers address these questions, Part I of this book includes three chapters providing detailed approaches to identifying and understanding creative travellers. Diana Zuluaga and Diana Guerra (Chapter 2) explain how three key variables help 5Bogota determine the interests of prospective creative travellers: traveller status (i.e. whether the traveller will have company), age, and factors that influence their creative tourism purchase decisions. Elena Paschinger (Chapter 3) points to key motivational traits such as surprise seeking, social adventurousness, and curiosity, as well as the importance of whether creative travellers 'experience creative travel individually, or as a group' (p. 26). Jaime Serra *et al.* (Chapter 4) discuss the findings of a survey of creative tourists participating in CREATOUR® pilot projects, segmented by place of residence (i.e. domestic or international visitors), and how these findings can inform practitioners' strategies. A subsection of Part III, entitled 'Marketing – reaching and engaging creative travellers', complements these chapters with a comprehensive chapter on marketing and market readiness (see Delisle, Chapter 19). Additional insights about creative travellers are woven throughout the book.

This book

The book is divided into six sections: Part I – Types of creative travellers; Part II – Trends in creative tourism; Part III – Designing and managing creative tourism products; Part IV – Embedding creative tourism activities within a community; Part V – Addressing challenges of impact and sustainability; and Part IV – Concluding remarks. Parts I through V include longer 'anchor' chapters as well as complementary shorter contributions to provide a wide variety of voices, perspectives, examples, and words of advice. Here we present an overview of each part; summaries of each chapter are found in an introduction in each section.

Part I: Types of creative travellers

Part I brings together knowledge and insights from both practitioners and researchers on identifying and understanding different types of creative travellers. The first three chapters provide profiles and insights about creative travellers from three geographic perspectives: Colombia, Austria, and Portugal. The first two are from practitioners' experiences (7 and 10 years, respectively), and the third is an outcome of a 3.5-year national research-and-practice project in which participants in creative tourism pilot activities completed a common questionnaire. These 'anchor' chapters are complemented by two shorter contributions providing, first, the perspective of an artist who practices community-engaged art projects in her travels and, secondly, insights on developing a creative tourism product for families. Altogether, the contributions in Part I aim to provide a variety of perspectives and approaches for thinking about and identifying different types of creative tourists and travellers for specific initiatives. Part I is complemented by chapters about creative tourism marketing within Part III, which provide further tools and resources for identifying, reaching, and engaging these creative travellers.

Part II: Trends in creative tourism

Part II explores the dynamically evolving trajectories of creative tourism, looking at both trends and tensions in the expanding and diversifying field. It aims to set out a 'big picture' landscape that highlights some of the emerging hot areas to watch, as background to Parts III and IV which focus on approaches to the design, development, and evolution of creative tourism initiatives on the ground. The first chapter in Part II provides an overview of 12 major trajectories within the diverse field of creative tourism, followed by observations about developments and challenges in the field from the perspective of the international Creative Tourism Network®. Three specific areas within creative tourism are then highlighted, with contributions providing insights on the development and implementation of creative gastronomy offers, which brings together creative tourism and food tourism; the appeal of mini-apprenticeships with master artisans and artists; and the varied relations between artistic residencies, digital nomads, and tourism.

Part III: Designing and managing creative tourism products

This central and largest part of the book is divided into three sections focusing on different dimensions of designing, launching, adjusting, and propelling a creative tourism initiative forward. The first section, 'The cycle of product design, implementation, and iterative adjustments', focuses on conceiving, designing, testing, and launching a creative tourism initiative. The second section, 'Marketing – reaching and engaging creative travellers', examines the multifaceted aspects related to marketing, promotion and communications, commercialization, and market readiness, and complements the chapters in Part I on identifying and understanding different types of creative travellers. The third section, 'Managing and innovating creative tourism products', looks at the longer trajectory of managing and evolving creative tourism initiatives and portfolios of creative tourism experiences over time.

Part IV: Embedding creative tourism activities within a community

Part IV presents contributions from a variety of contexts on how creative tourism organizers can promote resident engagement and work with their broader community. The chapters provide insights from different geographical perspectives and both academics and practitioners. The two 'core' chapters (the longest contributions) are based on the authors' experiences in Brazil and Namibia, with three shorter chapters featuring contributions based on situations in Spain, Denmark, and Portugal. A community-embedded approach to creative tourism can provide the 'link to locals' that the new generation of travellers is craving. It helps foster meaningful community involvement in conceiving, designing, and implementing creative tourism initiatives, and ensures a wider distribution of the benefits of creative tourism in the community. More broadly, community-engaged creative tourism initiatives can be applied as a strategy for inclusion, participation, and decolonization.

Part V: Addressing challenges of impact and sustainability

Part V addresses the potential greater impacts of creative tourism initiatives, from the well-being of the local community in which it is embedded, to addressing society's existential need to address our relationship with the planet and each other in more sustainable manners. It closes with a series of policy recommendations for the development of a locally embedded and networked creative tourism sector that brings together culture, tourism, and local/regional development. The three chapters present different pathways to think about the longer horizon of the impacts of creative tourism initiatives, in a holistic manner incorporating social, economic, environmental, and cultural dimensions.

Part VI: Concluding remarks

Finally, Part VI presents the book's closing chapter, written by the book's editors, which reflects on the volume as a whole and pulls together a series of key resonating points in the practice of creative tourism development. Informed by the book's contributors as well as the UN Sustainable Development Goals, it considers the potential trajectories and impacts that creative tourism can have in and for communities and, more broadly, reflects on the future of creative tourism in a context of sustainable development.

Creative tourism for ...

We believe this book will be valuable for many types of creative tourism entrepreneurs, developers, and facilitating agencies. We also hope to plant seeds of possibility that creative tourism development can be more than a business venture, and can spark broader local and regional initiatives and impacts. The development of culture-based creative tourism activities can foster pride in local cultural expressions and traditions, contributing to their vitality and sustainability; build local capacities, economic opportunities, and networks; and even lead to local empowerment.

From a cultural stewardship perspective, as Jutamas (Jan) Wisansing points out, creative tourism can play an important role:

> [D]iverse ranges of local culture and assets are unique to a specific place and ... unique forms of creativity are attractive because they are linked to the pride of a locality. Creative tourism activities can therefore not only provide an authentic taste of Thai local culture for visitors but can also create strong senses of ownership which lead to the desire to protect the value of such assets. (See Chapter 12, p. 85; see also Tan and Tan, 2019)

Participatory and collaborative creative practices driven by local communities can also promote counter narratives to stereotypical representations. In turn, this can help shape new and alternative stories related to place and local practices (see

Sarantou and Miettinen, Chapter 27; Marques, Chapter 23). As Catharina Sligting notes, the 'experience of art works created together adds to the public memory' (see Chapter 16, p. 120).

Closely integrating creative tourism with local development can encourage direct and indirect benefits to be shared among community residents, 'including those who might not "traditionally" be involved in the tourism sector' (see Gonçalves *et al.,* Chapter 31, p. 256). Lénia Marques (Chapter 23) and Larissa Almeida (Chapter 24) explicitly consider how innovation in creative tourism can 'be introduced in a successful and sustainable manner that contributes not only to the economic welfare of people, but also to their social and cultural empowerment' (see Marques, p. 173). In all communities, creative tourism can embrace principles of humanistic management, promoting human flourishing, engaging the other in journeys of mutual discovery, honouring the dignity of stakeholders involved, and contributing to the common good (Duxbury and Bakas, 2021).

The goal of participatory cultural mapping and creative tourism initiatives can go beyond a discovery of the features of the local culture to also 'expand the potential of that culture' through facilitating exchanges that can increase 'collective knowledge, skills, and understanding' (see Wisansing, Chapter 12, p. 86). Collaborative processes during mapping and planning exercises can serve to 'encourage networks and partnerships in designing and offering cultural and creative tourism activities', among other possibilities (see Scherf, Chapter 15, p. 117; see also L. Almeida, Chapter 24). These networks and partnerships can form the platforms for generative capacity building and community empowerment processes more widely (see, e.g. Musarò and Moralli, 2021).

During the COVID-19 pandemic, many agents in the tourism sector have been imagining new integrated and locally sensitive approaches to travel going forward (see, e.g. Holowack, 2020). Creative tourism entrepreneurs should focus on building and strengthening partnerships in their locale and region (Carvalho, 2020) to become 'more visible, more connected, and consequently more recognized as valuable and vital resources in their community' (see Delisle, Chapter 19, p. 147). With the connection to the local community becoming ever more central for visitors, creative tourism can serve as 'a strategy for initiating encounters with local communities' (see Sarantou and Miettinen, Chapter 27, p. 207).

Fostering 'coopetition' – cooperation among 'competitors' – can enable different actors to collectively 'bring out the distinctiveness of a place' as well as lead to inventive joint packaging ideas (see Delisle, Chapter 19, p. 139). Collaborative tourism arrangements and methods that feature 'shared values of hospitality, co-operation, and humanity' can generate new ecosystems of local partners and a kind of community-based tourism (Delisle, p. 147). This approach can align with growing interest internationally in local initiatives based on cultural and creative resources and networks that are 'guided by an ethics of care, social and environmental justice' (Benjamin *et al.*, 2020, p. 479).

When strategically linking creative tourism with community engagement, the associated dimensions of community decision making and empowerment should be considered. Local level decision making leaves the community 'empowered and enabled, at least to some extent, to influence their own futures' (see Sarantou and Miettinen, Chapter 27, p. 212). Taking a community-based approach to creative tourism, as Jutamas (Jan) Wisansing elaborates, means the goals for creative tourism development are derived from, and integrated into, 'the overall shared community visions and aspirations' and follow a principle of enabling the community's input through active participation in creative tourism development processes (see Chapter 12, p. 86).

Acknowledgements

This chapter was developed within the CREATOUR® project (no. 16437), which is funded by the Portuguese Foundation for Science and Technology (FCT/MEC) through national funds and co-funded by FEDER through the Joint Activities Programme of COMPETE 2020 and the Regional Operational Programmes of Lisbon and Algarve.

Notes

1. As creative tourism has evolved since the early 2000s, a wide variety of definitions highlighting different perspectives and contexts have been developed (see Duxbury and Richards, 2019). Today, a 'landscape of interpretations' of creative tourism is characterized by a fluidity, flexibility, and adaptability that enables it to be localized and to fit into a variety of contexts (see Marques, Chapter 23, p. 172).

2. Creative tourism entrepreneurship can encompass an array of organizational approaches, strategies, and models. For examples, see Duxbury *et al.* (2020); Henche *et al.* (2020); and Richards (2020).

References

Bakas, F.E., Duxbury, N., Silva, S., and Vinagre de Castro, T. (2020) Connecting to place through creative tourism. In Cabeça, S.M., Gonçalves, A.R., Marques, J.F., and Tavares, M. (eds.) *Creative Tourism Dynamics: Connecting Travellers, Communities, Cultures, and Places*. Editora Grácio, Coimbra, Portugal, pp. 119–139.

Benjamin, S., Dillette, A. and Alderman, D.H. (2020) "We can't return to normal": Committing to tourism equity in the post-pandemic age. *Tourism Geographies* 22(3), 476–483. doi: 10.1080/14616688.2020.1759130.

Carvalho, R. (2020) The need for creative actions in tourism in the Post-COVID 19 pandemic world. In: Figueira, L.M. and Oosterbeek, L. (eds.) *Turismo Mundial, Crise Sanitária Futuro: Visões Globais Partilhadas*. Instituto Politécnico de Tomar, Tomar, Portugal, pp. 177–190.

de Munck, V.C. and Bennardo, G. (2019) Disciplining culture: A sociocognitive approach. *Current Anthropology* 60(2), 174–193.

Duxbury, N. (2021). (Re)articulating culture, tourism, community, and place: Closing remarks. In: Duxbury, N. (ed.) *Cultural Sustainability, Tourism and Development: (Re)articulations in Tourism Contexts.* Routledge, London, UK, pp. 197–212.

Duxbury, N. and Bakas, F.E. (2021) Creative tourism: A humanistic paradigm in practice. In: Della Lucia, M. and Giudici, E. (eds.) *Humanistic Management and Sustainable Tourism: Human, Social and Environmental Challenges.* Routledge, London, UK, pp. 111–131.

Duxbury, N. and Richards, G. (2019) Towards a research agenda for creative tourism: Developments, diversity, and dynamics. In: Duxbury, N. and Richards, G. (eds.) *A Research Agenda for Creative Tourism*. Edward Elgar Publishing, Cheltenham, UK, pp. 1–14.

Duxbury, N. and Silva, S. (2020) Impulsionar o turismo criativo em cidades de pequena dimensão e zonas rurais de Portugal: a abordagem investigação-ação do projeto CREATOUR (Activating creative tourism in small cities and rural areas in Portugal: The CREATOUR research-and-application approach). In: Duxbury, N. and Silva, S. (eds.) *CREATOUR: Catalisando Turismo Criativo em Cidades de Pequena Dimensão e em Áreas Rurais.* Imprensa da Universidade de Coimbra, Coimbra, Portugal, pp. 11–37.

Duxbury, N., Bakas, F.E., Vinagre de Castro, T., and Silva, S. (2020) Creative tourism development models towards sustainable and regenerative tourism. *Sustainability* 13(2), 1–17. doi: 10.3390/su13010002.

Eräranta, K., Leino, T., Seppälä, T., Viña, S., and Timonen, E. (2016) Mapping the pig tale journey: A multidisciplinary design framework for cultural mapping in an old abattoir. *City, Culture and Society* 7(1), 25–33. doi: 10.1016/j.ccs.2015.07.006.

Gonçalves, A.R., Marques, J.F., Tavares, M., and Cabeça, S.M. (eds.) (2020) *Creative Tourism: The CREATOUR Recipe Book.* CINTURS, University of the Algarve, Faro, Portugal.

Henche, B.G., Salvaj, E., and Cuesta-Valiño, P. (2020) A sustainable management model for cultural creative tourism ecosystems. *Sustainability* 12, 9554. doi:10.3390/su12229554.

Holowack, D. (2020) Keynote address: Vital steps to building a local-first tourism economy. *Skift.com,* 23 November. Available at: https://skift.com/2020/11/23/crowdriff-keynote-vital-steps-building-local-first-tourism-economy (accessed 2 December 2020).

Isar, Y.R. (2013) *Creative Economy Report 2013 Special Edition: Widening Local Development Pathways.* UNESCO and UNDP, Paris, France.

Musarò, P. and Moralli, M. (2021) What is the role of responsible tourism in building stronger and intercultural communities? Two case studies from Italy. In: Duxbury, N. (ed.) *Cultural Sustainability, Tourism and Development: (Re)articulations in Tourism Contexts.* Routledge, London, UK, pp. 21–35.

Richards, G. (2020) Designing creative places: The role of creative tourism. *Annals of Tourism Research* 85, 102922. doi: 10.1016/j.annals.2020.102922.

Richards, G. and Raymond, C. (2000) Creative tourism. *ATLAS News, no.* 23, 16–20.

Richards, G. and Wilson, J. (2006) Developing creativity in tourist experiences: A solution to the serial reproduction of culture? *Tourism Management* 27(6), 1209–1223.

Richards, G., Wisansing, J., and Paschinger, E. (2019) *Creating Creative Tourism Toolkit*. DASTA, Bangkok, Thailand. Available at: https://perfectlink.co.th/wp-content/uploads/2019/01/Creating-Creative-Tourism-Toolkit_Version-Eng.pdf (accessed 23 July 2020).

Scherf, K. (ed.) (2021) *Creative Tourism in Smaller Communities: Place, Culture, and Local Representation.* University of Calgary Press, Calgary, Alberta, Canada.

Tan, S.-K. and Tan, S.-H. (2019) Nurturing the creative tourist in Malaysia. In: Duxbury, N. and Richards, G. (eds.) *A Research Agenda for Creative Tourism*. Edward Elgar Publishing, Cheltenham, UK, pp. 17–27.

Vinagre de Castro, T., Duxbury, N., Silva, S., Bakas, F., Carvalho, C., Sancho Querol, L., Borges, R., Albino, S., Gonçalves, A., Remoaldo, P. and Matos, O. (2020) *Creative Tourism: Guide for Practitioners / Turismo criativo: guia para profissionais.* (Published in English and Portuguese editions.) CREATOUR Project, Centre for Social Studies, University of Coimbra, Coimbra, Portugal. Available at: https://creatour.pt/en/publications/creative-tourism-guide-for-practitioners/ (accessed 12 January 2021).

World Tourism Organization (2016) *Global Report on the Transformative Power of Tourism: A Paradigm Shift towards a More Responsible Traveller.* Available at: www.e-unwto.org/doi/book/10.18111/9789284417834 (accessed 21 April 2020).

Part I: Types of creative travellers

Part I brings together knowledge and insights from both practitioners and researchers on identifying and understanding different types of creative travellers. The first three chapters provide profiles and insights about creative travellers from three geographic perspectives: Colombia, Austria, and Portugal. The first two are from practitioners' experiences (7 and 10 years, respectively), and the third is an outcome of a 3.5-year national research-and-practice project in which participants in creative tourism pilot activities completed a common questionnaire. These 'anchor' chapters are complemented by two shorter contributions providing, firstly, the perspective of an artist who practices community-engaged art projects in her travels and, secondly, insights on developing a creative tourism product for families.

The section begins with Diana Zuluaga and Diana Guerra's chapter, 'A typology of creative travellers in Bogota, Colombia: The experience of 5Bogota' (Chapter 2) which examines, reflects on, and classifies the different types of travellers they have encountered over the last 7 years at 5Bogota. The authors identify three variables that have assisted them to determine the interests of prospective creative travellers: traveller status, age, and factors that influence their creative tourism purchase decisions. While these categories derive from their specific experiences and may not be universally applicable, they provide keen insights and guidance for other creative tourism entrepreneurs. Towards this end, the chapter presents a tool that has been designed with the aim of helping creative tourism entrepreneurs to identify and profile prospective clients, or to use as a source of information to improve the services offered to existing customers. They also present a customer profiling exercise to develop a typology of creative travellers, a process that brings to light the key factors that play into travellers' decision-making processes.

In chapter 3, 'Creative travellers in Austria', Elena Paschinger of the association Kreativ Reisen Österreich (Creative Tourism Austria) presents a typology of creative travellers, providing a general perspective on the types of people she observes that are travelling to be creative, calling them the passion seeker, the family and friends type traveller, the urban explorer, the human adventurer, and the curious solo traveller. The chapter also provides a brief overview of the nature of creative tourism in Austria and reflections on a growing, diverse, and rich trend for creative tourism in the country.

In 'Domestic and international creative tourists in Portugal: Insights for practitioners', Jaime Serra, Noémi Marujo, Nancy Duxbury, Alexandra R. Gonçalves, Sónia Moreira Cabeça, Paula Remoaldo, and Juliana Alves present insights about the participants in the creative tourism pilot activities designed and implemented in small cities and rural areas within the CREATOUR® project. Chapter 4 presents the results of a detailed questionnaire completed by these visitor-participants, which was applied by the 40 participating organizations as they conducted their pilot activities in 2017, 2018, and 2019. In this brief chapter, the highlights of the questionnaire results are presented, segmented by place of residence (i.e. domestic or international visitors). The chapter concludes by suggesting some implications of these findings for practitioners.

Complementing these 'big picture' analyses and observations, two smaller contributions provide first-hand insights about two types of creative travellers: an artist who practices community-engaged art projects in her travels and a tourism animation entrepreneur who has developed a creative tourism product for families.

In chapter 5, 'Collective invention: A travelling artist's perspective', artist Margaret Feeney of Māpura Studios, Aotearoa New Zealand, writes about two projects she has organized with local communities while travelling, *The Stencil Project* and *Water Systems*. She discusses the value of such activities in her work and how they inspire her thinking on creative tourism as a platform that can bring together communities and their visitors through supporting artists to share their ideas. Based on her extensive experiences, the chapter also includes an insightful list of points on what makes a workshop work well, that is, 'strategies that encourage people to get involved and share ideas' (p. 42). Feeney proposes that the creative tourism approach, which privileges direct relations between makers and visitors/consumers, can be applied as a dynamic and flexible model to invigorate organizations that support artists and as a local and regional development tool in her home country.

Closing Part I, we present an interview with Mélanie Wolfram, of Vagar Walkingtours, Portugal, on connecting to family travel with creative tourism (Chapter 6). She emphasizes how, for many families, holidays are 'the means through which they strengthen their ties and build memories and experiences together' (p. 46). In this context, creative tourism products such as the 'Play Évora' kit provide carefully designed interactive and creative experiences for families to do together as they explore the city. Beyond this, she notes how tourism entrepreneurs must carefully promote tourism without destroying their destination, developing strategies that enable embedded experiences and interactions that support and nurture local businesses and other initiatives.

Altogether, the contributions in Part I aim to provide a variety of perspectives and approaches for thinking about and identifying different types of creative tourists and travellers for specific initiatives. They bring together insights from both research and practice to jointly inform this work. Part I is complemented by the chapters about creative tourism marketing within Part III, which provide further tools and resources for identifying, reaching, and engaging these creative travellers.

—N.D.

2 A typology of creative travellers in Bogota, Colombia: The experience of 5Bogota

DIANA ZULUAGA* AND DIANA GUERRA

5Bogota, Colombia

The goal of this chapter is to examine and classify the different types of travellers we have encountered throughout the last seven years at 5Bogota. To this end, we have created a tool that can be either a starting point for profiling prospective clients, or a source of information to improve the services offered to existing customers.

The general characteristics of travellers interested in creative tourism

In our experience, we have found that all possible traveller classifications (to be introduced later) share a common feature: they are all looking for a space to share intimately with a local host, a place that allows the traveller to feel as immersed as possible in the culture of their chosen destination. In fact, we have found that the personal bond with the host and the space matters to travellers more than the topic of the experience.

It is also worth noting that there is an important number of travellers who choose creative tourism allured by the trend of posting on their social media feeds about atypical experiences happening away from the usual touristic sites in real time, as a fresher and more genuine travel concept. This means there is a group of people who are more interested in filling their social media feeds than they are in immersing themselves in our creative tourism experiences. It behoves us to replace these ideas and gently lead the traveller to the essence of the experience we have created!

Continuing with the objective of this chapter, we have identified three variables that facilitate an analysis of the universe of travellers interested in creative tourism, allowing us to clearly determine the interests of prospective creative travellers. These variables are traveller status, age, and factors that influence their creative tourism purchase decisions.

Traveller status

Traveller status refers to whether the traveller will have company, in which case we must establish a relation between the traveller and their companions. We have identified five categories of traveller status: solo travellers, couples, families, groups of friends, and work groups.

Solo travellers

In most cases, this means people who travel for academic or labour reasons who decide to fill their free time with activities that include sharing with others. These travellers are sociable, open to new experiences, and willing to take advantage of each of their few available moments. They are also always on the lookout for recommendations in order to fully explore the destination. The price of the experiences is not a determining factor for these travellers; consequently, they do not put much effort into choosing an experience and opt for one based on the ease of purchase.

*Corresponding author: dianazuluagag@gmail.com

 Creative Tourism: Activating Cultural Resources and Engaging Creative Travellers (eds. N. Duxbury, S. Albino, and C. Pato de Carvalho)
DOI: 10.1079/9781789243536.0002

On the other hand, a considerable number of solo travellers undertake long voyages covering several countries during what is commonly known as a sabbatical year. This kind of traveller might be more concerned with money, making advance research about the experiences of paramount importance prior to their purchasing of an experience. This means these travellers will only invest in experiences with a stellar reputation.

Couples

Most couples take brief trips, often during long weekends or short vacations. These are travellers for whom knowing as many places and living as many experiences as possible is fundamental. They may sometimes be hard to contact since they are focused on living the experience as a couple and are not looking for cultural experiences, but rather for pastimes. They socialize more easily with other couples in a group. Regarding the experiences, they do not care much about the price, but they do care about the time as they seek to fully live in the day.

It is also common to find couples who travel for several months or even a year as a means of closing a cycle and starting another. Many of these travellers have quit their jobs and decided to turn their lives around; for them it is essential to have personal interactions with hosts as they are looking for inspiration for their new life projects. Having spent a long time away from home, these travellers also place a high value on the human qualities and the warmth of the host as they want to feel comfortable and welcome.

Families

Their travel itineraries usually include a number of destinations in the same country, and they travel during the holiday seasons. Families are normally looking for experiences for every member of their group, considering the differences in age and preferences. The age of the children determines how many of the activities are carried out, as a family with young children or teenagers will function differently than a family where all members are adults. The goal is always to create an environment where all members will have a task to complete, and where they will bond with each other. These travellers research extensively and focus exclusively on the best from each destination. To them, the quality and the level of detailed information available are decisive at the moment of purchase. For this reason, these travellers may often demand products tailored to their specifications, which may delay the sale by weeks until the service offer meets their expectations.

Groups of friends

Groups of friends share a reason or common interest to travel, for example, birthdays, special occasions, bachelor or bachelorette parties, group vacations, weddings, and so forth. There is usually one or two people who establish the itinerary, with everyone else following their lead. The number of participants must be taken into account when planning and executing activities for these travellers: if there are more than six participants, the interaction with the host is basically non-existent. The travellers will take control over the activity and it will be impossible for the host to take part. On the contrary, if there are fewer than six travellers, they will welcome the host into the group and every member will be more willing to enjoy the experience beyond the group setting. This will also make it easier for the group to interact with other travellers also participating in the experience. The price is paramount for these travellers, who are usually looking to take advantage of group rates or discounts.

Work groups

These travellers are professional colleagues who are in many cases attending events or training sessions. The experience is often chosen by the organizer of the corporate event, which results in the participants not bonding with the host or the experience.

Age

Age is a determining factor both for what travellers wish to encounter, and for the experience itself. Its importance stems from our need to meet the specific expectations of each participant in an experience, since in many cases they will belong to different age groups.

Children aged 12 and under

Though they did not make the final purchase decision, their preferences have informed their parents' choice. They are usually active, are interested in arts and crafts, and belong to families who are used to the complexities of travelling with children, so usually they are accustomed to getting to know other cultures and countries. Bonding with the children is key for positive group dynamics since the adults will always seek amusement and enjoyment for them. It is essential to ensure children participate in the experience, and to listen to their opinions.

Teenagers

These travellers have more power to sway their parents or guardians in favour of a decision as the latter look for experiences that match the interests of the youngsters. These travellers may have difficulty bonding with the host due to the difficult nature of their stage of life, which may cause them to become isolated, often even from their families. This makes it especially relevant to figure out their personality in order to decide whether to involve them in the experience.

Adults aged 20–29

These travellers make their own travel decisions and that means they are eager to bond with the host, but their lack of experience keeps them from doing so. They are also usually travelling away from home for the first time. The price of the experience is paramount to these travellers as their budget is limited. Travellers of this age are heavily engaged in social networks, which we must use in our favour by filling the experience with vivid and memorable images. But that same heavy social engagement can also enable relentless complaining, which means that creating an emotional bond based on the social reality of the destination is absolutely crucial. Finally, it is best not to schedule activities with limited and fixed schedules for these travellers as that may interfere with their quest for complete freedom.

Adults aged 30–45

These are mature, veteran travellers who are consciously looking for the most creative, best-ranked experiences in every destination. They invest considerable time and effort researching their chosen destination, which is why reviews are so important for them when selecting an experience. They rely heavily on friends' recommendations to fill their itinerary with activities, and even to pick a destination. They truly want to bond with the locals, and in some cases are even visiting the destination for a second time. They subscribe to a *slow* way of travelling and always want to take the best from each destination with them.

Adults aged 46–60

These are people who plan their travels carefully. They devote their time to define every detail of their trip. For them, keeping established schedules and respecting the duration of each activity are of the utmost importance, and to that end they demand absolute punctuality and quality from their hosts. They focus on comfort and ease, for which they invest a considerable budget. They are drawn to local culture and to bonding with the locals. They want to know about the country and about their host's personal life.

Adults aged over 60

These are quiet travellers, profoundly interested in bonding with their hosts. In most cases, their itineraries have been organized for them, which means they can be easily surprised. Occasionally, they require personalized assistance. It is best not to place them in groups with travellers unknown to them.

Factors influencing creative tourism purchase decisions

Having defined the general types of creative travellers by traveller status and age, the factors that influence their creative tourism purchase decisions need to be established. Table 2.1 lists and describes such factors, based on our experiences in interacting with creative tourists in Bogota.

Tool to categorize creative travellers

The matrix in Table 2.2 is a tool to chart the affinity the types of travellers described previously have with different factors that that influence their creative tourism purchase decisions. The column to the left lists the factors that travellers consider in making a decision (defined in Table 2.1). These are cross-referenced with each type of traveller, based on their status and age group. The aim of this

Table 2.1. Factors that influence creative tourism purchase decisions

Factor	Definition
Open to new experiences	Open-mindedness to participate in new situations, to try unknown things, to explore new destinations
Rapport with fellow travellers	Ease to quickly and effectively strike up conversation with different, unknown travellers
Travelling experience	Number of previously visited destinations and experiences lived there
Reviews	Relevance and personal appreciation for the comments posted on social networks by other travellers
Budget	High spending power and willingness to invest in experiences that will be worth the money in the traveller's opinion
Research about the destination	Previous research on social networks and search engines about the destination and the best things to do
Experience reputation	Personal consideration of travel experts' recognition (specialized books or magazines, blogs, etc.)
Bond with the local host	Desire to create a special and profound connection with the local hosts, showing interest in their way of life, the destination, and the culture
Detailed information	Need to receive every detail of the travel itinerary, and to know it beyond its basic features
Activity type	Influence of personal preferences or interests in connection with the activity at hand
Destination culture	Interest in the underlying factors of the local culture and way of life, in the ingredients of the traditional local dishes, and in the places frequented by locals
Unplugging from social networks	Unplugging from social networks and electronic appliances during the experience
Recommendation by family and friends	Personal validation of advice and recommendations from friends or relatives about the destination and the possible activities

process is to reveal how important each factor is to each type of traveller. When filled in, each cell should display a value between 1 and 3 which represents how influential each aspect is to each type of traveller, with 1 being *irrelevant* and 3 being *very important*.

This tool has been designed with the aim of helping creative tourism entrepreneurs carry out their own exercise based on their experience, their type of product, their geographic location, and the maturity of their company. Once the proposed exercise has been completed, the entrepreneur will have the necessary elements to profile different types of customers, which will be useful when creating grounded marketing and communication strategies.

All this requires special understanding of the activity you are offering and its attractiveness to different types of travellers – commercializing bicycle rides is different from offering culinary experiences, just as promoting arts and crafts workshops differs from selling urban art sightseeing tours. Likewise, communicating with a solo traveller will be different from doing so with a family. The approach will have to be specific for each type of traveller.

Categorizing creative travellers: the case of 5Bogota

With the results from the above matrix, and seven years of professional experience, we have produced the graph presented in Fig. 2.1. This analysis shows the types of travellers with whom we have found believe the most in the creative tourism offer available at 5Bogota. In the image, the types of travellers who most support creative tourism occupy the largest areas. We find it is a good idea to represent the results from the matrix above in a readable graph. This will become an essential tool for the categorization of client profiles.

Figure 2.1 illustrates that adults aged 46–60, adults aged 30–45, and families are the three types of travellers who believe the most in the creative tourism offer available at 5Bogota. In contrast, groups of friends and the work groups have not shown great affinity with our creative tourism offer.

As a means to showcase the versatility of the analysis, we have created an additional graph (Fig. 2.2) showing the predominance of the factors that influence travellers' decisions and their association with the three types of travellers who have

Table 2.2. Typology of creative travellers: a matrix

Type of traveller Factors	Solo traveller	Couple	Family	Group of friends	Work group	Children aged 12 and under	Teenagers	Adults aged 20–29	Adults aged 30–45	Adults aged 46–60	Adults aged over 60
Open to new experiences											
Rapport with fellow travellers											
Travelling experience											
Reviews											
Budget											
Research about the destination											
Experience reputation											
Bond with the local host											
Detailed information											
Activity type											
Destination culture											
Unplugging from social networks											
Recommendation by family and friends											
Total											

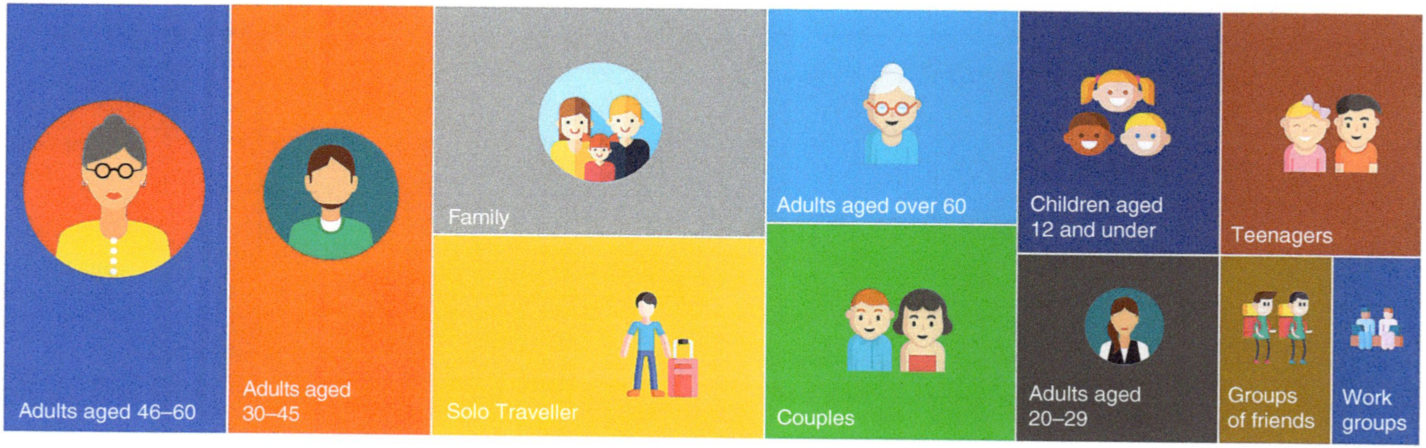

Fig. 2.1. Types of travellers and their belief in creative tourism: the case of 5Bogota.

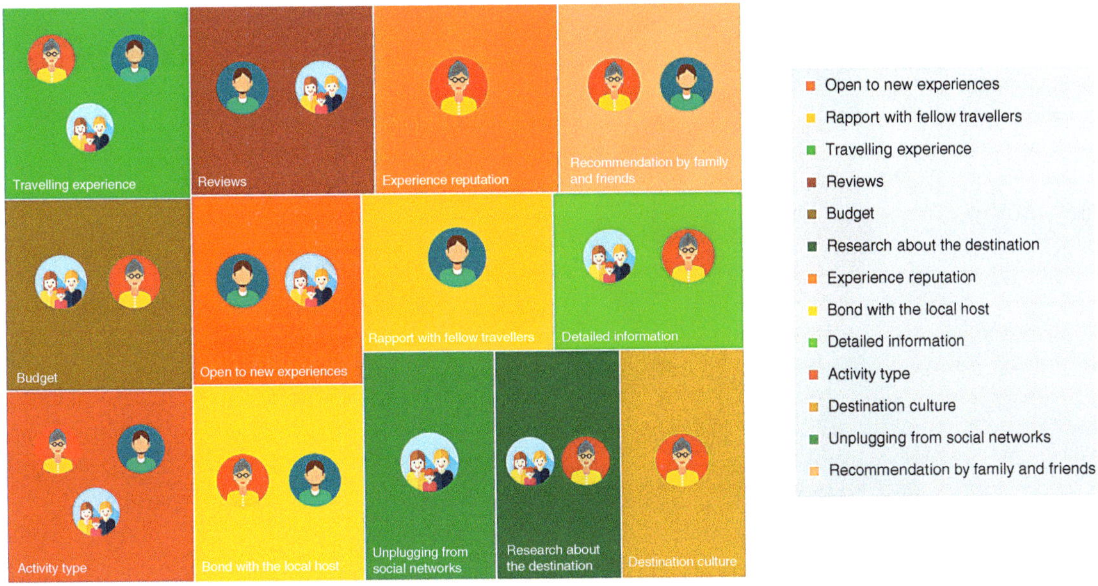

Fig. 2.2. The relation between the three types of travellers most inclined towards creative tourism and the factors influencing travellers' decisions: the case of 5Bogota.

favoured 5Bogota the most, that is: adults aged 46–60, adults aged 30–45, and families.

Based on the analysis illustrated in Fig. 2.2, we can see that travellers aged 46–60 regard the greatest number of factors as 'very important'. We also found that the travelling experience, the absence of budget limitations, and the type of activity are the most relevant factors when selecting one of the creative tourism experiences available at 5Bogota. It is worth mentioning that while some factors matter more to individual travellers, they consider all of them when purchasing a creative tourism experience, and therefore creative tourism entrepreneurs should consider them all when profiling their customers.

Figure 2.3 presents the results of a customer profiling exercise conducted for the three types of travellers with the most affinity for the creative tourism offer at 5Bogota. It is easier to build these profiles when they have names, ages, preferences, and other information that allows for their identification as a real person.

Remarks and reflections

The results presented in this chapter are the product of a 7-year empirical process at 5Bogota. During this time, we have had the chance to meet and serve over 8000 travellers. We have also been able to fraternize with other entrepreneurs in the tourism industry, which has enabled us to enrich our proposal and broaden our outlook through the experiences of our colleagues.

One of the most important takeaways from this typology of creative travellers is that it brings to light the key factors that play into the decision-making process of travellers. Such factors are usually on the minds of creative tourism entrepreneurs, yet not specifically evaluated for product design, marketing, or communication planning purposes.

The tool presented here for the categorization of creative travellers is intended to support entrepreneurs when conceiving or improving their business plans. Since our process was empirical, it took us years of experience to create and consolidate the tool, but those who are just beginning to ideate their own creative tourism portfolio could benefit from a solid starting point, which we hope the tool can provide.

It is worth noting that when we asked our colleagues about their most common types of travellers as well as about the factors they believed were most relevant, their answers were different from the results produced by the matrix used at 5Bogota. This leads us to think that each entrepreneur might find their own answers to the same

PEG

SENIOR TRAVELLER
(ADULTS AGED 46–60)

AGE: **57 YEARS OLD**

PROFESSION: **RETIRED GASTRONOMIC CRITIC**

FROM: **CALIFORNIA, UNITED STATES**

She is a traveller who loves cooking, and makes it the main activity in all her traveis.
What matters most to her during her travels is finding local hosts to have unique experiences in unconventional places, far from the usual tourist circuit.
Peg travels with her husband and their friends at least 3 times a year, and enjoys researching her chosen destination.
She plans her vacations ahead of time, and ensures her itineraries are full of experiences. She is an active person who wants to enjoy the local culture. Peg has visited many countries as her passion for travelling started at a very young age.
She always includes visits to local marketplaces, cooking classes, and pop-up dinners in her itineraries, especially if they have been recommended by friends or relatives. Trying new flavours motivates her to travei and to step out of her comfort zone. She wants to have all the details of her itineraries and travels, though her budget is not a concern.
She is willing to invest in unique experiences that change her life.

Fig. 2.3. Customer profiles of three types of travellers with the most affinity for the creative tourism offer at 5Bogota.

François

YOUNG ADULT TRAVELLER
(ADULTS AGED 30–45)

AGE: **33 YEARS OLD**

PROFESSION: **AIR TRAFFIC CONTROLLER**

FROM: **PARIS, FRANCE**

He is a traveller on a quest for local experiences that allow him to bond with the culture of his chosen destination. He loves trying local flavours and visiting exotic locations. He also tries to learn a bit of the local language and practice it with the hosts.

Due to his stressful job he takes long vacations every year during which he visits more than one country. He travels often with a romantic partner, though other times he is accompanied by relatives.

François does not plan his itineraries, preferring instead to be surprised. Once in the destination, he scouts out the best experiences or those recommended to him by his friends. One of his main motivators to travel is learning new things that open his mind and perspectives to other aspects of the world. For this reason he loves to strike up conversation and ask about the local culture.

Fig. 2.3. Continued.

This travelling family includes a dad, a mom, and two children aged 11 and 12. They love exploring new destinations and countries together. Mom enjoys planning the trip to make sure they get as many activities and experiences in as possible, usually beyond the tourist circuit. The children are used to the long journeys the family takes every year. They love trying new flavours and meeting new people. Whenever they don't understand the language they just ask their parents for a translation because they relish being actively involved in all the experiences. Travelling is a path to knowing new realities and exploring the world for this family. These parents want their children to be open to new experiences and enjoy meeting new people and cultures. Bonding with the hosts is essential for every member of the family. Every one of them has a positive and proactive attitude, and they love getting tips about their chosen destination.

ROTT

FAMILY

FROM: HAMBURG, GERMANY

Fig. 2.3. Continued.

questions, and therefore will be able to adjust their portfolio to better meet the needs of their customers. This also reasserts our intention that the tool should serve as a framework for the conception and design of a personalized portfolio for other companies.

On the other hand, it is fundamental to understand that the tourism industry must take into account people's natural dynamism. External situations, such as the ongoing pandemic, can change or otherwise weigh on travellers' behaviour, and will force us to rethink both the factors that influence their decision-making processes, and their reactions as customers. Matters such as biosafety practices, health indexes, the ability of governments to respond to environmental or sanitary emergencies, and the distance between countries, among others, may become new considerable factors for travellers and their decision-making process.

Finally, while this framework allows for the general categorization of travellers, we must remember that each person is unique, and that the objective of creative tourism is to create bonds between travellers and local hosts through memorable experiences. Creative tourism is not about designing numerous products. Rather, it is about responding to the needs of a traveller who trusts in creative tourism to discover a destination.

3 Creative travellers in Austria

Elena Paschinger

Creative Tourism Austria, Austria

Creative tourism has emerged as a global trend in cultural tourism, replacing the previous notion of cultural tourism as being a mostly passive, attraction- and museum-based activity. Creative travellers are actively seeking out opportunities to become engaged in the local cultural fabric of the place they visit, be it during a cultural festival, an opportunity to meet the locals through hands-on workshops, or by participating in an activity offered at formal institutions, such as cookery schools.

This chapter presents insights about types of creative travellers in Austria. Given the author's background as project manager in the association Kreativ Reisen Österreich (Creative Tourism Austria),[1] the remarks made here are observations based on 10 years of experience in managing, networking, and promoting creative tourism to both national and international travellers. This has been done in co-operation with such high-profile partners as the Austrian National Tourism Board as well as several regional destination management organizations across Austria.

Before addressing Austria as a case study, however, let us look at a more general picture of just who travels in order to become creative. Who, generally speaking, are some of the typical personae involved in creative tourism? In my working experiences and broader travels, I have observed the following types:

- *The passion seeker:* creative travellers actively looking for moments of surprise on their trip. They are ready to become absorbed in the creative experience, and typically have no fear of stepping out of their individual comfort zones.
- *The friends and family type traveller:* this type of traveller cares most about learning new things with and about the people who matter most to them.
- *The urban explorer:* this type of traveller is always on the lookout for new, unusual things about the city they visit. Creative breaks mean a route to escapism of the 'usual' city trip experience.
- *The human adventurer:* this type of traveller loves to meet and mingle with the locals. He or she is eager to learn about the differences and commonalities among all people.
- *The curious solo traveller:* this type of traveller has a huge potential for growth in the (creative tourism) market. The curious solo traveller, typically female, is looking to share a special travel moment with other like-minded individuals. They relish the new community thereby found, often well beyond the actual travel experience.

Another characterization to be made about the nature of creative travellers is whether they experience creative travel individually, or as a group. To this end, I refer you to a chart originally published in the book *The Creative Traveler's Handbook* (Paschinger, 2015), which is presented here as Table 3.1. In it, I address the various steps of the creative planning process for both individuals and group travellers for creative tourism.

The nature of creative travel in Austria

Austria is a country rich in cultural heritage, both tangible and intangible. UNESCO currently has recognized seven cultural practices as intangible cultural heritage of worldwide importance and recognition (UNESCO, 2020), some of which visitors

Corresponding author: reachme@creativelena.com

 Creative Tourism: Activating Cultural Resources and Engaging Creative Travellers (eds. N. Duxbury, S. Albino, and C. Pato de Carvalho)
DOI: 10.1079/9781789243536.0003

Table 3.1. The creative planning process for both individuals and group travellers for creative tourism. Adapted from Paschinger (2015)

Process	Creative solo travellers	Creative group travellers
Planning	Plan their trip when, where, and for how long they wish, depending on the desired intensity of the experience sought Focus on their specific interest Avoid major vacation periods or high season travel Take into account festival dates or creative events at the destination Maximize the flexibility they have	Take into account the availability, ability, and interests of the people they are travelling with Sign up for a creative travel experience that as a group can facilitate a flexible, sometimes even exclusive course date. Most tutors are happy to look into that option
Budgeting	Creative solo travellers join like-minded groups as it helps them reduce the cost of individual travel They typically ask accommodation or tourism staff or a local travel guide as to which interactive cultural experiences still have an open spot	Creative group travellers may easily cash in on group discounts They can spread individual travel costs among the group, such as hiring a car or sharing an apartment They may also maximize age or other discounts available to them
Packing	Packing for solo travellers is usually done lightly	Travel items may be more easily spread among the group
Creating	The ultimate bonus for creative solo travellers: In the creative travel activity, they are virtually guaranteed to meet travellers with the same values and interests, enriching their individual experience In this way, the shared experience is a wonderful opportunity for inspiration that solo travellers would not normally have on their own	On the road, creative group travellers feed off the inspiration from their peers and other individuals present in the creative exchange They can more easily overcome their perceived or actual level of inhibition by following trusted members of their group

may experience in the form of creative workshops and activities across Austria. Among them are:

- 'Blaudruck' block printing and indigo dyeing in Burgenland, a European cultural practice on the verge of disappearance;
- regional centres for craftsmanship, particularly in the Alpine regions of Austria;
- classical horsemanship and falconry; and
- the 'Schemenlaufen' carnival experience in Imst, Tirol.

The 'Blaudruck' process of indigo dyeing, as an example, may still be experienced in southern Burgenland, open to the public in the form of small workshops at the original place of manufacturing, yet only after previous enquiry and registration.[2]

Other types of craft experiences include restoring old furniture, wood carving and turning, and iconography workshops at monasteries, among many more. It is important to note that these activities are at different levels of formal organization and communication to the public. Many run their workshops with merely artistic goals and are poorly set up to accommodate international visitors who do not speak the local language.

On the other hand regional funds such as LEADER issued in co-operation with provincial governments as well as networking and knowledge exchange meetings have often helped stimulate the economic process of getting small-scale artisans off the ground in order to become ready to embrace creative tourism as a new market opportunity, both nationally and internationally.

Besides craft experiences, Austria has long focused on a strong culinary heritage and tradition for its international marketing activities. This, too, is reflected in the nature of Creative Tourism Austria, grouping many different gastronomic experiences such as wine tastings, harvest experiences, beer brewing workshops, cooking experiences, and cake decoration workshops, among many more.

Creative travel in Austria is equally attractive to national and international visitors, although for different motives. Whereas national visitors are typically attracted to creative experiences as repeat visitors to certain places ('I have visited place X more than once and would now like to experience it in a new light, at another time in the year, etc.'), international travellers are still largely attracted by activities centred around cultural clichés (e.g. learning how to bake your own apple strudel, participating in a Waltz dance lesson, renting a Dirndl whilst exploring Austria's mountain areas, etc.). To them, creative travel offers a more intimate, personal experience over other types of travel activities.

Creative travellers in Austria

According to an interview with Klaudia Zortea, managing director of the Altenmarkt-Zauchensee regional tourism board in the province of Salzburg and founder of the Natura.Kreativ creative summer festival,[3]

> ... the creative tourist to our mountain destination during the summer is all of the following: He or she travels alone, as a couple, as a family or in a group, either individually or through the organization we provide. Some also come following a long-standing bond they have with our local accommodation providers. Creative activities we provide, such as courses on how to yodel, how to bake your own bread, how to carve your own sculpture, are the primary motivation for the trip. They either want to perfect practices that they have already carried out at home or use their holiday in order to approach new experiences. What we have observed is that they are not looking for 'monumentalism', but rather something more subtle, a local gesture, the warmth of a smile, the vibe of the group. They are very selective, too: Once they have experienced creative tourism in our destination, many come back the next year and the year after, bringing more family members and friends with them. I have a feeling they can no longer adapt to conventional or standard tourism. Also, they have a significant budget to spend for the experience they are interested in, and often like to combine different types of tourism during their stay: creative, gastronomic, eco-tourism, slow tourism ...

Key points that may be drawn from this statement reflecting on several years of the Natura.Kreativ taking place in the province of Salzburg, and that are likened to the general nature of creative travellers across Austria, include:

- The socio-demographic characteristics of creative travellers in Austria are varied. They are not just the typical 50+ female special-interest traveller, but also family travellers, (young) couples, and groups bound together by shared interests, such as singing, painting, or wood carving.
- Many travellers value personal service and long-standing traditions, that is, choosing the same accommodation provider for when they make a repeat or return visit.
- Workshops and courses that tie in with the local cultural fabric are among the most popular, but not only that: workshops that tie in with the mega trends in society, such as health, individual pursuits, work–life balance, etc., are also very popular (see, for instance, the popularity of the workshop on how to make your own natural pharmacy or learn how to bake your own bread).
- Creative travellers in Austria come with high expectations, but also a high level of disposable income – and the willingness to spend it on the activities they love.
- They are also driven by other travel styles similar to creative tourism, such as slow tourism, food tourism, and (mountain) eco-tourism.
- Certain creative tourism providers attract a large volume of international travellers, for example, the wood carving school in Elbigenalp, Tyrol, which is very popular with US visitors. Others, in contrast, offer activities that are especially popular among local and regional creative travellers, such as the many creative ceramic courses offered by the Kunstmuseum Schrems in the Lower Austrian Waldviertel region.

Conclusions

Summing up, we can observe a growing trend for creative tourism in a country as rich and diverse in cultural heritage and tradition as are the nine provinces of the Alpine republic. Especially in the light of the current crisis situation caused by the Coronavirus pandemic, innovation and creativity are considered key skills for established tourism products and destinations to survive. The existing potential and know-how of local craftspeople is something that Austria has already very cleverly started to market, both nationally and internationally. This is, in part, due to the fact that the tourism industry is at the forefront of the national economy. Among the many efforts in regional, national, and international destination marketing, creative tourism has helped to sustain many (alpine) traditions in Austria that might otherwise have already disappeared. Creative tourism, as a trend, is set to grow even further, attracting an equally growing market segment of creative travellers to and within Austria. Furthermore, creative tourism overlaps with some of the major trends in society, such as sustainability, social connectivity, and physical and emotional health – potentially fuelling new iterations of creative tourism that intersect with these other interest areas.

Notes

1. Creative Tourism Austria was launched in 2011. For further information, visit www.kreativreisen.at (accessed 27 April 2020).
2. See http://www.originalblaudruck.at (accessed 27 April 2020) for further information.
3. The interview was conducted in August 2014. For further information, see: www.altenmarkt-zauchensee.at (accessed 27 April 2020)

References

Creative Tourism Austria (n.d.) Available at: http://www.kreativreisen.at (accessed 27 April 2020)

Paschinger, E. (2015) *The Creative Traveler's Handbook: The Art of World Travel—Life-seeing instead of Sightseeing*. Full Flight Press, Toronto, Canada.

UNESCO (2020) Austria and the 2003 Convention. Available at: https://ich.unesco.org/en/state/austria-AT (accessed 23 April 2020).

4 Domestic and international creative tourists in Portugal: Insights for practitioners

Jaime Serra,[a*] Noémi Marujo,[a] Nancy Duxbury,[b] Alexandra R. Gonçalves,[c] Sónia Moreira Cabeça,[c] Paula Remoaldo,[d] and Juliana Alves[d]

[a]*Interdisciplinary Centre for History, Culture and Societies (CIDEHUS), University of Évora, Portugal;* [b]*Centre for Social Studies (CES), University of Coimbra, Portugal;* [c]*Research Center for Tourism Sustainability and Well-being (CinTurs), University of Algarve, Portugal;* [d]*Landscape, Heritage and Territory Laboratory (Lab2PT), University of Minho, Portugal*

A great diversity of definitions of creative tourists exist, ranging from those who refer to visitors of dance, art, or handicraft workshops, to those who include people who take up temporary artistic residences to practice their creative expression and develop their art forms. In recent decades, we have observed the emergence of a new generation of travellers.[1] These tourists are increasingly seeking co-creation processes, leading to more relational forms of cultural tourism, and active participation in creative experiences (Richards, 2020).

And yet, pinning down the diverse, niche-oriented creative tourist has been an ongoing challenge. Internationally, a number of studies have been conducted to profile the creative tourist (see Remoaldo *et al.*, 2020) but no overarching comparative framework has yet been developed, and in Portugal no study on the creative tourist had previously been conducted. Within the project CREATOUR® (Creative Tourism Destination Development in Small Cities and Rural Areas), we developed a detailed questionnaire for participants in the creative tourism pilot activities organized within the project, which was applied by the 40 participating organizations as they conducted their pilot activities in 2017, 2018, and 2019.

These activities were developed and situated in small cities and rural areas in the Norte, Centro, Alentejo, and Algarve regions of the Portugal mainland. They ranged from small-scale participatory cultural festivals to gastronomy workshops to handicraft, mosaic-making, and other hands-on workshops. The common thread was an aspiration to develop creative tourism activities adhering to the CREATOUR® approach to creative tourism, which incorporates active participation, learning, opportunities for creative self-expression, and connections to the local community.

The questionnaire enabled us to gain insights on socio-demographic characteristics, motivations, behaviours, experiences, and perceptions of the activities – providing a rich source of insights on the creative tourist in Portugal. In this chapter, we present highlights of these questionnaire results, segmented by place of residence (i.e. domestic or international visitors) and suggest some implications of these findings. A motivational analysis of creative tourist participants can be found in Remoaldo *et al.* (2020). Based on socio-demographic, travel behaviour, and motivation-based criteria, three clusters were found: novelty seekers, knowledge and skills learners, and leisure creative seekers.

*Corresponding author: jserra@uevora.pt

 Creative Tourism: Activating Cultural Resources and Engaging Creative Travellers (eds. N. Duxbury, S. Albino, and C. Pato de Carvalho)
DOI: 10.1079/9781789243536.0004

Developing a profile of creative tourists in Portugal

Questionnaire development and setting

In 2017, a detailed questionnaire was developed by CREATOUR® researchers based on the key dimensions in research on tourists and a literature review of creative tourism, tourist motivations, and cultural tourist profiles. The questionnaire consisted of 31 closed questions oriented to the participant's profile, motivations, perceptions, and evaluation of activities, as well as the impacts on the local economy (e.g. accommodation, meals, and local commerce). It was prepared in English, Portuguese, Spanish, and French, and distributed and applied by the 40 participating organizations as they conducted their creative tourism pilot activities in 2017, 2018, and 2019. A total of 814 individuals completed these questionnaires after they participated in one of the CREATOUR® creative tourism pilot activities. Completed questionnaires were returned to the in-region research centre involved in CREATOUR® to be entered into a central database and subsequently analysed.

Data analysis procedure

To analyse the data, a descriptive analysis was adopted, divided into two groups of analysis: (i) socio-demographic profile and (ii) travel behaviour and motivations. The criteria 'place of residence' was used to split the data into two groups in order to compare international and domestic participants. In order to identify differences among the two groups, T-tests were adopted to validate the statistical differences.

Who are the creative tourists in Portugal?

Both domestic and foreign tourists participated in the CREATOUR® creative tourism pilot activities and completed the questionnaires. Overall, 75% of the questionnaires were completed by domestic travellers. Both groups included all age segments, from children through to seniors (65 years and over) (see Table 4.1). In both groups, the largest age segment was 35–54 years. Domestic (Portuguese) creative tourists included most of the children reported, likely indicating families (see also the interview with Wolfram in Chapter 6). Overall,

Table 4.1. Sociodemographic profile of domestic and international creative tourists in Portugal. Source: CREATOUR (2019)

Variables	Categories	Domestic (%)	International (%)
Age	<12 years old	8.4	0.5
	13–19 years	7.4	7.4
	20–34 years	21.2	27.5
	35–54 years	**42.6**	**37.3**
	55–64 years	11.7	19.1
	≥65 years	8.6	8.3
Marital status	Single	**53.8**	**45.6**
	Married	36.2	44.0
	Divorced	7.4	10.4
	Widow	2.7	0.0
Education	Primary education (up to 6 years of schooling)	10.6	1.5
	Secondary education (up to 9 years)	10.3	3.0
	Tertiary education (up to 12 years)	8.4	8.5
	Professional training (up to 12 years)	9.2	12.0
	Bachelor's degree	**34.1**	**39.0**
	Post-graduation/master's degree	22.4	27.0
	Doctoral degree	4.8	8.5
	Other	0.1	0.5
Socio-professional situation	Employed	**62.2**	**67.5**
	Student	20.4	17.0
	Retired	9.8	9.0
	Unemployed	5.9	4.5
	Domestic	1.7	2.0

women accounted for about two-thirds of all participants (Remoaldo *et al.*, 2020). About 61% of domestic tourists and 75% of international participants held a university or post-graduate degree. About two-thirds of both groups were employed, with students representing the next largest group (17–20%), then retired individuals (9–10%).

Most participants were accompanied, with friend(s) and spouse/partner most commonly mentioned (see Fig. 4.1). Children formed part of the participating group for 21% of domestic participants and 26% of international participants. These findings suggest that the designer of creative tourism activities should have in mind how to attract and facilitate the participation of small groups, and – if it can be accommodated – the involvement of children. Solitary travellers made up only 20% of domestic participants and 16% of international participants.

Awareness and decision to participate: How did they find out? What are they looking for?

Participants became aware of the activity/experience mainly through friends or family, and secondarily through social networks (see Fig. 4.2). For 17.5% of the international visitors, they became aware of the experience when they were passing by the site; this was the case for only 8.4% of the domestic participants.

Based on a selection of possible reasons listed in the questionnaire, the highest-ranked reasons for choosing a creative tourism activity, for both domestic and international travellers, was that it is 'culturally motivating' and 'original' (Table 4.2). There are statistically significant differences by place of residence, for the reasons 'For being fun', 'For allowing to know and interact with the local community', and 'For knowing the promoter of the activity'. The 'fun' factor was more important for the international travellers, while interaction with the local community was more important for domestic travellers. The (less important) 'knowing the promoter' factor is understandably more likely to be relevant for domestic rather than international tourists.

Characterizing and evaluating the experience

The questionnaire included a list of emotions to describe the creative tourism experience (based on

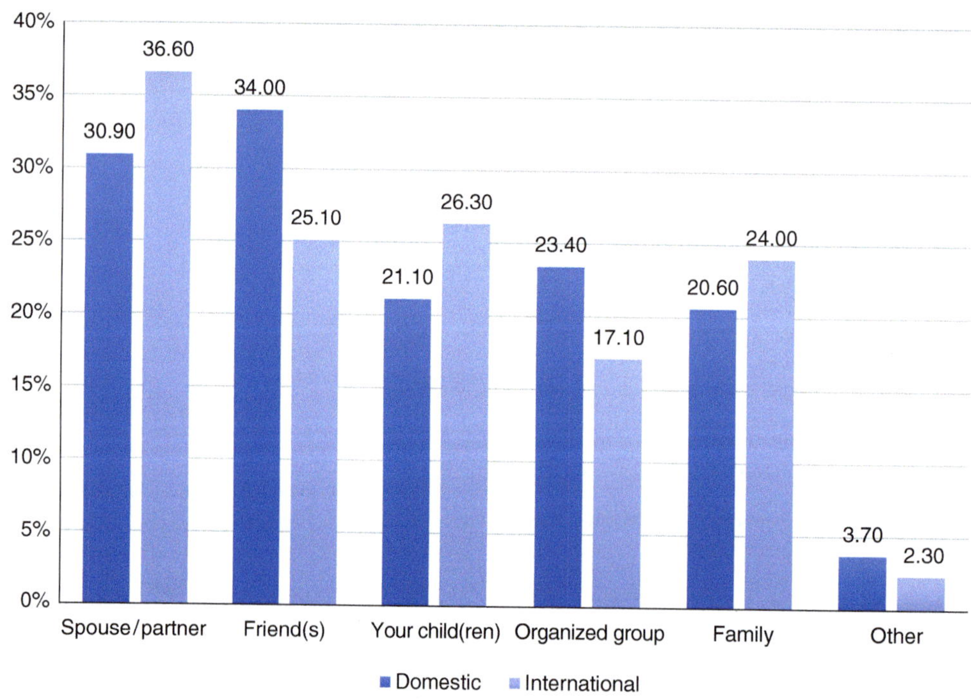

Fig. 4.1. Travelling companions of domestic and international creative tourists in Portugal (%).

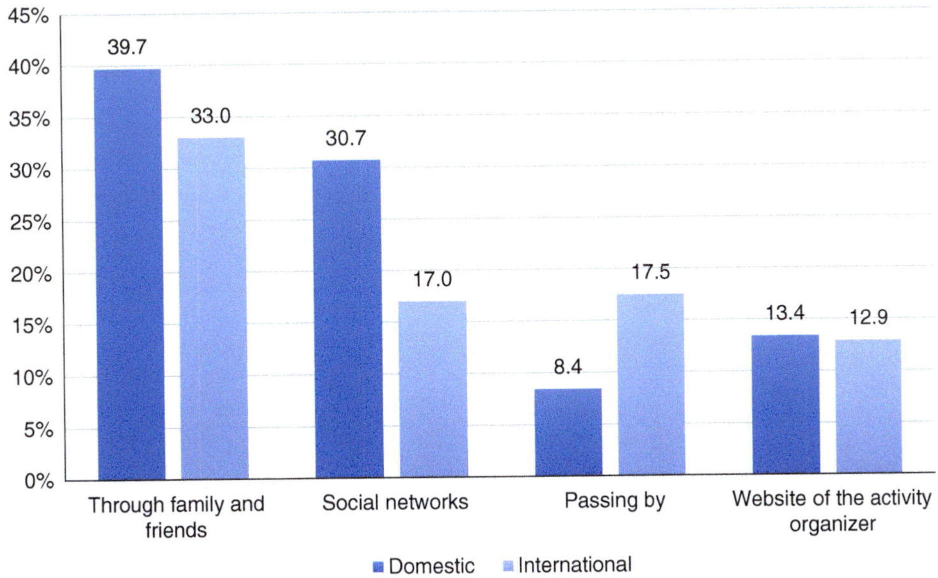

Fig. 4.2. How did you learn about the experience?

Table 4.2. Reasons to choose a tourism creative activity, by place of residence. Source: CREATOUR (2019)

	Domestic	International	T-test	
	Mean	Mean	Value	*p-value*
It is culturally motivating	**4.7**	**4.7**	0.153	0.878
It is original	**4.5**	**4.5**	−0.959	0.338
For being fun	4.3	4.4	−2.760	**0.006**
To stimulate my creativity	4.3	4.3	−0.429	0.668
It enables me to meet and interact with the local community	4.3	4.1	2.315	**0.021**
It permitted interaction with other participants	4.1	4.1	0.340	0.734
It is suitable for the whole family	4.0	3.9	0.868	0.386
Because of its location	3.9	3.9	−0.036	0.972
To accompany someone	3.2	3.3	−0.683	0.495
I know the promoter of the activity	2.7	2.4	2.093	**0.037**

*significant at the 5% level
**likert scale: 1 - disagree completely, 5 - agree completely

a review of relevant literature). For both groups of tourists, all positive emotions (i.e. creative, enriching, original, stimulating, memorable, absorbent) attained a high average level of agreement, while negative emotions (i.e. tired, frustrating, annoyed) had a low average level of agreement (see Table 4.3). Only the characteristics 'original' and 'absorbent' reveal statistically significant differences, the former being more relevant for international tourists and the latter for domestic tourists.

Regarding the participants' evaluation of the experience, the ordering of the array of statements on the questionnaire was similar for both domestic and international participants, with 'I had fun', 'I met interesting people', and 'I tried a new activity' at the top. Only the statement 'I contributed to the local community' revealed a lower level of agreement; while still positive (3.6/5), this indicates that the design of the pilot creative tourism activities did not emphasize this dimension. Only the statement 'I met interesting people' shows a statistically significant difference between the two groups, with this characteristic of the experience being more valued by international tourists (Table 4.3).

Table 4.3. Characterizing and evaluating the creative tourism experience, by place of residence. Source: CREATOUR (2019)

	Domestic	International	T-test	
	Mean	Mean	Value	p-value
Characterization of the experience				
Creative	**4.6**	4.6	−1.595	0.112
Enriching	**4.6**	4.5	0.841	0.401
Original	4.5	**4.7**	−2.530	**0.012**
Stimulating	4.4	4.5	−0.193	0.847
Memorable	4.3	4.4	−1.808	0.071
Absorbing	4.2	4.0	2.300	**0.022**
Emotive	4.0	4.1	−1.261	0.208
Tiring	1.3	1.3	−0.349	0.727
Frustrating	1.2	1.3	−1.147	0.252
Annoying	1.1	1.2	−0.946	0.345
Evaluation of the experience				
I had fun	**4.6**	**4.7**	−1.312	0.190
I met interesting people	4.3	4.5	−2.986	**0.003**
I tried a new activity	4.3	4.4	−1.344	0.179
I learned more about the local culture	4.2	4.2	−0.514	0.607
I learned to do something	4.2	4.1	1.214	0.225
I interacted with the local community	4.0	4.0	−0.453	0.651
I acquired new skills	4.0	3.9	1.303	0.193
I contributed to the local community	3.6	3.6	0.217	0.828

*significant at the 5% level
**likert scale: 1 - disagree completely, 5 - agree completely

These two scales may be useful in terms of structuring more in-depth discussions with visitor participants about their creative tourism experiences. Creative tourism activities can provide, on the one hand, individual-focused experiential take-aways and skills development opportunities and, on the other hand, social interactions among the visitor participants as well as with the local community. Furthermore, the activities may be structured to allow visitors to contribute to the local community in an explicit way. Realizing each dimension requires strategic planning and careful implementation.

Creative tourism attracts travellers, with local economic benefits – and they can become serial participants

We found that creative tourism activities can attract travellers to a location: both domestic (76%) and international tourists (64%) indicated that the primary reason for visiting the locale was to participate in the creative tourism experience. This was an important finding for us given that the locations were small cities and rural areas – and some were quite remote.

The questionnaire findings also revealed that these visitors generated wider economic impacts in the locale. Almost 70% of domestic participants were staying in the creative tourism locale overnight, for an average of about three nights, with 42% of these staying in a hotel or similar establishment (see Table 4.4). Although only 32% of international participants stayed overnight, they stayed for a longer period of time, on average five nights. In keeping with the longer period of staying in the locale, international participants were much more likely to indicate having meals in the locale, reporting about twice as many meals compared to domestic travellers. In regard to other local purchases, domestic participants were more likely to purchase something from a local store (68%) than were international participants (42%).

For most participants (68.7% Portuguese and 72.6% international tourists), the activity was their first experience with creative tourism. For those who had participated in creative tourism previously, Portuguese tourists indicated a higher number of experiences (on average) than the international tourists (2.42 versus 1.67).

The majority of the participating tourists (over 92%) indicated they were interested in repeating, recommending, or participating in more activities with the same organization. Over 85% of participating tourists indicated they would be interested in participating in creative tourism activities elsewhere in Portugal (see Table 4.5). These findings suggest the potential for repeat visitors – and the need to continually evolve activities to re-attract their interest – as well as the value of operating in a network where travellers' participation in a creative tourism activity at one location can lead to participation in other activities elsewhere. This potentially virtuous cycle can strengthen awareness and connectedness among organizers of creative tourism activities in smaller places and rural areas.

A few closing remarks

The results of this initial research on the creative tourist in Portugal revealed variety in the profile of individuals participating in the CREATOUR® pilot project activities, providing data on their demographic, socio-economic, behaviour, and motivational characteristics, and thus can support efforts of market segmentation and product design. Marketing strategies are needed in order to match specific creative tourism activities with the needs and the expectations of target participants (see Chapter 19 by Delisle). This research is a starting point for further analysis of the profile of participants for specific types of creative activities to enable organizers to adapt their creative tourism activities to the motivations and preferences of particular participant groups.

Based on this initial research, we can say that creative tourism has an enormous potential outside large city centres because it can successfully attract a great diversity of people and enable new and

Table 4.4. Accommodation, meals, and shopping behaviour by place of residence. Source: CREATOUR (2019)

	Domestic	International
Did you stay or will you stay (overnight)? Yes:	**69.5%**	31.5%
If yes, how many nights? (mean)	3.32	5.23
Where did you stay?		
Hotel or similar establishments	42.2%	35.1%
Accommodation provided free by family/friends	19.6%	20.6%
Other private accommodation	16.1%	21.4%
House rented	15.1%	19.1%
Second residence (includes own home)	8.0%	5.3%
Room rented in private home	2.0%	4.6%
Outside this activity, did you have any meals on site? Yes:	45.9%	**73.4%**
If yes, how many meals: (mean)	2.25	4.43
Did you shop at a local store? Yes:	**67.7%**	41.9%
If yes, type of product:		
Other local products	48.9%	50.5%
Artisan crafts	33.9%	41.2%
Souvenirs	23.0%	37.1%
Guides	15.5%	6.2%

Table 4.5. Degree of interest to participate in creative tourism activities in future or to recommend, by place of residence. Source: CREATOUR (2019)

	Domestic		International	
Degree of interest to ...	% less interest	% higher interest	% less interest	% higher interest
Repeat the experience	2.1%	**92.5%**	1.9%	**93.2%**
Recommend the experience	1.1%	**94.0%**	1.0%	**98.6%**
To participate in more activities of this organization	1.4%	**93.7%**	0.5%	**90.6%**
Would you be interested in participating in creative tourism activities in other locations in Portugal?				
Yes	**87.1%**		**85.5%**	

different activities to be developed based on endogenous resources and local ingenuity. But it is challenging and requires better co-operation between a range of different stakeholders so that organizers can successfully reach and attract interested travellers. Importantly, providing meaningful connections with the local community, which the 'new traveller' craves, requires local co-operation, strategic planning, and an ethos and practice of collaboration (Bakas *et al.*, 2019).

Caution should be applied when adopting the results of this research, however, due to the non-random sample and the data collection process, which was not balanced among the regions where the survey was applied nor adjusted for the type of creative tourism activity involved. In the current analysis, we analysed all the questionnaires that were completed by participants in the CREATOUR® pilot projects. Some organizations carried out more activities than others and provided a greater number of completed questionnaires than others. Also, activities had varying numbers of participants, with some activities designed for smaller numbers of visitors while others were open to larger groups. A further limitation is concerned with the non-inclusion of seasonal effects in this analysis. Further analysis using this dataset can examine the participants of different event types more closely and segment the activities by the time of year they were organized.

Take-aways for practitioners

1. The characteristics of creative tourism participants vary widely, but they usually travel accompanied, with one in four or five participants accompanied by their children. Give attention to the design of the activities to enable these small groups to interact with one another (as well as others), and for the possibility of engaging and involving children in the activities.
2. Creative tourism activities are not yet widely experienced, and activity organizers should plan for 'first time' creative tourism participants.
3. Creative tourism activities provide opportunities for both individual-focused experiences and skills development, as well as social interactions among visitor participants and with the local community. Furthermore, the activities may be structured to allow visitors to contribute to the local community in an explicit way. To realize each of these dimensions, strategic planning and careful implementation are required.
4. Creative tourism activities can attract both domestic and international travellers to small cities and rural locations. In the current COVID-19 era, attention to domestic tourism is rising, and creative tourism activities can be an engaging component of packages to attract domestic travellers to extra-metropolitan locales.
5. Organizers should use a diverse range of channels of communication to publicise their activities. They also need to establish partnerships with local entities and institutions (tourist offices, local accommodation, etc.) so that tourists are more likely to encounter information about the activities.
6. Creative tourism activities can be combined with hotel and other experiences to create packages that may keep travellers in a locale longer. For example, strategically inserting creative tourism activities within the programming of established cultural events is a good strategy to consolidate the creative activities within local promotional and organizational networks.
7. Tourists are not generally willing to repeat the same activity, but they can be lured with related or evolved offers in areas in which they are passionate. They also want to try other creative tourism activities. Our findings suggest the potential for repeat visitors – and the need to continually evolve activities to re-attract their interest.
8. Operating in a network may create collective benefits for multiple agents and locales, where travellers' participation in a creative tourism activity at one location can lead to participation in other activities elsewhere. This potentially virtuous cycle can strengthen awareness and connectedness among organizers of creative tourism activities in smaller places and rural areas.

Acknowledgements

The chapter is an outcome of research conducted within the CREATOUR® project (project 16437), which is funded by the Portuguese Foundation for Science and Technology (FCT/MEC) through national funds and co-funded by FEDER through the Joint Activities Programme of COMPETE 2020 and the Regional Operational Programmes of Lisbon and Algarve. The authors also wish to acknowledge a grant from the FCT – National Funding Agency for Science, Research and Technology, under the project UIDB/00057/2020 – CIDEHUS.

Note

1. These 'new' tourists have been called 'creative consumers' (Richards, 2009; Richards and Wilson, 2006; Pappalepore *et al.*, 2014; Zhang and Yu, 2018), creative tourists (Raymond, 2003; Tan *et al.*, 2014; Smith, 2016; Zhang and Yu, 2018; Remoaldo *et al.*, 2020), and postmodern travellers (O'Dell, 2007; Tan *et al.*, 2014; Pappalepore *et al.*, 2014).

References

Bakas, F.E., Duxbury, N., Remoaldo, P.C., and Matos, O. (2019) The social utility of small-scale art festivals with creative tourism in Portugal. *International Journal of Event and Festival Management* 10(3), 248–266. doi: 10.1108/IJEFM-02-2019-0009.

CREATOUR (2019) Análise dos inquéritos aos turistas. Relatório não publicado. Available upon request to creatour@ces.uc.pt.

O'Dell, T. (2007) Tourist experiences and academic junctures. *Scandinavian Journal of Hospitality and Tourism* 7(1), 34–45. doi:10.1080/15022250701224001.

Pappalepore, I., Maitland, R., and Smith, A. (2014) Prosuming creative urban areas: Evidence from East London. *Annals of Tourism Research* 44, 227–240.

Raymond, C. (2003) *Cultural Renewal + Tourism: Case Study – Creative Tourism*. New Zealand. Available at: www.creativenz.govt.nz/assets/paperclip/publication_documents/documents/97/original/casestudy-creative-tourism-new-zealand.pdf?1322079829 (accessed 29 January 2019).

Remoaldo, P., Serra, J., Marujo, N., Alves, J., Gonçalves, A., Cabeça, S., and Duxbury, N. (2020) Profiling the participants in creative tourism activities: Case studies from small and medium-sized cities and rural areas of continental Portugal. *Tourism Management Perspectives* 36, 100746, 1–12. doi: 10.1016/j.tmp.2020.100746.

Richards, G. (2009) Creative tourism and local development. In: Wurzburger, R., Aageson, T., Pattakos, A. and Pratt, S. (eds.) *Creative Tourism: A Global Conversation*. SunstonePress, Santa Fe, New Mexico, USA, pp. 70–90.

Richards, G. (2020) What is creative tourism? A retrospective view. In Gonçalves, A.R., Marques, J.F., Tavares, M. and Cabeça, S.M. (eds.) *Creative Tourism: The CREATOUR Recipe Book*. CinTurs Research Centre, University of Algarve, Faro, Portugal, pp. 39–10.

Richards, G. and Wilson, J. (2006) Developing creativity in tourist experiences: A solution to the serial reproduction of culture? *Tourism Management* 27, 1209–1223.

Smith, M.K. (2016) *Issues in Cultural Tourism Studies* (3rd ed.). Routledge, New York, New York, USA.

Tan, S.-K., Luh, D.-B., and Kung, S.-F. (2014) A taxonomy of creative tourists in creative tourism. *Tourism Management* 42, 248–259.

Zhang, Y. and Yu, X. (2018) Urban tourism and the politic of creative class: A study of the chefs in Macao. *International Journal of Tourism Sciences* 18(2), 139–151.

5 Collective invention: A travelling artist's perspective

MARGARET FEENEY

Māpura Studios, New Zealand

People travel because they are searching. They are not content to download experience through mediated internet platforms, or to find themselves curated into a vat of preserved content, steeped in academic salt. Travelling is free of the dreary imperatives of everyday life – we step into the world to enlarge our frame of reference, not to be poked into another grid of curated meaning. Tourist hotspots everywhere have been heating up as fast as the climate, and unique cultural sensibilities and practices are perishing in the glare of the demands of the modern tourist. Busy with the business of experience, I believe many tourists are missing the very thing they are searching for. Experiences are preformed and performed, according to commercial expectations. Yet the commercial imperative is a blunt instrument when it comes to negotiating culture, whereas art is a complex, living initiative that adapts to need, capability, and circumstance and is better suited to lead meaningful engagement between communities and their tourists. The money can follow us.

I am a visual artist with a commitment to collective invention that intertwines my fine art practice with my community art practice. I work from my studio in Auckland, Aotearoa New Zealand (ANZ), and travel to work on projects with unique and remote workshop groups. In this chapter, two of these projects, *The Stencil Project* and *Water Systems,* inspire my thinking on creative tourism. These projects highlight the creative and social strength of collaboration and diversity, and demonstrate how the creative tourism model can unify communities and their visitors by supporting artists to share their ideas. I have included creative practitioners of all fields – artisans, crafts people, and artists – under the rubric of artist for this discussion.

I have chosen to discuss *Water Systems* and *The Stencil Project* because climate change and artificial intelligence are two of the greatest threats we face.[1] They are so serious they can appear overpowering, and often trigger disenchantment and disengagement. But through creative community collaboration, we can approach these threatening realities together in a warm and inventive way. And through the collective invention generated we feel less powerless and may even find solutions! Furthermore, co-ordinated creative tourism initiatives generate international networks that encourage extended sharing and discussion of the ideas that are important to people. I seek out unique community groups to provoke discussion and discover fresh thinking and solutions to these issues.

Water Systems[2]

Water Systems is concerned with ecology, and brings people together to make installations of small, connected sculptures to provide water for birds and insects during times of drought. The core *kaupapa*[3] is inclusion and invention and through this *mahi*[4] people are brought closer to each other and to the environment and its creatures. The project aims to form a point of mediation and meditation between the natural world and the human-built world.

The *Water Systems* workshops bring diverse clusters of people together, giving them an opportunity for direct creative action: sharing ideas and inspiring each other. Artists, non-artists, locals, and travellers all work together – grandparents, parents,

Corresponding author: margaret@mapurastudios.org.nz

 Creative Tourism: Activating Cultural Resources and Engaging Creative Travellers (eds. N. Duxbury, S. Albino, and C. Pato de Carvalho)
DOI: 10.1079/9781789243536.0005

children, cool young people, and seasoned travellers all hunch over worktables making tiny sculptural water vessels for birds and insects in the dry season. They discuss form and function, and the past and the future, in a multi-media exploration of humans and the natural world (Fig. 5.1).

The Stencil Project[5]

In 2018 I enjoyed a residency at the Buinho Creative Hub in Messejana, Portugal, run by artist-academics Sara Albino and Carlos Alcobia. During my stay I developed *The Stencil Project*, which I then shared with communities back in ANZ. *The Stencil Project* is an investigation of artificial intelligence. The project looks at how people's thinking and feeling is mediated by algorithms through digital platforms. It is an investigation of the expanding influence of form over content in the human-built world. I use pattern as the main investigative tool for this project, as pattern recognition is central to machine learning and to human learning (something we have in common). Stencils are a good tool to explore artificial intelligence because the core design has to follow the rules of another binary system: positive and negative space. The stencils form a template that directs the marks made by the user in the same way our thoughts are directed by algorithms.

While I was absorbed at my Buinho workstation designing the cold, analytic patterns of Deep Learning, I was surrounded by the vivacious reality of humanness at the residency: with Sara, Carlos, and Jaime; the other Buinho artists; and the townsfolk of Messejana all around me. During my time there I interacted with artists, non-artists, and children who, through the workshops, gave the project the human response to pattern making which was warm and complex. The challenges posed by language made the workshop process dynamic and unpredictable which, as intrinsic human properties, suited my project well. On my return to ANZ I ran more workshops modelled on the Buinho workshops and gathered more community designs for the stencils.

The contrast between artificial intelligence and humanness was highlighted at the subsequent exhibition.[6] I displayed the aloof, invisible face of artificial intelligence (machine-cut transparent stencils from my Buinho designs) formally on the gallery wall. The hand-cut workshop stencils and a multitude of coloured pencils were set up at an interactive station for the public to play with. Tangential ideas in the form of drawings and writings sprang inspired from the workshop stencils and breached the designated space for interaction. By the end of the exhibition, all four walls displayed the collective vitality of the visitors to the gallery. People are creative; they just want the chance to be involved (Fig. 5.2).

Community collaboration and the artist

Both *Water Systems* and *The Stencil Project* use collaborative workshops and emphasize experimentation.

Fig. 5.1. Māngere Arts Centre – Nga Tohu o Uenuku *Water Systems* workshop, 2020. Photo by Grace Thresher, used with permission.

Fig. 5.2. *The Stencil Project*, Waiheke Art Gallery, after, 2018. Photo by Margaret Feeney, used with permission.

We start with spontaneous, volatile materials such as charcoal, ink, assemblage items, and clay. Because we use warm and accessible materials, with unpredictable properties, the participants cannot control the outcome, so they relax and shift their focus from achieving a product to enjoying the process. As a result, they engage directly with the ideas as they are manifested in the materials. The group is supported by the collective energy generated by working closely together which, in turn, supports a culture of experimentation and adventure. This acts as an analogy of the difference between cultural tourism and creative tourism, the difference between viewing a product and being involved in the creative process of making.

Both projects also embrace diversity. Groups of people from disparate demographic clusters meet up and focus on the ideas through the workshop *mahi*. Established research has shown that higher rates of diversity and a flat structure engender higher rates of innovation (Kastelle, 2013; Gotkin, 2012). Greater numbers of clusters of tangentially connected and diverse content in a system allow for more connections, and more surprising connections, to be made. Scientists now apply computational network tools to look at the semantic memory network of creative people because their minds are more associative than others (Kaufman, 2014). This is true of any system, and when applied to groups of people it allows for greater intellectual surface area.

These properties – experimentation and diversity – make community collaboration a wonderful research tool for the artist. The artist can test and extend their ideas, and share them with other people in a more direct and dynamic forum than that of the more detached space of the gallery exhibition. Community collaboration also gives the artist direct contact with the stakeholders and consumers of art and culture, enriching their lives in the process.

In 2019, I was the creative lead of a collaborative initiative for the Waiheke Community Art Gallery – Te Whare Taonga o Waiheke to make an interactive installation with the community and visitors to 'Sculpture on the Gulf 2019' as a tangential part of my ongoing project *Water Systems*. I worked with disparate groups on Waiheke Island, unifying them with our common purpose. After establishing the *kaupapa* and some basic design parameters I opened the initiative up to the participants and encouraged them to lead from their own diverse fields of expertise: weaving, construction, and pottery – or to be brave and experiment if they had never tried art before! Every item made was included in the final installation, whether made by a professional artist or a complete novice. This made the overall work warm and sweet, with a lovely collective energy (Fig. 5.3).

Fig. 5.3. Waiheke Community Art Gallery *Water Systems* installation, 2019. Photo by Margaret Feeney, used with permission.

While the Waiheke project was concept-led, the dynamic was completely different with the Buinho Creative Hub project in Messejana, where I worked with the children of Técnicas de Acção Educativa (Fig. 5.4). Because of the language barrier, I used the materials to tell the story of my idea. To develop the designs for the artificial intelligence stencils, we first used charcoal and ink to explore humanness, as charcoal and ink are so unpredictable and exuberant. To explore machines, we began a second work using random, pre-cut stickers of generic forms. The stickers were machine cut with hard edges, hard colours, and tight design parameters. The children layered the stickers into more and more complex arrangements and I introduced vividly coloured gouache paints to support the creative path they had taken. I had expected the 'human' images to be rich and complex but thought the children would respond to the coldness of the machine-cut stickers and make more rigid compositions, reflecting the coldness of artificial intelligence. But by repetition the sticker images achieved an anarchic level of complexity equal to that of the 'human' composition. This made sense in terms of philosopher Daniel Dennett's crane theory of evolution (i.e. natural selection is a mindless, mechanical, and algorithmic process), and opened up a new path of discovery for my own art practice (Dennett, 1995). I am now using clay, an ancient technology, to look at artificial intelligence struggling to make a form for itself in the world of matter, by layering multiple iterations of the same motifs again and again.

What makes a workshop work?

What are the elements that make one project take off like a rocket and another wobble around and fizzle out? Which platforms can inspire collective invention and allow people to take the initiative? What are the strategies that encourage people to get involved and share ideas, and what causes other initiatives to be cautiously skirted and politely ignored?

- While all good research embraces risk, accident, and failure, it is good to strengthen and support your project with planning and networking as well. A successful workshop experience is created spontaneously by the group but planning and networking is essential to support this success. A brilliant concept and high-end skill do not guarantee success.
- Visibility and reliability are important. It is good to speak with people either in person or on the phone when seeking support and involvement. Emails get lost. Put time into composing a document with images and text that gives the reader an immediate sense of your project and includes links to social platforms that feature your ideas. After a good conversation, this document can be sent out with a brief bio to assure people the project is viable. Always follow up. People are busy and they might love an idea but lose it in the blizzard of communication in which we all live.
- Sit down and interrogate your idea. What does it need to be realized? What venues, collaborators, materials, timeframe, participants, and promotion does it need? Find alternatives for everything so the project doesn't stall when something doesn't work out.
- You may need to run a volunteer project initially to raise the profile and visibility of your idea. Document this process well and use the images to apply for funding or to use for promotion. This proves you are committed and can organize a successful event – according to your own criteria. It is also a brilliant opportunity to test your ideas and procedures so they can be refined for when you start charging people to be involved or, if you want to keep the project free, when you apply for funding. I am interested in working with people who are outside the mainstream arts industry, or who live off the beaten tourist tracks and contexts, so I tend to apply for funding to keep it accessible. (Keep all actions consistent with the *kaupapa* of your project – see below!)
- An initial free project also creates contacts and starts to spread the word. When I began the *Water Systems* workshops with the Waiheke Community Art Gallery, I collaborated with the Waiheke Adult Learning group and ran a series of free workshops with them to test the ideas. This also meant that when the public project began, the worktables featured clay sculptures and weavings already made by the group. This created a stream of energy and interest that made the project accessible to newcomers and inspired them to join in. (As opposed to a bare table and verbal guidelines.) This process also set broad community connections in motion. The adult learning artists invested in the project and shared with their networks – this is how we found new collaborators, including wonderful local artist Angela Newtown who taught us all how to harvest the pest plant ginger to make rope for the installation.
- Word of mouth is powerful. People love ideas and new things. Every conversation you have gives your project visibility and by explaining the project to people, your thoughts are clarified and your ideas refined.
- Believe in what you are doing and let your *kaupapa* guide the process. For example, I believe in collective invention and outsider thinking, so my process involves consultation and collaboration with outsider groups.
- And don't wait until everything is perfect! Failure is not in binary opposition to success, it is part of the process, so be brave and adapt your method with every interaction to refine your ideas and improve their communicability.
- Creativity can happen anywhere. It doesn't have to take place in an official space but accessing the network and communication systems of established organizations can really help artists to share their ideas and can help people to find them.

Fig. 5.4. *The Stencil Project*, Buinho, Messejana, Técnicas de Acção Educativa workshop, 2018. Photo by Margaret Feeney, used with permission.

The artist and infrastructure

Embedded in the Te Reo Māori word *ako* is the acknowledgement that the sharing of information and experience is always reciprocal. Reciprocity makes interactions rich and innovative, allowing everyone to benefit. The more complex and connected the relationships are between the artist, their community, and its visitors, the more rewarding the creative, social, and economic gains are for all. Forming social partnerships with the many organs of the local and regional councils, and of central government, as well as the private sector provides further breadth of access and engagement for the artist's project. In turn, the collaborations allow these organizations to benefit through becoming better connected to the communities they serve.

I believe the creative tourism model, with its direct interface between maker and consumer, is a dynamic and flexible model that can invigorate organizations as those organizations support artists in return. Many community art galleries in ANZ work closely and directly with artists. I have been lucky to work with some with amazing places including the Waiheke Community Art Gallery – Te Whare Taonga o Waiheke, Māngere Arts Centre – Nga Tohu o Uenuku, and the Village Arts Gallery, Hokianga. Working with diverse groups in diverse places, from fringe city sub-culture sites to remote rural areas, makes me realize the resilience and portability of art as a communicative technology. Fewer and fewer artists can afford to live in cities, and many choose to form satellite cultural centres in the regions alongside extant practitioners of the traditional arts. Potential sites are dotted throughout the regions: community halls, regional galleries and libraries, charitable trust centres, community galleries and halls, artist-run galleries, residency sites, homeless shelters, learning centres, and community art studios. The national and metropolitan cultural institutions could operate as communication hubs – connecting people and activating remote, under-utilized spaces – linking people, ideas, and spaces in real time.

By setting up outreach sites in their sprawling regions, public institutions could draw visitors from the more generic city centres and into direct engagement with art and artists. Galleries and other arts institutions can gain social traction by embracing the creative tourism model as it is socially nimble and can connect them with tourists and with everyday people in a more direct and accessible way. This can foster a more energized and connected dialogue with culture, which would in turn enrich their other organizational systems. The institutions have the networks and infrastructure to support artists in their work and in return they will be revitalized, and more properly represent the lived reality of a live culture.

Display and interaction

The tourist model in general, and the cultural tourist model in particular, has always been about the purchasing of experience as a measurable, knowable commodity. But people now want more genuine engagement. Creative tourism is experiential and immersive, it is a direct transmission of information, not mediated or controlled by a culture of display. The creative space, whether an artist's studio or a public workshop, is open to interaction. It is anarchic and tangential. Discoveries and experiences are made rather than directed. The cultural model of display is the performance of didacticism, a delivery system of intellectually and economically ameliorated notions of what art is. These are the structures talking, not the art. And they are in the service of the shareholders of those structures. The viewers, however, are not shareholders: they are the fuel. But fuel is consumed in the process, thus creating the need for more and more fuel. Can we refer to *ako* and instil a more sustainable and reciprocal mode of exchange? I believe this would improve the rigour of large cultural institutions where chance and discovery are manufactured and outcomes are predicted and fulfilled.

In the light of the COVID-19 pandemic, the principles of creative tourism can still apply. And they don't have to only be applied internationally but can apply at any level: national, city, town, or even street and apartment building level. They could also be engaged with on virtual platforms constructed by a dedicated, cooperative body of practitioners and institutions working together. Virtual workshops and interactions could provide quality creative engagement to alleviate the loneliness, boredom, and anxiety of isolation. People can still be meaningfully and creatively connected across continents and oceans, and in their own apartment buildings. I think the critical factor here is to keep the structure of such a venture flat and non-hierarchical, to ensure the content remains the focus – the interaction of the practitioner and the travelling mind.

I believe sharing ideas with artists allows the tourist to expand their worldview and to become agents of positive change. Equally, the travelling artist enters new worlds of thought through interaction with communities in new places. I have learned so much working with different groups of cool people. The collective invention workshops have formed a research model that is now integral to my art practice. My art practice is challenged and revitalized with each workshop.

Creative tourism has the potential to form an international community of connected thinkers, and of connected actors. Art used to be the dominant technology of mass media and had great relevancy as the purveyor of reality. Perhaps by speaking directly to people through creative tourism it can once again be an agent of change. Creative tourism can respond to the swift adaptions culture needs to make in an increasingly volatile world. It is decentralized and nimble and can share ideas quickly, thus being better adapted to the modern world and its challenges.

Postscript: A note to Aotearoa New Zealand

I am excited about the potential for creative tourism in ANZ. ANZ artists are a well-connected group and form close clusters all over the country, but we would benefit from centralized coordination and support from public and private infrastructure. National institutes such as museums and galleries, regional and local council galleries, libraries and community facilities, conference centres, corporate and small business hubs, and transport hubs would all benefit from co-ordinated networking and the monetization of art and craft that creative tourism offers. We also have extensive national digital infrastructure and platforms to facilitate these initiatives.

Creative tourism in ANZ has the potential to:

- relieve unemployment in the regions as creative tourism employs many more people than cultural tourism;
- maintain Indigenous skills and arts;
- take pressure off the living *taonga*[7] of our national parks and beaches;
- help the housing crisis, as rental accommodation is currently being repurposed as Airbnb accommodations in the big centres and the tourist hotspots such as Queenstown;
- encourage New Zealanders to visit and then to perhaps move to the regions to live and thrive,

providing some balance to our cultural and intellectual landscape, and a sharing of sensibility to dissolve the urban–rural divide;
- establish contacts for ANZ artists and artisans all over the world; and
- accelerate innovation through the sharing of ideas through a non-hierarchical, flat structure.

ANZ creative tourism could be extended to so many ideas. Here are a few:

- farming practices that embrace organic, eco, or technological innovations;
- coastal Māori history tours visiting famous archaeological Pa and war sites by waka;[8]
- specialist artisan skills such as boat builders, craft beer brewing, and wine making;
- scientific research projects such as earthquake, volcanic, marine, and waterways research; and
- sanctuary, planting, and pest eradication projects.

By purposefully connecting and capacitating artists to develop and implement creative tourism initiatives, we could guide the wave of tourism towards positive and beneficial outcomes for more New Zealanders in a way that properly represents our quality of thinking and practice across a broad spectrum of activities. We are more than the sports stadium, the scenic vista, and the public art space. We are part of a living culture that we need to share amongst ourselves as much as we need to share it with the world.

Notes

1. Pandemics such as COVID-19 are cyclic, recurring threats that visit humanity periodically. I believe climate change and AI are different threats as they will not wax and wane – if they take hold, they will stay.

2. I have run four *Water Systems* workshops at Waiheke Community Art Gallery for travellers and visitors to Sculpture on the Gulf, Māngere Arts Centre, and Māpura Studios in Tāmaki-makau-rau, ANZ. *Water Systems* is an ongoing project that has taken many forms: a gallery installation, a symposium paper, and a subsequent chapter in Stupples (2018).

3. *Kaupapa* is an ANZ Te Reo Māori term broadly meaning a set of values, principles, and plans agreed upon as a foundation for action.

4. *Mahi* is the ANZ Te Reo Māori word for work.

5. I have run five *The Stencil Project* workshops at Buinho Creative Hub, Messejana, Portugal, and at Village Arts Gallery, Hokianga, Waiheke Adult Learning, Waiheke Island and Māpura Studios, and Tāmaki-makau-rau, Aotearoa New Zealand.

6. *It Started with A Kiss: A Low Resolution Show about High Technology*, Waiheke Community Art Gallery – Te Whare Taonga o Waiheke, Waiheke Island, ANZ, 2018.

7. *Taonga* is the ANZ Te Reo Māori word for treasure, anything prized. It is applied to anything considered to be of value, including socially or culturally valuable objects, resources, phenomenon, ideas, and techniques.

8. *Waka* is the ANZ Te Reo Māori word for canoe, vehicle, conveyance, spirit medium, or medium.

References

Dennett, D. (1995) Darwin's dangerous idea. *The Sciences* May/June, p. 37.

Gotkin, Z. (2012) America's innovative companies are going flat. *Huffington Post*, 17 December. Available at: http://www.huffingtonpost.com/zev-gotkin/corporate-hierarchy-work_b_1962345.html (accessed 23 April 2020).

Kastelle, T. (2013) Hierarchy is overrated. *Harvard Business Review*, 20 November. Available at: https://hbr.org/2013/11/hierarchy-is-overrated (accessed 23 April 2020).

Kaufman, S.B. (2014) The controlled chaos of creativity. *Scientific American*, 25 June. Available at: https://blogs.scientificamerican.com/beautiful-minds/the-controlled-chaos-of-creativity/ (accessed 23 April 2020).

Stupples, P. (ed.) (2018) *Art and Future: Energy, Climate, Cultures*. Cambridge Scholars Publishing, Newcastle Upon Tyne, UK.

6 Interview with Mélanie Wolfram, Vagar Walkingtours: Connecting to family travel with creative tourism

Mélanie Wolfram*
Interview by Sara Albino

Vagar Walkingtours, Portugal

Sara Albino (SA) – Thank you for participating in this interview (and book), Mélanie. To begin, please tell us about Vagar Walkingtours.[1]

Mélanie Wolfram (MW) – Vagar Walkingtours is a small tourism company (registered as a tourism 'animation' company in Évora) that promotes 'slow touristic visits', conducting these cultural visits in all of the Alentejo region and, more particularly, focusing on the North Alentejo. We do essentially private tours and try to adapt to the will of our clients, providing heritage tours, wine tours, and arts and crafts tours, among other options. We have a deep interest in archaeology, built heritage, and local practices of intangible cultural heritage, such as horse riding and gastronomy.

The name of this tourism microventure, Vagar Walkingtours, was inspired by the Portuguese word *vagar,* which has a regional symbolic meaning, 'to do things without rushing' (*fazer coisas com vagar*). The concept of time is something specific in the culture of Alentejo. Compared to other local tourism companies that frame each tourism experience within a pre-determined schedule and timeframe of a few hours, Vagar Walkingtours organizes its activities without fixed end times.

SA – Why do you think that Vagar Walkingtours' approach is important?

MW – It is important because this region needs this kind of initiative: bringing small numbers of tourists and showing them all the beauty of it. It gives also strength to the local activities, shops, wineries, and craftsmen, who can earn a little bit of money and visibility thanks to these visits. Also, we believe that leisure time between family members is becoming ever more important in the face of the growing pressures of work and everyday routines. Holidays, for many families, are the means through which they strengthen their ties and build memories and experiences together.

SA – What do you think are its main contributions?

MW – The main contribution of Vagar Walkingtours is to promote a whole region – the Alentejo – to the rest of the world, since the visits are essentially given to foreigners who speak several languages. Through our visits, people from the United States, Germany, France, Belgium, Swiss, the UK and other countries take a little of our culture to their home and, often, come back to the Alentejo. And, as I said before, our activities provide opportunities for families to reconnect through leisure time together. Family tourism is becoming more and more prominent. The experiences of family tourism that we create provide families opportunities to participate actively in cultural activities in different environments that can connect them to the place they are visiting. With our family offers we try to create activities that are carried out at the pace of each family.

SA – What do you think are the main challenges for Vagar Walkingtours and the creative tourism activities it organizes?

*Corresponding author: melanie@vagarwalkingtours.com

DOI: 10.1079/9781789243536.0006

MW – The main challenge for all tourism animation companies is to promote tourism without destroying the destination. The activities organized by Vagar Walkingtours are inspired by and represent local cultural heritage, contemporary artistic expressions, and the community – 'place' is the primary content of the creative tourism offer. Tourists' experiences are embedded in the geographical, spatial, and historical–cultural contexts of the city. We try to deepen their knowledge and appreciation of the place and its elements, rather than creating a separate 'tourism experience'. One example is the kit 'Play Évora', where activities involve interactions with a local ceramist who owns the museum *Aldeia da Terra* and Queens traditional pastry shop. Play Évora presents the main traits of regional Alentejo culture in an attempt to characterize and transmit knowledge about the traditional and modern 'spirit of place'. The Play Évora approach also fosters interdependencies with local people, not only in the promotion and distribution of kits but also in the implementation of the tourist experience, in which tourists are encouraged to interact with people from local businesses in an everyday activity.

SA – What are the most popular offers you have at Vagar Walkingtours?

MW – The three main principles underlying the creation of the tourism offers at Vagar Walkingtours are time, personalization of activities, and cultural values associated with Alentejo – all three form the foundation for the establishment of local partnerships. That is why we have created the tailor-made visits: offering exactly what the client wishes, mostly in their native language. I highlight specifically the offer of activities that we have for families, in particular the Play Évora creative tourism approach, which combines cultural and nature tourism activities, an educational component, and immersion in local culture.

SA – Can you tell us a bit more about the family programme, Play Évora?

MW – We thought there was a gap in the city of Évora, that is, no specific programme for families. How can you discover a city with kids without boring them? So we created an educational kit, a kind of game for kids with their parents, called Play Évora, that unfolds through the streets of Évora: they have to answer questions, observe monuments, discover and draw details, and during almost 2 hours they walk, play, and learn at the same time!

The Play Évora kit, which is purchased from local trade and hospitality partners, consists of a backpack with a map, the three notebooks, and writing and drawing materials. In addition, a sealed envelope with quiz answers and gifts for children made of regional materials is included.

Play Évora has been designed as a game to encourage interplay between the users and elements of the landscape, which leads to the discovery of the city (Fig. 6.1). The activity is structured in a flexible way that aims to respect the rhythm of each family and to extend their stay in Évora.

Play Évora includes three activities entirely designed by the Vagar Walkingtours team: Peddy-paper for children from 5 to 12 years old; street art for teenagers (photographic collection of existing street art or creation of city sketches by the participant); and Alentejo gastronomy, asking participants to interact with the local community and its uses and customs. For instance, Play Évora focuses on urban street art through a co-operation with the Cultural Association *100 Pavor*, which re-interprets elements of the history of Évora and re-expresses them as graffiti interventions on urban elements such as public electricity boxes. We thought of finding a place where the users could make their own street art as part of their creative visiting experience.

Overall, the users of Play Évora are invited to discover and draw the urban art elements related to local legends and the history of Évora, to draw in the notebooks of the activity kit, and to actively participate and share their own creativity through social media. The game features creative activities regarding the interpretation of heritage sites and curiosities of the city and it is structured into thematic books (local culture and gastronomy) that are directed to various age ranges, both teenagers and younger children. A third book for the parents explains historical details and provides answers to the quizzes, enabling them to actively contribute and participate in the explorations.

SA – What do you perceive is the impact of these creative experiences on travellers?

MW – Very positive. We felt that parents often do not know how to make their kids and themselves happy at the same time during touristic tours. With Play Évora, they can do it! We also try to offer

Fig. 6.1. Visiting family participating in Play Évora activity, Évora, Portugal. Photo by Mélanie Wolfram, used with permission.

activities that are shaped by the participants' contributions, so they have greater impact on them. We believe the activities also create new routes and personal perspectives of the city. When carrying out the activities, families initiate and experience different 'appropriations' of the places and spaces where they interact, building their own perceptions and memories of the city of Évora and its heritage.

SA – What can you tell us about the artistic and financial impact of this project locally?

MW – We have to be honest: it has a small impact locally. It has more impact for the family. The kids have to draw the Roman temple and to observe architectural details on the monuments (artistic impact) and, at the end, they have a snack, which is financially good for the café, but it is a small input regionally speaking. However, their relationship with some small local businesses and associations is relevant, and it encourages partnerships among various organizations and business in the territory.

SA – What kind of strategies do you use to disseminate awareness of Vagar Walkingtours?

MW – We use a lot of social media, very regularly (at least before the pandemic): Facebook, Instagram, LinkedIn, and Twitter. We produced a newsletter for a couple of months. We also promote Play Évora through the hotels in the region.

SA – What are your plans for the future of Vagar Walkingtours? What would you like to do in the next few years?

MW – In 2020, we are not able to answer this question. The pandemic is making everything complicated. We would like to continue to promote Alentejo and all its secrets in engaging and creative ways.

SA – What are the key challenges for creative tourism?

MW – Getting the tourists involved with the locals, with the real traditions, and not fall into a 'touristic product' that has nothing to do with what locals really do. Or, like Play Évora, creating dynamic solutions to visit a place without spoiling it. We think that Vagar Walkingtours has the ability to create new activities and encourage new perspectives on places, and that this can generate new tourism demands and publics – alternatives to mass tourism. These types of tourists include smaller groups that are interested in embedding local experiences constructed at their own pace, and families with young and/or teenage children. We think that creating an array of creative tourism options for different family configurations and interests is a valuable challenge to pursue.

SA – What are your expectations on the future development and impact of the activities that you have, in particular the one for families? (Related passions, aspirations, etc.)

MW – Visitors led by the Play Évora kit of activities can help to shape the future tourism profile of our city. We expect that Play Évora may contribute, in a small way due to its micro-scale, to profile Évora as a family friendly creative tourism destination, thus differentiating itself from Évora's predominant image as a mass tourism destination. The idea of seeing the city as a place for family time and bonding not only provides new positive dimensions to the city's tourism profile but also enhances the interaction between tourists and the local community, extends visitors' length of stay, and increases the number of repeat visitors who engage with Vagar Walkingtours' activities. We expect that in the near future this activity can help change the image of the city of Évora.

Note

1. Some of this information is also published in Albino *et al*. (2020) and Bakas *et al*. (2021).

References

Albino, S., Marujo, N., and Wolfram, M. (2020) O projeto 'Play Évora' da Vagar Walkingtours. In: Duxbury, N. and Silva, S. (eds.) *CREATOUR: Desenvolver Destinos de Turismo Criativo em Cidades de Pequena Dimensão e em Áreas Rurais.* Imprensa da Universidade de Coimbra, Coimbra, Portugal.

Bakas, F.E., Duxbury, N., and Albino, S. (2021) Creative tourism microentrepreneurs in Portugal. In: Morais, D.B. (ed.) *Tourism Microentrepreneurship:* Vol. 12. 'Bridging Tourism Theory and Practice' series. Emerald Publishing, Bingley, UK.

Part II: Trends in creative tourism

Part II explores the dynamically evolving trajectories of creative tourism, looking at both trends and tensions in the expanding and diversifying field. It aims to set out a 'big picture' landscape that highlights some of the emerging hot areas to watch, as background to Parts III and IV which focus on approaches to the design, development, and evolution of creative tourism initiatives on the ground. The first chapter in Part II provides an overview of 12 major trajectories within the diverse field of creative tourism, followed by observations about developments and challenges in the field from the perspective of the international Creative Tourism Network®. Three specific areas within creative tourism are then highlighted, with contributions providing insights on the development and implementation of creative gastronomy offers, which brings together creative tourism and food tourism; the appeal of mini-apprenticeships with master artisans and artists; and the varied relations between artistic residencies, digital nomads, and tourism.

In the opening chapter of Part II, 'Trajectories and trends in creative tourism: Where are we headed?' (Chapter 7), Greg Richards and Nancy Duxbury provide an overview of observed trends among the international array of creative tourism offerings today, aligning them with socio-cultural changes and travellers' emerging aspirations. Illustrating the contemporary scope of creative tourism, they briefly outline 12 major trajectories in creative tourism, organized into four general categories: taking home creative skills as well as souvenirs, finding space for creativity, cultivating meaningful travel, and connecting with creative networks and hubs. They also speak to the potential for small-scale creative tourism approaches in the post-pandemic tourism context.

Next, in Chapter 8, we present an interview with Caroline Couret, Executive Director of the Creative Tourism Network®, on her observations of recent trends and dynamics – and the future – of creative tourism internationally. Couret remarks on how creative tourism has quickly evolved, responding to demands that reflect social changes, and can play an important role as a lever for local and regional sustainable development through developing linked value chains of different actors. She also stresses how fostering bottom-up collaborations in which cross-sectoral co-creation involving very different actors has become increasingly essential.

In Chapter 9, 'Gastronomic creative tourism: Experiences in the Algarve region of Portugal', Sónia Moreira Cabeça, Alexandra R. Gonçalves, João Filipe Marques, and Mirian Tavares, researchers at the University of Algarve, explain how the area of creative gastronomy, one of the identified trends in the opening chapter, is developing in the Algarve region as an alternative to its traditional 'sun and sea' tourism. In the context of an international explosion of interest in all things related to food, they examine how gastronomic creative tourism, or 'creative gastronomy', strives to develop 'co-creative experiences, making use of the spirit of the place and boosting the participants' creativity in an interactive process' (p. 63). Based on an array of creative gastronomy pilot initiatives that were designed and piloted in the region within the CREATOUR® project, the chapter shows how these initiatives are demonstrating strong links with the history and heritage of the territory where they take place, involving community members in the experiences, and integrating learning processes that promote healthier lifestyles and the conviviality of making and sharing food together.

Next, an interview with Geetika Agrawal, the founder and director of Vacation with An Artist, which is working with artists in 27 countries, provides an entrepreneurial perspective on the evolution of creative tourism and its future. In Chapter 10, she describes the transformational experience of being in the company of master artists, 'the creative lifelines of a culture' (p. 68), and her passion to enable and share such experiences with others. Agrawal explains how, on one hand, creative tourism meets the growing desire for personal growth-enhancing experiences and connections through the hand-made, but that an ongoing challenge is connected to people thinking that they are not creative. On the other hand, she also explains how creative tourism can serve as an inspirational experience for both the host and visiting creators and can play an important role in intergenerational knowledge transfer – and survival – of traditional crafts and techniques.

Extending the discussion on visiting artistic residencies further, and closing Part II, Sara Albino and Carlos Alcobia's chapter, '*I need you, but I don't want you*—the ambiguity of artistic residencies and tourism' (Chapter 11), delves further into the diverse scenes of artistic and creative residencies, exposing some of the tensions that have been emerging in recent years. They observe a growing convergence among digitally oriented fabrication spaces (such as FabLabs, makerspaces, and hackerspaces), co-working spaces, and process-oriented supports such as Artists in Residency programmes that is fuelling an expansion of concepts and categories. These configurations are attracting creative nomads and creative tourists and altering the terrain for artistic residences for professionals. They elaborate on how these developments expose an ambiguous relationship between the worlds of creative tourism and of professional artist development that crosses intellectual, cultural, and economic dimensions.

—N.D.

7 Trajectories and trends in creative tourism: Where are we headed?

Greg Richards,[a*] and Nancy Duxbury[b]

[a]Breda University of Applied Sciences and Tilburg University, The Netherlands; [b]Centre for Social Studies, University of Coimbra, Portugal

Creative tourism is part of a number of general trends towards more creative, relational, and locally focused styles of travel. For example, the 'live like a local' trend encourages travellers to reject the 'role' of tourist and to integrate into everyday life and culture at the destination, reciprocated by local recognition of 'temporary residents' (Richards and Marques, 2018). The rise of the experience economy (Pine and Gilmore, 1999) brought forth 'experiential tourism' (King and Zhang, 2017) and more active views of tourists as well as choices for their consumption. The 'prosumer revolution' and rise of social networking platforms meant a growing desire for customized travel itineraries and opportunities for expanding one's network and co-creation (Binkhorst and den Dekker, 2009). The desire for more-than-accumulating-tourism-experiences spawned more meaningful travel (Wilson and Harris, 2006) and 'transformational tourism' (Reisinger, 2013). As these tendencies pile on one another, propelled by broader societal changes and trends, they are inter-mingling and spawning new developments. In this crowded sea of interactive, inspiring, meaningful, and (big picture) creative activities, how is creative tourism envisioned? From the roots of creative tourism activities and strategies, what is emerging now?

In this chapter, we briefly outline what we envision as 12 major trajectories in creative tourism, organized into four general categories: taking home creative skills as well as souvenirs, finding space for creativity, cultivating meaningful travel, and connecting with creative networks and hubs.

Taking home creative skills as well as souvenirs

One of the major differences between creative tourism and cultural tourism is that the creative tourist seeks to expand not just their knowledge of the places they visit, but also their own creative skills. Many creative tourism experiences are based on the exchange of skills between hosts and visitors, which also increases the relational dimension of the experience.

In many creative experiences there is also a tangible end product, such as the production of art, photos, or craft objects. This can provide a direct link between the destination culture and the tourist and give them something to take home as an I-made-it-myself souvenir.

Creative learning and making experiences can help individuals tap into and realize their latent creative abilities (see the interview with Geetika Agrawal, Chapter 10). The learning that takes place during a creative tourism experience can develop the personal skills of the traveller, enabling them to enhance their own creative capabilities and potential. This is the case with gastronomic creative experiences (learning to cook the local food) as well as art and crafts (learning specific techniques). The depth of immersion in these experiences can vary widely, from time-limited 'snack culture' activities to immersive mini-apprenticeships and programmes – and we expect both ends of this spectrum to grow.

*Corresponding author: richards.g@buas.nl

 Creative Tourism: Activating Cultural Resources and Engaging Creative Travellers (eds. N. Duxbury, S. Albino, and C. Pato de Carvalho)
DOI: 10.1079/9781789243536.0007

Trajectory 1: Creative gastronomy

Gastronomy has become a key element of the cultural tourism market, and there are now a wide range of courses and workshops enabling people to hone their culinary skills. Experiences that develop skills as well as provide a distinct link to local culture – learning to make pasta in Italy, *fois gras* in France, or cataplana in Portugal's Algarve region – are likely to continue growing. This can also extend to a knowledge of ingredients and how to obtain them, including foraging for wild foods. As Sónia Moreira Cabeça *et al.* (see Chapter 9) point out, these creative gastronomy experiences extend beyond specific skills to also incorporate narratives and histories of the culture and place linked to the food being prepared.

Trajectory 2: Snack culture creative tourism

In our mobile societies, the emergence of 'snack culture', the quick habit of consuming information and cultural resources, is ever more ubiquitous. Linked to the growing prevalence of consuming online digital content via smartphones, digital viewers typically consume online cultural content for about 10 minutes and there is a sense that individuals 'get addicted to the short segments' (Chung, 2014). This practice has spawned a wide range of online cultural contents; in South Korea, for example, 'snack culture' is becoming representative of the Korean cultural scene. Transferring this to the creative tourism context, to enable sampling of a variety of skill-expanding creative experiences, these new experiences are offered in time-limited 'snack'-sized pieces, providing 'snack culture creative tourism' (Seo, 2019).

Trajectory 3: Mini-apprenticeships

Alternately, individuals are seeking to develop and/or deepen their knowledge and skills through workshops and short-term immersion experiences in artistic residency-type programmes. Opportunities are emerging for travellers to work and learn with artists and artisans, such as Creative Iceland's wood carving, Icelandic knitting, and knife-making workshops, among other topics (https://creativeiceland.is; accessed 28 April 2021). Vacation With An Artist (https://vawaa.com; accessed 28 April 2021) offers a range of 'mini-apprenticeships' with master artists in different parts of the world. For example, you can learn Japanese calligraphy with Chikako in Kyoto, street art with Pum in Buenos Aires, or monochrome drawing with Yoshi in Barcelona. These apprenticeships usually last between 3 and 7 days. During the COVID-19 pandemic, VAWAA turned to online workshops and classes delivered by members of its artistic network, a practice they are expecting to continue (see Agrawal, Chapter 10).

Finding space for creativity

The development of creative spaces has been one of the main development trends in recent decades (Richards and Wilson, 2006). Creative spaces can be formal, such as those provided by creative clusters and districts around the world (Marques and Richards, 2014), or they can be more informal, integrated into the everyday life and placemaking of the destination. Creative spaces in both urban and rural areas provide opportunities for consumers and producers to gather, to exchange ideas, and to co-create experiences. Jutamas (Jan) Wisansing (Chapter 12) points to the importance of these creative cultural spaces as a connector between local creators, the broader community, and visitors. Three types of 'spaces for creativity' are prevalent among travellers: spaces to conduct their creative work 'on the road', creative residencies, and more personal 'escape' spaces where one can retreat from the busyness and pressures of urban life and re-connect personally.

Trajectory 4: Engaging in creative work on the road

Working in the creative industries has become more popular in recent years as people try and combine their work and their passion. Co-working spaces and living labs are becoming more common destinations for travellers. The WYSE Travel Confederation (2018) *New Horizons* survey indicated that around 8% of young travellers had used a co-working space while travelling in 2017. Those who said they were also 'developing their creativity' during travel were much more likely to use a co-working space (15%). This trend is supported by the drive towards remote working and the growing number of 'digital nomads' (Richards, 2015). With the COVID-19 pandemic accelerating remote working among a wide range of office workers, some travel firms and destinations have begun catering to a new breed of (wealthier) nomadic worker, 'the rise of the "half-tourist" who combines work with a change of scene' (Turner, 2020).

Trajectory 5: Creative residencies

Artistic residencies have been an important part of professional development and career trajectories for artists for some time. In recent years, this professional segment has been augmented by a stream of residency offers for non-professional creators that have proliferated internationally, with a variety of training and other programmes offered in their areas. As Sara Albino and Carlos Alcobia discuss (Chapter 11), there is some tension between these two components, accompanied by processes of distinction and diversification within this movement, enabling specialties to emerge within a broader array of sites (see also Falick, 2019). This connects closely with the skills-development trend discussed previously.

Trajectory 6: Creative personal space

For people living in big cities, life has become increasingly pressured and busy. Many try to 'get away from it all' during their holidays, only to find that many destinations have also become more crowded and stressful through 'over tourism'. This has produced a growing demand for retreats and other ways of finding peaceful surroundings during travel for recharging and renewal. This is often linked to a search for (new age) spirituality, including yoga experiences, meditation, or modern-day pilgrimage. On top of this, the COVID-19 pandemic has drawn public attention to managing personal well-being and, informed by both popular and scientific writings about connections between creativity and well-being, has highlighted actively participating in creation processes as part of nurturing 'self-care' and personal wellness (see, for example, the Creating for Wellbeing website; Davidson and Tahsin, 2019). The combination of creative activities and landscapes that allow one to 'commune with nature' will also become more important as people seek to also 're-connect' and experience self-in-nature.

Cultivating meaningful travel

Travellers are increasingly seeking to extract meaning from the experiences they have. Creativity is one important strategy for giving experiences meaning and making them memorable (Tan *et al.*, 2013). Meaningful experiences also link to the final stage of the experience economy evolution suggested by Pine and Gilmore (1999), that of 'transformations'.

Trajectory 7: Transformative travel

'Travellers want, more than ever, an experience that will transform them as human beings and make them become better versions of themselves' (Miron, 2017). Creative tourism provides experiences that can change a traveller's thinking about the world and their place in it, and contribute to their own identity construction (Sano, 2016, referencing Krajnovic and Gortan-Carlin, 2007). This potential is part of a growing societal *Zeitgeist* in which 'analogue arts' (hands-on creation) and other forms of personal participation in creative activity and aesthetic self-expression are viewed as an integral dimension of personal self-development and well-being (Duxbury *et al.*, 2019; Connor *et al.*, 2018). Personal self-expression and cultivating one's 'authentic' distinctiveness are stimulated by one's cultural identities mixed with personal trajectories – and one's travel choices become part of this process.

Trajectory 8: Living like a (creative) local

The local has become the new 'authentic' (Richards, 2014), and living 'like a local' and 'immersion' are the new mottos for meaningful tourism experiences. Travel trends organization Skift has emphasized the drive for authentic, local experiences: 'People, especially millennials, want to have an authentic experience and rub shoulders with people who live in the area' (Miron, 2017). Living like a local begins with people staying in sharing economy accommodation such as Airbnb rather than a hotel (Russo and Richards, 2016). Now, Airbnb is enlisting its hosts to provide local experiences, which can also include creative elements. The ability of the 'local' to act as a space for creative transformation lies in the authenticity it can offer (see Chapter 14 by Diana Zuluaga and Diana Guerra on 5Bogota). Close local–visitor relationships can foster an extended community around creative organizations following the experiences, whose members feel connected in an ongoing way.

Trajectory 9: Social connectivity and impact

Creative tourism is an important means of bringing people together. The relational dimension of creative tourism will increase in the future as people become more concerned about social issues and developing their own social capital. Many of those who volunteer while travelling also see themselves as creative (WYSE Travel Confederation, 2018). This means there is potential to develop creative tourism experiences with a more explicitly social dimension. This might include volunteer programmes directed towards particular forms of creativity (for example, artistic skills) that can be useful in community development.

Connecting with creative networks and hubs

Creativity thrives on connections. Creative tourism has been supported by the growth of networks linking producers and consumers across regions, countries, and continents. We have been seeing increased mobility of creative people and activities as the search for new creative experiences expands. In this context, destinations have been strategically positioning themselves to attract these mobile creatives through fostering hubs and platforms for creative activities. We are also seeing creative placemaking initiatives being circulated through grassroots networks and individual creatives who travel (Chatzinakos, 2017).

Trajectory 10: Holidays for the creative class

Much has been written about the rise of the creative class (Florida, 2002). Whether one believes them to be a real social class or not, there is no doubt that creative occupations are growing and that creative people often want to continue their creative consumption and inspiration gathering on holiday. This is a specific trend that has generated programmes such as Creative Tourism Austria, which links creatives with leading edge creative experiences in Austrian cities (see Paschinger, Chapter 3). Although these tend to be largely concentrated in urban areas, there is also potential for developing smaller creative clusters in rural areas. The important point is that many of these experiences are produced and curated by members of the creative class for their peers, and this requires the presence of 'trendsetters' and 'coolhunters' who can identify and promote new creative experiences.

Trajectory 11: Destinations as the hub of creative networks

Many successful creative destinations have managed to place themselves at the hub of international networks. By positioning themselves as the 'place to be' for specific creative activities and/or knowledge, they can attract people from around the globe. The development of creative networks will continue in future, as people turn their virtual contacts into physical ones. Even though we can be virtually connected anywhere, anytime, the need for physical contact is still paramount. Particularly for creative collaboration, we need to know the people we are working with personally in order to develop trust and to be able to exchange creative ideas more effectively.

Trajectory 12: Destinations as creative platforms

Cities worldwide are rapidly discovering the power of big data and open data to enable creative activities. By providing access to archives and databases, places can provide the raw material for creative activity that can spawn a range of products (such as new apps) and experiences (based on storytelling, for example). In the future, destinations will have a more important enabling role for creative tourism development by acting as platforms on which creative content can be assembled, processed, and developed. Such platforms will generate value for users and gain importance as they link greater numbers of producers and consumers, enabling the 'co-creation' of tourism products. The Welcome City Lab (https://welcomecitylab.parisandco.com; accessed 28 April 2021) in Paris provides a programme to stimulate innovation in the tourist sector including the world's first incubator that is dedicated to this sector, evolving and advancing the trends presented at the beginning of this chapter.

Creative tourism as a post-COVID-19 form of tourism?

With travel being disrupted by the COVID-19 pandemic (Burini, 2020), destinations are searching for new models of tourism for the future. The Creative

Tourism Network® is positioning creative tourism as a more sustainable and small-scale form of cultural tourism that can help local communities reconnect with the global space of flows. Although in some circumstances there may be problems of intimate contact versus the need for social distance, creative tourism is well positioned to significantly contribute to post-pandemic tourism. It is designed for small groups (e.g. families and social bubbles) and aligns well with domestic tourism and longer stays in one place – 'exploring one's city, region, or country with new perspectives on its diversities; pursuing personal interests and curiosities; and developing new skills' (Remoaldo *et al.*, 2020, p. 10). Furthermore, with the wide diversity of activities within creative tourism, its transversality is a strength, complementing and extending other types of tourism (Gonçalves *et al.*, 2020).

Creative tourism sits among an array of allied modes of tourism, such as relational tourism, conscious travel, responsible tourism, geotourism, situated tourism, and community tourism. These labels highlight 'an incremental materialization of tourism alternatives that direct attention to local benefits (and impacts) from tourism activities, emphasize conscientiousness in travel decisions and behaviours, strive to reshape host–guest relations, and prioritize local distinctiveness and diversity' (Duxbury, 2021, p. 201). These emerging alternatives to mass tourism will contribute to shaping future trajectories of travel and tourism, and it is strategic to envision creative tourism embedded within this web.

Acknowledgement

This chapter was developed within the CREATOUR® project (no. 16437), which is funded by the Portuguese Foundation for Science and Technology (FCT/MEC) through national funds and co-funded by FEDER through the Joint Activities Programme of COMPETE 2020 and the Regional Operational Programmes of Lisbon and Algarve.

References

Binkhorst, E. and den Dekker, T. (2009) Agenda for co-creation tourism experience research. *Journal of Hospitality Marketing and Management* 18, 311–327.

Burini, F. (2020) *Tourism Facing a Pandemic: From Crisis to Recovery*. University of Bergamo, Bergamo, Italy.

Chatzinakos, G. (2017) Urban experiments in times of crisis: The case of Svolou's Neighbourhood Initiative in Thessaloniki/Greece. Paper presented at the Conference Heteropolitics International Workshop: Refiguring the Common and the Political, Aristotle University of Thessaloniki, September 2017.

Chung, A. (2014) Fast-changing web-based content spawns new pop sub-genres. *The Korea Times*, February 2. Available at: www.koreatimes.co.kr/www/news/culture/2014/02/386_150813.html (accessed 26 September 2020).

Connor, T.S., DeYoung, C.G., and Silvia, P.J. (2018) Everyday creative activity as a path to flourishing. *Journal of Positive Psychology* 13(2), 181–189.

Creating for Wellbeing (n.d.) Available at: https://creatingforwellbeing.com (accessed 28 April 2021).

Davidson, R. and Tahsin, A. (2019) *Craftfulness: Mend Yourself by Making Things*. Harper Wave, New York, New York, USA.

Duxbury, N. (2021) (Re)articulating culture, tourism, community, and place: Closing remarks. In: Duxbury, N. (ed.) *Cultural Sustainability, Tourism and Development: (Re)articulations in Tourism Contexts*. Routledge, London, UK, pp. 197–212.

Duxbury, N., Kastenholz, E., and Cunha, C. (2019) Co-producing cultural heritage experiences through creative tourism. In: Gronau, W., Bonadei, R., Kastenholz, E., and Pashkevich, A. (eds.) *E-CuL-Tours: Enhancing Networks in Heritage Tourism*. tab edizioni, Rome, Italy, pp. 189–206.

Falick, M. (2019) *Making a Life: Working by Hand and Discovering the Life you are Meant to Live*. Artisan, New York, New York, USA.

Florida, R. (2002) *The Rise of the Creative Class*. Basic Books, New York, New York, USA.

Gonçalves, A.R., Borges, R., Duxbury, N., Carvalho, C.P., and Costa, P. (2020) *Policy Recommendations on Creative Tourism Development in Small Cities and Rural Areas*. Coimbra: CREATOUR project, Centre for Social Studies, University of Coimbra, Portugal.

King, B. and Zhang, H.Q. (2017) Experiential tourism and hospitality learning: Principles and practice. In: Benckendorff, P. and Zehrer, A. (eds.) *Handbook of Teaching and Learning in Tourism*. Edward Elgar Publishing, Northampton, Massachusetts, USA, pp. 207–217.

Krajnovic, A. and Gortan-Carlin, I.P. (2007) Possibilities of music heritage tourist valorization–an example of Istria. *Tourism and Hospitality Management* 13(2), 469–482.

Marques, L. and Richards, G. (eds.) (2014) *Creative Districts Around the World: Celebrating the 500th Anniversary of Bairro Alto*. NHTV, Breda University of Applied Sciences, Breda, The Netherlands.

Miron, K. (2017) 8 Key travel trends in tourism for 2018 at Skift Forum 2017 New York City. *Medium.com*. Available at: https://medium.com/@KarineMiron/8-key-travel-trends-in-tourism-for-2018-at-skift-forum-2017-new-york-city-1c255379c68e (accessed 28 July 2020).

Pine, B.J., II and Gilmore, J.H. (1999) *The Experience Economy: Work is Theatre and Every Business a Stage*. Harvard Business School Press, Boston, Massachusetts, USA.

Reisinger, Y. (2013) *Transformational Tourism: Tourist Perspectives*. CABI, Wallingford, UK.

Remoaldo, P., Serra, J. Marujo, N., Alves, J., Gonçalves, A., Cabeça, S., and Duxbury, N. (2020) Profiling the participants in creative tourism activities: Case studies from small and medium-sized cities and rural areas of Continental Portugal. *Tourism Management Perspectives* 36, art. 100746, 1–12. doi: 10.1016/j.tmp.2020.100746.

Richards, G. (2014) Creativity and tourism in the city. *Current Issues in Tourism* 17(2), 119–144.

Richards, G. (2015) The new global nomads: Youth travel in a globalizing world. *Tourism Recreation Research* 40(3), 340–352.

Richards, G. and Marques, L. (2018) *Creating Synergies between Cultural Policy and Tourism for Permanent and Temporary Citizens*. UCLG/ICUB, Barcelona, Spain.

Richards, G. and Wilson, J. (2006) Developing creativity in tourist experiences: A solution to the serial reproduction of culture? *Tourism Management* 27(6), 1209–1223.

Russo, A.P. and Richards, G. (2016) *Reinventing the Local in Tourism: Producing, Consuming and Negotiating Place*. Channel View Publications, Bristol, UK.

Sano, H. (2016) Theoretical consideration on creative tourism. *Journal of Global Tourism Research* 1(2), 127–132.

Seo, U.-S. (2019) Coffee tourism as creative tourism: Implications from Gangneung's experiences. In: Duxbury, N. and Richards, G. (eds.) *A Research Agenda for Creative Tourism*. Edward Elgar Publishing, Cheltenham, UK, pp. 84–96.

Tan, S.-K., Kung, S.-F., and Luh, D.-B. (2013) A model of 'creative experience' in creative tourism. *Annals of Tourism Research* 41, 153–174.

Turner, A. (2020) The rise of the 'half-tourist' who combines work with a change of scene. *The Guardian*, 25 September. Available at: www.theguardian.com/travel/2020/sep/25/the-rise-of-the-half-tourist-who-combines-work-with-a-change-of-scene (accessed 30 September 2020).

Wilson, E. and Harris, C. (2006) Meaningful travel: Women, independent travel and the search for self and meaning. *Turizam* 54(2), 161–172.

WYSE Travel Confederation (2018) *New Horizons VI Youth Travel Survey*. WYSETC, Amsterdam, The Netherlands.

8 Interview with Caroline Couret about the Creative Tourism Network®

CAROLINE COURET*
INTERVIEW BY CLÁUDIA PATO DE CARVALHO

Creative Tourism Network®, Spain

Cláudia Pato de Carvalho (CPC) – To begin, please tell us about your work in the field of creative tourism. How did the Creative Tourism Network® begin?

Caroline Couret (CC) – I could say that the Creative Tourism Network® (CTN) was born 'naturally' and was part of our DNA. Thanks to Professor Greg Richards' works on creative tourism and his valuable collaboration, we had already created the first city-wide platform of its kind, in Barcelona, to provide creative experiences for these new travellers. The idea of networking with other destinations that shared the same 'philosophy' and goals – to make creative tourism a lever for sustainable development – seemed obvious. That's why we created the network in 2011, with the support of the European Union, although it quickly extended to other continents.

CPC – What are the CTN's main functions?

CC – Our first mission is to identify the quickly evolving needs of this tourism in order to provide destinations of all kinds with solutions and services that enable them to position themselves in this booming market, while creating value chains that benefit their local territory. Then, we support them in the development of their creative tourism programme management though training and consultancy and, of course, we promote them internationally, labelled as CreativeFriendlyDestinations.

CPC – Why do you think creative tourism is important? What is the value that creative tourism can add to communities and to regional and local development? What local impacts have you observed?

CC – Creative tourism is important in two main ways: it meets a growing and irreversible demand that reflects social changes more than a marketing gimmick, and it is, above all, perceived by the territories as a lever for sustainable development. Indeed, far from being considered as a danger for local communities – as mainstream tourism used to be – creative tourism is valued for its capacity to create a value chain at the local and regional scale, by turning weaknesses and negative externalities into a virtuous ecosystem. In addition to this, assets such as communities' empowerment, inclusiveness, seasonality, resilience, intangible heritage recovery, social cohesion, and governance (just to mention a few) are applicable to any kind of destination and are particularly beneficial for the rural areas that cannot rely on other industrial sectors. The destinations that are part of the Creative Tourism Network® are labelled 'CreativeFriendly' for guaranteeing both the quality and authenticity of the creative experiences they are promoting, as well as for the value chain generated through the project design.

For instance, territories that did not have any tourist vocation *a priori*, once converting their traditional resources and know-how into creative tourism experiences, are given the opportunity to attract tourists. This relies on a bottom-up model at the local level that enables the involvement of a wide array of stakeholders, from farmers to artists,

*Corresponding author: carolinecouretv@gmail.com

 Creative Tourism: Activating Cultural Resources and Engaging Creative Travellers (eds. N. Duxbury, S. Albino, and C. Pato de Carvalho)
DOI: 10.1079/9781789243536.0008

going through festival organizers and hospitality entrepreneurs, gathered around public bodies. In addition to this, creative tourism tends to be concentrated in the low season, which allows these stakeholders to incorporate it as a complement to their main activity, without distorting it. It is also, at the local scale, a natural way to recreate the DNA of the territories, relying on the perception the communities have of themselves while also favouring social cohesion. This is certainly the most efficient way to generate economic and sustainable growth.

CPC – How do you see the future of creative tourism around the world? What are the key trends you observe?

CC – Since we started working in this field, in early 2000, we have observed a very fast evolution, that really reflected the evolution of the broader society. We could say that the mass tourism that corresponded to the Fordist model of the economy had to adapt itself to the new paradigm, characterized by the emergence of 'new' economies like the experience economy, the circular economy, the sharing economy, and, of course, the creative economy.

Such a disruption empowered the tourists or, better said, the travellers. The result is that while the creative tourists were still a minority in the early 2000s, represented by the budding artist who travelled on his own, we can say that creative tourists are now the new generations of travellers whose profile can be very diverse, which is also merged with existing segments, like MICE, singles, seniors, linguistic, and so on.

More and more tour operators and travel agencies are also including creative tourism offers in their catalogues, which makes the role of in-between organizations like CREATOUR® or the Creative Tourism Network® even more essential to guarantee the quality and authenticity of the experiences through standardized marketing channels.

The adaptability that has characterized creative tourism lets us infer its high versatility accompanying the evolution of society. For instance, the current pandemic has stressed the relevance of creative tourism for staycations. Local people seized the creative tourism experiences proposed by their own destination, to re-discover their own culture by participating in a wide array of proposals, whose segmentation enabled reaching the most diverse – local – tourists in reduced groups (e.g. female communities, multi-generational groups, sketchers, foodies, etc.).

The future of creative tourism is thus optimistic as, on the one hand, the demand evolves organically with the society and, on the other hand, it deals with virtuous tourism, which creates value in its development.

CPC – What do you think are the main challenges to the development of creative tourism?

CC – The first challenge for the destinations and stakeholders is to be aware that they can no longer propose the same standardized tourism and they must meet the new demands, be they creative tourism or other thematic ones. This implies the need to change their model, from a top-down one to a bottom-up one in which co-creation between very different actors will be essential, but not easy! And yet, they will have to co-design experiences that are, at the same time, authentic, creative, and profitable by integrating intangible resources from sectors such as crafts, creative industries, or agriculture. Despite the fact that our 'dream makers' as I like to call them – who are these craftsmen, these artists, these fishermen at the centre of the activity – are very generous when it comes to sharing their knowledge, the experience also has to meet other criteria related to the protection of intangible heritage and sustainability as well as, of course, profitability, marketing, and what is more prominent today, health and security. Such a challenge finds solutions in the intervention of in-between organizations, like the ones mentioned before, that not only give support to such a process, but also contribute to foster new synergies.

Let's say it can be as simple and as complicated as human relationships are! But it is certain that people and creativity are resources which can only be enriched during such a process. It is such a worthy challenge!

CPC – How do you evaluate the impact of creative experiences on travellers?

CC – As we are dealing with experiences, that are, by definition, unique, it is difficult to establish standardized indicators. Of course, the perception of authenticity or sustainability are common denominators but, on the other hand, the new generations of tourists are the best judges to evaluate their own experiences as they are more trained and active as prosumers. That's why there is direct feedback between the 'providers' of the experiences and the travellers themselves. This is part of the co-creation. At an upper level, we, at the Creative Tourism Network®, work together with the destinations in order to identify the good practices that can be applied to any kind of experience or territory.

CPC – What kind of strategies do you use to disseminate awareness of creative tourism?

CC – Actually, in addition to the promotion of the CreativeFriendlyDestinations, the global awareness of creative tourism has been rising quite naturally, just by following the changes in the values of society. But it is important to stress the role of the academic research as well as programmes like CREATOUR® and other initiatives, to structure and optimize it. Disruptions like the pandemic also sadly contribute to making the world more aware of the urgency of betting on a more sustainable model of tourism development like creative tourism.

9 Gastronomic creative tourism: Experiences in the Algarve region of Portugal

SÓNIA MOREIRA CABEÇA, ALEXANDRA R. GONÇALVES,* JOÃO FILIPE MARQUES, AND MIRIAN TAVARES

University of Algarve, Portugal

Gastronomy tourism is more than consuming good quality and exquisite food while travelling. Food is a vital resource: to share a meal can be a form of socialization and relaxation, it can be a way of knowing the culture of the other, and it can be a ritual. Gastronomy has always been part of the tourist consumption experience but, according to the World Tourism Organization's *Global Report on Food Tourism*, gastronomic tourism – defined as when travellers meet their destinations led by gastronomic curiosity and the will to know new eating habits and lifestyles (dos Santos and Antonini, 2004) – has:

> grown considerably and has become one of the most dynamic and creative segments of tourism. Both destinations and tourism companies are aware of the importance of gastronomy in order to diversify tourism and stimulate local, regional and national economic development. Furthermore, Food Tourism includes in its discourse ethical and sustainable values based on the territory, the landscape, the sea, local culture, local products, authenticity. (WTO, 2012, p. 5)

Today, gastronomy is an integral part of the cultural experience, the new cultural capital of a destination (Nistoreanu *et al.*, 2018). It is a value proposition that can influence travel destinations for tourists who seek to experience traditional recipes and homemade food, and to socialize through food: food consumption implies acquiring cultural capital, exploring the local culture(s), and seeking authentic experiences (Mak, 2018).

Food is a very popular product. On Instagram, food photos break records, the media covers festivals and events that show food, and, throughout the world, destinations organize culinary tourism programmes. The regions of Provence (France) and Tuscany (Italy) in Europe, and Canada and Australia are some of the most recognized destinations that promote local cuisine in their marketing strategies. Cooking classes, visits to markets and wineries, and tasting experiences are some of the most prominent offers internationally. Creative gastronomy workshops and cooking classes are among the most successful, providing more engaged and active experiences that guide tourists to collect ingredients, cook, and then experiment (Ilincic, 2014).

In Portugal, gastronomy is also seen as a valued asset that can enrich the tourist experience and add value to the territories, differentiating them (Turismo de Portugal, 2017). In a mass tourism destination like the Portuguese region of the Algarve, a major sea and sun destination, gastronomy can contribute to a more sustainable tourism, helping to diversify the Algarve's attraction base and to reduce tourism seasonality.

Conviviality at the table is a vital element of cultural identity and a tradition of communities across the Mediterranean basin, being a moment of social exchange and communication, of affirming and renewing the identity of the family, the group, or the community itself. For instance, the Mediterranean diet emphasizes and promotes values of hospitality,

*Corresponding author: cinturs@ualg.pt

DOI: 10.1079/9781789243536.0009

neighbourhood, intercultural dialogue, and creativity, playing a vital role in cultural spaces, festivals and celebrations, and other rituals and artistic expressions, bringing people together in an intergenerational way and promoting social cohesion and integration. A new Algarve has been arising from these new offers and trends, which bring together gastronomy, a healthy lifestyle, and tourism.

The satisfaction of tourists with the gastronomy experience in the Algarve was studied some years ago and the results show that it depends on three main determinants (Correia *et al.*, 2008): quality and price, gastronomy, and atmosphere. The evaluation showed that satisfaction 'was very much related with gastronomy related factors, including local courses, originality and exoticness, and the presentation of food and staff. ... quality, and price were also found to be an important influence on satisfaction although to a lesser extent than gastronomy' (Correia *et al.*, 2008, p. 174). Atmosphere was considered a less important factor.

More than mere good food consumption, gastronomy tourism is an experience. In fact, gastronomy can be considered a creative industry, stimulating innovation by involving tourists in co-creation processes (Cabeça *et al.*, 2020) and fostering links between global and local cultures and creating narratives around food (Richards, 2002). Through food, visitors have the opportunity to undertake creative and learning paths, participating in cooking lessons, harvesting, tasting local products, learning traditional techniques, etc. Through this learning and doing process, gastronomic creative activities are joint exercises of self-affirmation that express the creative potential of each participant. This is to say that they are co-created by the tourists themselves and by their hosts. A co-created experience involves tourists and residents as full participants and not passive subjects (Binkhorst, 2008) and contains all the people and things that are needed to provide the experience environment (Binkhorst and Dekker, 2009). These gastronomic experiences are, therefore, co-creative experiences, making use of the spirit of the place and boosting the participants' creativity in an interactive process.

Recently, some creative gastronomy projects have been developed in the region of Algarve. They are all working within CREATOUR® (www.creatour.pt, accessed 3 May 2021), the multidisciplinary research and incubation project devoted to the development of creative experiences in low population-density territories in Portugal (see box in Chapter 1). 'Algarve Cooking Vacations' is a culinary holiday programme developed by Tertúlia Algarvia in which participants learn how to make traditional recipes using ancestral kitchenware. Each cooking class is preceded by related activities, such as visiting a farm to pick the ingredients needed. The idea is to experience local culture through gastronomy and local heritage. Eating Algarve Tours is providing 'Food & Cultural Tours', gastronomic and cultural pedestrian tours accompanied by residents, in places from locals' daily lives such as food markets, restaurants, and heritage buildings, with tasting experiences and storytelling. 'Tasting Algarve' is an ongoing gastronomy route developed by the non-profit association In Loco that links food and landscape, in which participants learn traditional know-how related to fishing, agriculture, and regional products. Its emphasis is the Mediterranean diet, a healthy way of living. Espírito da Terra (which, in English, means Spirit of the Earth), a localized network, receives visitors in an organic farm, giving them the opportunity to pick their own ingredients and make fresh dishes such as soup. Odiana, a local development agency, offers traditional know-how workshops where participants can be a shepherd or a sea salt maker for a day, learn traditional recipes, or make traditional homemade bread in a community oven. All these initiatives seek to develop alternatives to the type of tourism that, in general, continues to be promoted in the Algarve.

Besides different strategies and offers, these experiences illustrate an approach to food consumption that testifies that gastronomy is not limited to a mere act of eating. An analysis of these activities (Table 9.1) provides some insightful and useful information about some key factors when considering developing a gastronomy creative offer.

Link to the territory

What immediately stands out in all the activities presented in Table 9.1 is that they share a strong link with the territory where they take place: they all make use of local resources and knowledge and they address local food consumption and the traditional know-how related to it (e.g. fishing techniques, products from a local farm, restaurants, etc.). The experience created is closely linked with the uniqueness of a given territory (Cabeça *et al.*,

Table 9.1. Gastronomy tourism programmes within CREATOUR® Algarve

Programme	Cooking vacations	Food and cultural tours	Tasting Algarve	Mornings at the farm	Traditional know-how
Approach	Provide participants with skills related to traditional recipes	Bring tourists to the community's daily life spaces	From nature to dish: learning the origin and transformation of products	Allow each participant to know and enjoy local products and the agricultural circle	Integrate tourists in local culture through activities characteristic of the region
Goals	Experience local culture through gastronomy and local heritage	Break seasonality and increase the time spent in the destination	Raise awareness of the Algarve's culture and history and promote the Mediterranean diet as a healthy lifestyle	Promote agriculture in a sustainable way, and recover mills and river site	Revitalize and promote local traditions involving the local community
Resources	Local: traditional recipes with local products	Local: partnership with local restaurants and monuments	Local: variety of each's place diet	Local: the farm's products, landscape, and surroundings	Local: natural and cultural resources that provide unique experiences
Promoters and providers	Local entrepreneurs and masters	Local entrepreneurs and residents	Local community	Local farmers	Local community
Creative action	To cook	To meet	To know	To harvest	To do

2018): local food, local recipes, and local spaces attach participants to the place and to its cultural landscape. Most of the experiences are provided by the community as well, extending the connection between gastronomy and culture and culture bearers, between tourists and the places and communities they visit: the shepherd (not a performer) teaches visitors how to milk a sheep; the local farmers (real farmers, not the promoters) help visitors pick the agricultural products, while explaining their properties and what can be done with them. In a visit through a city centre, local owners of restaurants and market sellers invite you to learn and taste traditional cuisine.

Deepen connections between food, heritage, and place

The goals of these creative programmes go beyond providing (good) food consumption. Through creative gastronomy, promoters aim to make known the history and heritage of their territories, to promote lifestyles, and to contribute to a more sustainable tourism by making use of local resources, eventually reducing tourism seasonality and increasing the number of days spent in the destination. It is also important to underline that there is a learning process in all these experiences: to cook, to meet, to know, to pick, to harvest and, of course, to taste. Whether the experience is related to food production or to tasting, from shop visits to food distribution channels to cooking classes, this is what a gastronomic experience means to our partners: a place-based touristic offer that invites visitors to immerse themselves into the local culture and involves a learning process, as seen in Fig. 9.1.

Interviews with the organizers of three of the gastronomic projects – 'Algarve Cooking Vacations', 'Tasting Algarve', and 'Food & Cultural Tours' – seem to reinforce these key ideas. 'The tourist must learn and know the origin of the ingredients and also the history of the utensils. And during the experiment itself, he learns the entire process of making a recipe and he makes it itself. Tourists create their own meal,' says Lélia Madeira from Tertúlia Algarvia (interviewed on May 22, 2019). In the case of 'Food & Cultural Tours', although the creation of an original recipe is not involved, the strong connection with local culture remains central: 'our customers want to know our culture, they

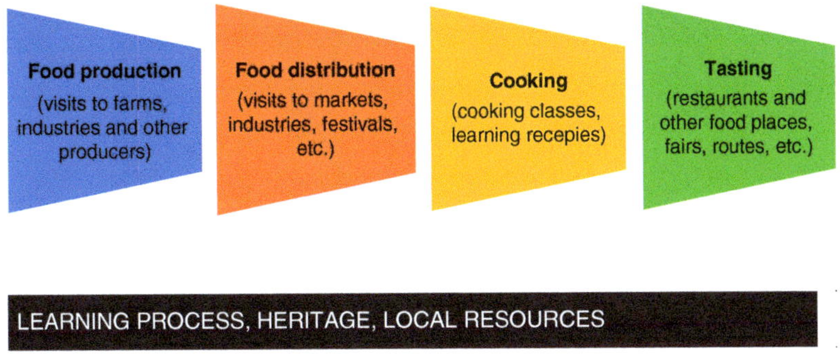

Fig. 9.1. Different types of actions in gastronomic creative tourism activities.

want to know the region and they want to eat,' says Joana Martins from Eating Algarve Tours (interviewed on June 19, 2019). To see, to smell, and to taste form the basis of the gastronomic experiences provided by the company, in which food consumption cannot be disconnected from the place where the experience occurs. It is not separable from the people who inhabit the places either. To make a visit with Eating Algarve Tours is to discover the alleys and narrow streets frequented by local residents, while listening to their stories; it is to learn about the traditions and the ways of life associated with them. What better hostess than passionate residents? A gastronomy experience 'is much more than letting the tourist know the unmatched beauty of this region and its typical gastronomy. It is necessary that the tourist/visitor feels connected with the landscape' and to create a link in the visitor's mind between the words gastronomy/food and landscape and territory, asserts Artur Gregório from In Loco (interviewed on June 12, 2019).

The process of self-expression, one of the key dimensions of creative tourism according to CREATOUR® (2016–2020), is also guaranteed: 'the recipes we make are traditional, but we do not strictly follow the recipe and often the chefs who accompany these experiences adapt it to the taste [of the participants],' says Lélia, meaning that chefs 'give tips to do it differently', especially since tourists, once back in their countries of origin, don't always have all the ingredients used in the Algarve region. By providing other suggestions and alternative ingredients, whoever makes the recipe can do it using their creativity and personal taste and then take it abroad.

We also invited our interviewees to characterize a gastronomic experience, considering the analysis dimension proposed by Correia *et al.* (2008). These researchers conducted an inquiry to better understand which assets are more important in a gastronomic experience: 'Quality and price' (i.e. quality, quality versus quantity, price of drinks and food), 'Gastronomy' (local courses/workshops, food presentation, originality and exoticness, staff presentation), 'Atmosphere' (ethnic decoration, place decoration, music, lighting), and 'Experience' (interaction with locals, creativity, exclusivity, authenticity, learning, emotional, entertainment, escape). The outcome from the CREATOUR® interviews reinforces the previous analysis about the key factors of gastronomic tourism. When asked about the importance of these dimensions, opinions differ depending on what type of tourism the organization has to offer but, remarkably, they all point out the importance of the fourth dimension at the expense of the others: 'Experience'. Furthermore, they all point out the same five key factors as being the most important (and none of them pointed out any other that was listed as 'Experience'): the creativity, the interaction with locals, the authenticity, the learning process, and the emotions that arise.

For Tertúlia Algarvia, which operates from its restaurant space, the existence of courses, the originality of the food, and the decoration are important factors as well. For Eating Algarve Tours, it is the way food is presented (not only visually but also using storytelling) that can make a difference. For In Loco, which seeks to establish a route from east to west Algarve, both the gastronomy and the atmosphere dimensions are of little importance and they do not characterize a gastronomic experience (they are merely important to the catering sector). Nevertheless, they all agree that it is the experience dimension that characterizes a creative gastronomic offer.

Fig. 9.2. Key factors of creative gastronomy offers.

A gastronomic creative tourism activity is a food-related, authentic experience that implies a creative learning path, promotes interaction with the destination's residents, and generates emotional links (Fig. 9.2). This statement meets Richards' and Maks' studies about gastronomic experiences, and also the CREATOUR® vision about creative tourism: 'creative activity encouraging personal self-expression and interaction between visitors and local residents, inspired by local endogenous resources (place and people), and designed and implemented by local residents' (Cabeça *et al.*, 2019, p. 4).

The creative gastronomic experiences briefly presented here are potentially sustainable tourism offers that might constitute an alternative not only to mass tourism (Cabeça *et al.*, 2018), but also to other types of gastronomy tourism, emphasizing the cooking and tasting experience (learn to do, to do, to taste), involving local resources and communities, valuing local products and knowledge, breaking food stereotypes, and stating that food sharing is cultural sharing.

Acknowledgements

This work was developed within the CREATOUR® project (no. 16437), which is funded by the Portuguese Foundation for Science and Technology (FCT/MEC) through national funds and co-funded by FEDER through the Joint Activities Programme of COMPETE 2020 and the Regional Operational Programmes of Lisbon and Algarve.

References

Binkhorst, E. (2008) Turismo de cocreación, valor añadido en escenarios turísticos. *Journal of Tourism Research / Revista de Investigación en Turismo* 1(1), 40–48.

Binkhorst, E. and Dekker, T.D. (2009) Agenda for co-creation tourism experience research. *Journal of Hospitality Marketing & Management* 18(2–3), 311–327.

Cabeça, S.M., Gonçalves, A.R., Marques, J.F., and Tavares, M. (2018) Contributions of creative tourism to sustainable development: The CREATOUR research and incubation project. In: Amoêda, R., Lira, S., Pinheiro, C., Zaragoza, J.M.S., Serrano, J.C.S., and Carrillo, F.G. (eds.) *Heritage 2018—Proceedings of the 6th International Conference on Heritage and Sustainable Development*. Editorial Universidad de Granada / Green Lines Institute for Sustainable Development, Granada, Spain, pp. 1671–1680.

Cabeça, S.M., Gonçalves, A.R., Marques, J.F., and Tavares, M. (eds.) (2019) *Book of Abstracts of the 3rd CREATOUR International Conference and Creative Tourism Showcase: Creative Tourism Dynamics: Connecting Travellers, Communities, Cultures, and Places*. CinTurs, Faro, Portugal. Available at: http://hdl.handle.net/10400.1/12848 (accessed 2 July 2020).

Cabeça, S.M., Gonçalves, A.R., Marques, J.F., and Tavares, M. (2020) Creative tourism as an inductor of co-creation experiences: The CREATOUR project in the Algarve. In: Pinto, P. and Guerreiro, M. (eds.) *Handbook of Research on Resident and Tourist Perspectives on Travel Destinations*. IGI Global, Hershey, Pennsylvania, USA.

Correia, A., Mortal, M. Costa, C., and Peres, R. (2008) The determinants of gastronomic tourists' satisfaction: A second-order factor analysis. *Journal of Food Service* 19, 164–176. doi: 10.1111/j.1745-4506.2008.00097.x.

dos Santos, R. and Antonini, O. (2004) La gastronomía típica de la isla de Santa Catarina, Brasil: su identidad como atractivo para el turismo cultural. *Estudios y Perspectivas en Turismo* 13(1–2), 89–110.

Ilincic, M. (2014) Benefits of creative tourism—the tourist perspective. In: Richards, G. and Russo, A.P. (eds.) *Alternative and Creative Tourism. ATLAS*, Arnhem, The Netherlands, pp. 99–113.

Mak, A.H.N. (2018) Motivations underlying tourist food consumption. In: Pforr, C. and Phau, I. (eds.) *Food, Wine and China: A Tourism Perspective*. Routledge, Abingdon, UK, pp. 72–95. doi: 10.4324/9781315188317.

Nistoreanu, B., Nicodin, L., and Diaconescu, D. (2018) Gastronomic tourism—stages and evolution. *Proceedings of the 12th International Conference on Business Excellence*, Sciendo, Warsaw, Poland, pp. 711–717. doi: 10.2478/picbe-2018-0063.

Richards, G. (2002) Gastronomy: An essential ingredient in tourism production and consumption? In: Hjalager, A. and Richards, G. (eds.) *Tourism and Gastronomy*. Routledge, London, UK, pp. 3–20.

Turismo de Portugal (2017) *Estratégia do Turismo 2027—Liderar o turismo do futuro*. Turismo de Portugal, Lisbon, Portugal.

World Tourism Organization (WTO) (2012) *Global Report on Food Tourism*. WTO, Madrid, Spain.

10 Interview with Geetika Agrawal about Vacation With An Artist

GEETIKA AGRAWAL*
INTERVIEW BY CLÁUDIA PATO DE CARVALHO

Vacation With An Artist, United States

Cláudia Pato de Carvalho (CPC) – Thank you for participating in this interview (and book), Geetika.

Geetika Agrawal (GA) – First of all, what a lovely project! This topic of creative tourism hasn't been documented very well in the world and is very fragmented. It is so nice to learn about this project, which contains valuable insights and stories on what people are doing so the space can grow. I am very grateful that you guys reached out and I am very excited and happy to share everything that can help this book and this space ... and I am very passionate about this space, so thank you for including me.

CPC – How and why did your company begin?

GA – My background is in architecture and design and as a student of Bauhaus philosophy I was exposed to this philosophy of combining multidisciplinary crafts and technology and seeing the impact of that in our everyday lives. To do that, I used to travel to different parts of India to work with different crafts people. I would spend weeks and months just working with them. I was a student and didn't know much. I realized how much it transformed me as a person. It was not so much about learning the craft ... and today you can learn about everything in the internet. What was transformational was being in the company of an artist. Artists are the creative lifelines of a culture. They shape a culture. They define it. And they hold, I feel, the wisdom of a culture. And to be in the presence of them, not just learning the crafts and the techniques, but also seeing the place through their eyes, because the crafts have so much influence of the place that they are in ... it was all so transformational ... and I didn't realize that. So, over the last 20 years I have been doing the same because it is just the only way I know how to travel.

Now I live in New York, with a very busy job, sitting in front of the computer all the time. Our hands are not even touching any materials ... I started craving it more and I saw that the younger generation was craving it too. Craving the opportunity to step outside their day to day, their routines, their screens and expand and grow. And I thought: this is such a transformational experience, what if I bring this to more people around the world?

People have become more conscious about experiences, not just materialism, not just buying things. They want to have experiences that make them become a better person, that help them grow. Now you have all these cultural programmes, mastering and teaching things, and people are spending more time with yoga teachers and meditation. And that is all to expand our minds.

And on the other hand, you have artists – you have that power and even ten times more and they have been kind of neglected in our society. So, my goal was to help people discover this amazing wisdom we are surrounded with in our cities and places and tap into that. That was the main reason why I started Vacation With An Artist (VAWAA). Over

*Corresponding author: geetika@vawaa.com

© CAB International 2021. *Creative Tourism: Activating Cultural Resources and Engaging Creative Travellers* (eds. N. Duxbury, S. Albino, and C. Pato de Carvalho)
DOI: 10.1079/9781789243536.0010

the years, I started to work with many masters. I also realized that, by doing this, we are contributing to knowledge transfer because a lot of crafts are not documented, the artists get old, they die, and the crafts die with them. If we can help people learn with these masters and help with knowledge transfer between generations, we can have this tiny impact on the preservation of global crafts. That has also become a big part of our mission.

CPC – Why do you think this project is important? From the experience of people who come to visit these artists, what do you see happening that is important?

GA – There are a bunch of things. The project touches on a few trends that we are seeing in the world. First of all, we as a society, we are tired of mass consumption. We want things to be handmade and we want to know the story about how it was made, what were the materials, and what is the source. People are just tired of seeing things on Amazon and having no history, no heritage, nothing to it. So, I think it brings people closer to objects and things around them and, because of that, people feel connected to even themselves as human being and to others. So, that is the main thing. It gets us out of mass consumption and mass commoditization to connect with something that is handmade, that takes time to make, that is not perfect, that is imperfect.

There is a beauty in it and because we are so surrounded by technology, to see something that is so human makes it emotional, it gives it a soul. So this attracts people: it is this emotional aspect that people are craving so much. So, that is the second aspect.

The third aspect is that, as Picasso said, 'every child is born an artist. The problem is how to remain an artist when you grow up'. And I cannot tell you how many people who booked through VAWAA said 'I am not an artist!' and it is so sad to think that an engineer who is seated behind a computer is not creative. It is so sad to see that a lawyer is not creative because they do not have the 'title'. So, what happens? Our day-to-day work suffers because everything we do requires creativity. So, we aspire to help unlock the idea that 'You are creative, everyone is creative. You just need to find it and find those juices flowing.' Imagine what our world could be if we tapped into our creativity? We would not be seeing the world that we see today. It would be a completely different world.

So that is one struggle: people don't think that they are creative. But then they feel that they are stuck. Once you are in doubt, where do you go for this? You cannot go back to school. Back in the day we had access to these masters and you could do an internship. But today, where do you learn? You can go and take an hour-long class and that is it. As an adult, you don't have that the opportunity to find that creativity. VAWAA also provides a very safe and welcome environment in a studio of an artist to immerse yourself in it and experience it yourself. Every single VAWAA guest has left the studio after a few days completely transformed. Because they suddenly realize that the creativity is not in the perfect drawing. Creativity is feeling comfortable to get out of our comfort zone and play. We forget to play.

These are the three top things that really have impact on a person. And I can say a fourth thing: the experiences become a launching pad for people to do something more. If you are an artist, to work with another artist is a way to learn and improve and expand your skill. We have people who have started their own businesses after making an internship.

CPC – I was just going to ask you that: if you do know what happens to visitors who come to you. Do you have that kind of monitoring?

GA – We do but it is not formal. We stay in contact; we call it the VAWAA family and some people go to multiple VAWAAs and we see ... people sometimes discover an art form and end up doing a book on it or go to another artist to develop the craft even more. Because they interact with another artist, we see that this influences their new learnings. It is so beautiful and so powerful. And sometimes we have the artist and the guest collaborate too. They become friends and they do exhibitions together.

CPC – Do you have a way of evaluating the impact of these experiences on these travellers? Or it is a more person-to-person evaluation?

GA – Yes, it is more person-to-person because it is more story-based. We don't have a metric base to evaluate this impact. But the way we see it is that if we feel that before going to a VAWAA they were feeling stuck somewhere or they had an idea or goal to accomplish and afterwards they achieved that or if they felt unstuck, then our job is done. So, for us it is more about improving that opportunity for people to do whatever they are dreaming of, to provide guidance. Before each VAWAA, we ask people 'what is their goal?' and 'what are they looking to get out of this?' It's not as if you land in a studio and everything has a curriculum. No. We ask people: 'What do you want?' Because someone might say: 'I am interested in learning the glazing technic and how you apply to a specific material...' So, some people are very specific about what they are interested in learning, while some may say: 'I've lost my husband and I feel stuck in my life'. Or someone may say: 'I've just retired and my all life I was just busy after my corporate career. But I was an artist and I lost that artist in me and I want to rediscover it'. Sometimes it can be: 'I've always wanted to learn calligraphy'. And to learn from a grand master is Japan is so powerful! So, people have dreams and if they are able to achieve those dreams, that is impact.

CPC – Do you have a profile of the people who come to look for this type of experience?

GA – It's interesting. Our youngest guest has been 8 years old and the oldest one has been 78 years old. It's really wide, which proves that we are creative. But I would say that the core is usually 30 to 45 years old, and I feel that is because, at that time, we have been doing the things we have been doing, we have a little bit of money also, and now we are looking for what is next in my life. They are usually female – 70% of them are females. I don't know why that is, but I think women are more self-growth oriented. I am making a very stereotypical statement here, this is very generic. There is no research that is backing this; I am just saying this from observation. They seem to be more reflective and into self-growth. And they tend to be more into creative arts. But also, the way VAWAA is structured, it is a very safe way to travel for people who are travelling internationally. This gives them a framework, but in that framework they have a lot of space for adventure.

CPC – Can you tell us a little bit about that structure?

GA – Typically, most creative experiences are tightly scheduled: these workshops will happen from this to this date, on this morning we will do this and then this... and there is a group and everyone in the group does that. It is very rigid. On the other hand, with entirely independent travel, you go and do it on your own, just like no structure because you are left on your own. We are kind of in the middle. Over five days, roughly you will spend about five hours in the studio with the artist but there is no fixed itinerary and it is one-on-one. So, when you land there, it depends on your chemistry with the artist. The day can shift so there is a lot more flexibility. Let's say that in the third day, you make ceramic pieces and they are in the kitchen and they are firing, you can't do anything. And the artist says: "It is a great summer day, so let's go for a drive in Croatia!" So, it is more spontaneous. You have the artist as your point person who becomes your friend and that makes it very safe.

CPC – What about the challenges? What are the challenges of a project of this nature?

GA – The challenges are connected to the fact that people think that they are not creative, so they think that they don't need this. Over time, as people are learning this about themselves, it will grow. But when it's new ... people didn't grow up going for creative vacations, you know? They went for skiing vacations, they went for cooking vacations, but they never went for creative vacations. It is a very new idea for people, so they don't even realize that this is something that they could be doing or that it would lead to this magic. Even yoga vacations, wellness vacations – people never used to go for yoga vacations or to yoga retreats, or wellness vacations. But it is a common thing now because people now realize that yoga makes them feel amazing. So, the biggest challenges are, first, people realizing that this is something they need and, then, the power that this experience can have on them. Education to people, telling them what this experience can unlock – this is the biggest challenge.

CPC – What about the artists themselves? What can you tell us about how these experiences have impacted the artist, that changes the way he/she sees his/her work? What are the possibilities that happen after these encounters?

GA – The artists with whom we work are already masters. They have developed these wonderful skills, they have been doing this for years. So, you are right, they have a lot of stuff going on. However, as artists, you are always looking for inspiration and exchange of ideas. So, if you are in Slovenia, if you are in Vietnam, imagine someone coming from Colombia. You suddenly have fresh ideas in your studio. This person who speaks a different language, who looks different, wears clothes different, talks different, thinks different. And for an artist, that is a gold mine. And the biggest benefit to artists is that they have the world coming into their studio, they have the inspiration coming into their studio. The exchange of ideas between them and this person is something that they find very valuable. Another thing is that they also like the idea of teaching their craft because, as you know, artists are normally just busy making, making, making. But when you teach something, it makes you process the art in a different way. So, by teaching, by giving back, by seeing it learned by another, it highlights that their knowledge is useful for someone else. What they are doing, it brings joy. And revenue, of course. They have more money. So, I would say that these are the three biggest benefits: inspiration, validation through teaching that brings joy, and revenue.

CPC – What type of strategies do you use to disseminate this project? I imagine digital tools are the most common, but do you have others ways to reach and engage people?

GA – We produce a regular e-newsletter/communication that profiles different artists and opportunities. But right now, it has mainly been word of mouth. So, people do it and tell other people. We have been kind of relying on that right now. I am sure there are other ways. We are always looking for other partners, so that is another way we want to grow. But we want to keep it very thoughtful.

CPC – So, you also want to keep it very close to people and not massive because it is a very 'niche' kind of work, is that it?

GA – Yeah, it is niche right now, but our goal is that it is going to be big, to expand. Right now we are working with artists in 27 countries and there are so many amazing masters you can learn from. But we also do not want to be 'big too fast'.

CPC – And how do you look for the artists? Are you part of any networks that give you access to those artists?

GA – We have a full curation side, so we have artists who apply to the site. We have thousands of applications and we keep going through them. Also, the VAWAA artists tell other artists so it is word of mouth on this side too. We have been published in a lot of magazines, so the word spreads. And I am a designer so I have my own networks that I have established over the years. So, we tap into all of that.

CPC – How many people do you have working with you?

GA – I am the only full-time employee and have another four people in part-time positions.

CPC – What about the future? What are your plans for the future of the company?

GA – We have big plans. As you may have seen in our website, we have been offering online sessions and we have been very successful. We really like this idea because – imagine: you are in the Zoom, with an artist from Morocco and you are sitting in New York, and it feels so amazing. There is also some kind of intimacy ... and we have a group of people who join from around the world, so it is not just us. You meet people from Johannesburg, from the Philippines, from Spain, from Brazil and everyone is on the same Zoom, so it becomes a very global cultural exchange environment. So, we are going to continue to keep growing that to see what evolves. Even after the pandemic is over, we are going to continue doing that because it is still very powerful. While we love the handmade, there is something to be said about technology. And it makes it easier for people who wanted to go on a VAWAA experience and could not travel – this way, just from the home they can connect with the artist.

We are still continuing to add more artists, so we are going through applications right now. And our goal is to keep developing products that are useful for audiences, for the community, and for the artists. We are also doing an interview series because there is just so much amazing knowledge and there are so many ways we can use to keep transferring it.

CPC – Going back to beginning of our conversation, from your perspective, what are the key challenges for creative tourism in the future?

GA – One, is that it is very fragmented. There are things happening, individuals are doing it, small businesses are doing it, but there is not someone or one company that is leading the pack and setting the future of creative tourism. In that space we are probably the only one who is at the global level, at that scale, doing something. So, we hope to be the ones to lead the wave for what creative tourism could be in the future. Like I said earlier, education around creative tourism – that is a big challenge. People just don't know. I think that in people's minds, creative tourism still means: either you go to a cooking class or you go to a place where there is a stand and you make a handkerchief, you dip it in ink and you let it dry, and it's art, you know? Or you go for some quick painting class. What you see in the masses is very kitsch right now. Of course there are independent artists doing retreats and workshops and they are great. But how do you even find them? So, in summary, key issues are: educating people on what creative tourism can be and what are the opportunities, and building a way so that people can find it.

11 *I need you, but I don't want you*: The ambiguity of artistic residencies and tourism

Sara Albino* and Carlos Alcobia

CIDEHUS, University of Évora, Portugal

Introduction

This contribution comes from a double-sided experience of setting-up a creative residencies programme in a rural context and engaging as action-research academics in the fields of maker education and creative tourism. In 2015, Carlos Alcobia and Sara Albino founded Buinho, a private rural FabLab (digital fabrication laboratory) in Portugal (see Fig. 11.1), which has been a pioneer in the context of multidisciplinary artistic residencies with a focus on digital making. Back then, very few FabLabs in the world had an in-house residency programme that enabled artists and creatives to explore the potential of digital fabrication while being able to explore interconnections with local community as a basis of inspiration and creative place making. Furthermore, Buinho has the distinctive characteristic of being an entrepreneurial initiative located in a remote historic town in Baixo-Alentejo, since FabLabs and makerspaces originated from urban contexts and are mostly publicly funded and concentrated in those settings.

The artists in residency (AiR) programme attached to this project has been developed according to its mission of enabling a transient community for the FabLab while raising the profile of the place through creative interventions and community-led activities. Buinho's motto has emerged to be 'serious play', as the Buinho's AiR programme has been much inspired by Michael Resnik's idea of a 'lifelong kindergarten' (Resnik, 2018). Buinho acts as an enabling arena for both artistic research and technological experimentation. Creative practices are unleashed from formal methodological restraints through a programme that privileges an experimental approach.

Creative travel as a mean of inspiration and self-development

Travelling to be creative finds itself amid a poetry of geography, where inspiration is part of a purposeful process that can be enhanced by externalities such as nature, working with local communities, the social environment, intellectual, political and religious freedom, and openness for experimentation. As children we have a borderless mind and we are more prone to be creative, but the creative spiral is a process that requires training from our early school years (Robinson and Aronica, 2016). As such, creative professionals such as writers, architects, visual artists, designers, researchers, etc. train themselves to develop their critical and imaginative thinking and often choose enabling places to enhance their production fluency, exchange ideas, and network amongst peers (Helguera, 2011; Ptak, 2011).

The philosopher Michel Onfray (2009) talks of travellers as representatives of a clash between two worlds: the world of the cosmopolitanism of nomad travellers and the nationalism of sedentary peasants that spans history – from antiquity to the contemporary consciousness of a networked society. For this reason, artistic residency programmes have been seen, but not exclusively, as a shelter for those nomads who need places with a certain place-making

*Corresponding author: saralb@uevora.pt

 Creative Tourism: Activating Cultural Resources and Engaging Creative Travellers (eds. N. Duxbury, S. Albino, and C. Pato de Carvalho)
DOI: 10.1079/9781789243536.0011

Fig. 11.1. Buinho FabLab. Photo by authors, used with permission.

environment as well as peers to help unleash their creative spiral, which very often is persecuted in rigid working environments and established routines or, more seriously, in non-democratic societies.

The progression of AiR spaces

In the knowledge society, creative spaces have evolved into multiple configurations. Beyond the artistic residencies world, which has gained a significant expression and dimension over the last two decades, the widening of the creative industries with the appearance of new digitally oriented creative professions has stimulated the appearance of new typologies of creative spaces and support services. We observe growing convergence in the relation of digitally oriented fabrication spaces (FabLabs, makerspaces, hackerspaces) with co-working spaces and process-oriented support such as AiR programmes. As a new class of creative nomads has emerged, it has become increasingly popular to connect this new generation of spaces for those who work and live remotely to curated contents that enhance experimentation. Furthermore, municipalities and museums have increasingly entered the sphere of artist residencies both as promoters of socially engaged projects targeting secluded communities and intangible heritage practices, and also as providers of cultural equipment and facilities as an attempt to democratize the position of the arts and provide access to a wide public (Badham, 2017).

These twenty-first century spaces, born initially in urban contexts, also receive creative travellers and have challenged the nineteenth-century idea of utopian closed artist retreats and colonies sponsored by philanthropists (e.g. the Birdcliff Retreat in Woodstock, USA, a utopian self-sustained arts and crafts community founded by Ralph Radcliffe Whitehead in 1902). New digital spaces are evolving and even moving to rural settings, as secluded places can provide comparative advantages as creative retreats and opportunities for new approaches to place making.

A novel and hybrid approach to AiR programmes enables new opportunities not only for creative nomadism, but also for creative tourists. Nevertheless, creative tourism within AiRs or artist-led programmes exposes an ambiguous relationship between two worlds that cross the intellectual, cultural, and economic dimensions. Despite economic dependency, the arts sphere repulses tourism viewing it as analogous to an industrial and mass activity to which it pretends to be an alternative. Such a contradictory relationship seems to reinforce John

Ruskin's statement, 'Life without industry is guilt, and industry without art is brutality' (1870, sect. 95).

Artistic residencies under the shape of artist retreats and colonies emerged related to philanthropy and, although precarious, have been an opportunity for creatives to find good working conditions and position themselves in the art markets. Nowadays, an AiR can range from a programme inclusive to all creative areas to exclusive thematic areas and projects. In opposition to AiR programmes funded by private cultural institutions, philanthropists, and governments, self-paid AiRs have a wider flexibility in different aspects and do not compromise artists in completing and showing the results of their work at the end of the period of stay. Self-funded AiR programmes can provide a short-term focus for artists and writers who seek a season of retreat for a week-long workshop, a summer course, or a one- to three-months-long residency where life-practice is mixed with conviviality (Fig. 11.2).

Artistic residencies and tourism – bridging two conflicting worlds

At the first sight it may seem that the main differences between an AiR Call for Artists and a tourism-led artistic experience is in the discourse used, the curatorial selection board, and in its founding philosophy. Going deeper, these activities differ in their funding schemes, in economic activities legislation, and in the content design and structure of AiR programmes.

Artistic residences for professionals focus on offering a process-oriented programme that involves curatorial mentorship, and sometimes even proposes a common given topic for focus (e.g. post-colonialism, alterity). It also may provide production support and often includes the realization of activities, such as artist talks and peer presentations, that stimulate the creation process and offer advice in artistic research. These programmes accept artists upon application, based on a project to be developed and analysis of the artistic portfolio. The residency programmes that have support from funding schemes are usually stricter regarding the dimension of the portfolio and curriculum of the artist and prioritize the contribution of the residency experience to reinforcing and advancing the individual's professional development. Self-funded residency programmes tend to focus more on the quality of the proposed project due to a need for high-quality inputs for place making and visibility of the residency programme in the arts sphere.

Fig. 11.2. Discussion of the artistic research processes amongst artist in residence peers, Buinho artist in residence programme, 2019. Photo by authors, used with permission.

Finally, creative and cultural tourism within the artistic residency scene tends to not be as strict, regarding neither the artistic career dimension nor the proposed project during analysis of the applications for the AiR programmes. In these cases, the focus is on targeting travellers that look for embedded local culture and nature experiences while having a learning experience in arts, crafts, or digital technologies through participating in beginner courses and workshops. Some of the residencies described previously may also find reason to participate in this 'simulacrum' approach to an artistic life, which ultimately compromises the artistic identity of the AiR programme.

Clashing discourses

While the academy discusses the frontiers between the concepts behind cultural, creative, and educational tourism activities, most often practitioners direct themselves to target niche publics that mingle these three dimensions according to their needs and scope of action. As a result, they offer programmes that emphasize both the creative and process-oriented dimensions of the classic AiRs and the cultural landscapes and accommodation experience of cultural tourism.

Nevertheless, higher degree cultural tourism AiR programmes tend to go beyond such a compromise, promising to elevate the status of a common guest or tourist to a 'prosumer' who creates or co-creates an 'artistic' experience – an artist for a week. For instance, the Cerdeira Village Art & Craft Project (n.d.), located in Lousã, Portugal, explores the rural dimension as the main feature for inspiration for guests enrolling in its workshops, with a focus on short-term stays. There are also examples of city-based creative tourism programmes, such as ART-POLI Painting on Location International (https://rivet.es/orgs/art-poli-r, accessed 10 August 2020). Despite appearing in the directories of artistic residency networks such as Rivet, ART-POLI is an arts and humanities travel programme and academy that offers a culture-led programme with language classes, tours, culinary studies, gallery exhibits, critiques, and readings in order to 'inspire a creative, unique personal journey for each participant' (https://rivet.es, accessed 6 August 2020).

Such different spheres clash in discourse, although the arts and tourism are increasingly economically interconnected through aspects of consumption behaviours. For instance, Lithgow and Wall (2020) question the dynamics of AiR programmes that focus on hospitality, conviviality, and the personal connection between the guest artist and the host institution that configures the residency. The division between programmes with a creative tourism motivation and a professional practice focus are becoming more visible in the public offer and advertisement of programmes featured in the main AiR international networks, such as *Wooloo* (www.wooloo.org, accessed 6 August 2020), *Res Artis* (https://resartis.org, accessed 6 August 2020), *TransArtists* (https://www.transartists.org, accessed 6 August 2020), *The Alliance of Artists Communities* (www.artistcommunities.org, accessed 6 August 2020) and *Rivet* (https://rivet.es, accessed 6 August 2020). The presence of different types of spaces and programme offers varies according to the acceptance criteria of these networks and their target audience.

Challenges and learning outcomes

Since the valorization of an artistic career is related to international artistic experiences and works developed in AiRs, professional artists, whether being funded or self-funded, tend to apply to the best programmes that can add value to their learning path and artistic research. In parallel, digital nomadism related to creative and entrepreneurial professions has caused the appearance of spaces focusing on new modes of production, most of which are proposing functioning models inspired by the process-oriented AiR programmes. As such, the main challenges and opportunities facing the organization of international artistic residency programmes are related with the issue of defining its own identity while facing different target groups.

The first challenge is related to the existence of new creative professions that require new spaces for practicing. These new spaces of production offer new approaches and processes for learning and creation, while appropriating the operational model of the more traditional AiR programmes. The appearance of such new actors has pushed the AiR sector to become more competitive and dynamic to the emergent requests of a new class of creative practitioners.

The second challenge for AiR programmes brings us to a higher level of complexity related to AiR's own identity and boundaries, and it is a direct result of tourism-led stakeholders who compete in the same networks as professionally led programmes.

The fact that the AiR as a work practice born from the arts is no longer exclusive to this sector – it is also being reproduced by the cultural, creative, and educative tourism sectors as a business opportunity – is causing a defensive movement within the arts sector. These resistances are found within the cultural institutions, networks, career artists, arts scholars, and curators who use AiRs to enhance their career profiles and self-development.

Nevertheless, and in regard to both challenges, the resistant cultural organizations who opted to remain faithful to the original principles of AiR now find themselves in the ambiguous situation of needing to play with multiple stakeholders and to engage in open activities with the broader public in order to enlarge the scope of their self-funding opportunities (Bishop, 2005; Leban, 2017). The widely chronicled instability and economic precariousness of the cultural sector has pushed cultural organizations to see creative travel and tourism as a base for its stability. It is in this ambiguous relationship between the defence of the original and more pure identity regarding AiR programmes and the need for sustainability that, in part, creative tourism activity can provide, that a state of imbalance is sustained between opposing forces.

From our learning experience, we do not believe that pressure from tourism companies working in the field is causing a loss of identity in AiRs due to the existence of a growing specialization of spaces and positioning of networks where projects become featured. Firstly, because we are witnessing an expansion of concepts and categories, which generates a new equilibrium and alternative

Fig. 11.3. *Ver o Extraordinário no Ordinário*, 2019. Installation artwork by Luba Zygarewicz developed at Buinho's AiR Summer Residence Programme. Photo by authors, used with permission.

perspectives without compromising the identity of AiRs. Many of the new typologies of creative spaces focus on a new class of transient digital nomads and are able to function without strangling sectorial entropies.

Secondly, the 'borrowing' of the AiR framework within a logic of sustainable creative tourism practices, focusing on different forms of education in the arts field through experimentation and critical thinking-led processes, may actually contribute to a better public understanding of the contemporary arts. Furthermore, when located in secluded rural contexts, creative travel-led AiRs may contribute to the valorization of communities and their endangered tangible and intangible heritage (Fig. 11.3). As such, it is believed that both culture and tourism narratives can co-exist in the context of the new spaces of creative production and the smart specialization of places.

References

ART-POLI (n.d.) https://rivet.es/orgs/art-poli-r (accessed 10 August 2020).

Badham, M. (2017) The social life of artist residencies: Working with people and places not your own—part 2. *Seismopolite*, no. 19. Available at: www.seismopolite.com/the-social-life-of-artist-residencies-working-with-people-and-places-not-your-own-part-2 (accessed 2 September 2020).

Bishop, C. (2005) The social turn: Collaboration and its discontents. *Artforum* 44(6), 178–183.

Cerdeira Village (n.d.) www.cerdeirahomeforcreativity.com/art-village (accessed 12 August 2020).

Helguera, P. (2011) *Education for Socially Engaged Art: A Materials and Techniques Handbook*. Jorge Pinto Books, New York, New York, USA.

Leban, S. (2017) Art in residency: Precarity or opportunity? *Seismopolite*, no. 18. Available at: www.seismopolite.com/art-in-residency-precarity-or-opportunity (accessed 2 September 2020).

Lithgow, M. and Wall, K. (2017) Embedded aesthetics: Artist-in-residencies as sites of discursive struggle and social innovation. *Seismopolite*, no. 19. Available at: www.seismopolite.com/embedded-aesthetics-artist-in-residencies-as-sites-of-discursive-struggle-and-social-innovation (accessed 18 August 2021).

Onfray, M. (2009) *Teoria da Viagem. Uma Poética da Geografia*. Edições Quetzal, Lisbon, Portugal.

Ptak, A. (ed.) (2011) *Re-tooling Residencies: A Closer Look at the Mobility of Art Professionals.* CCA Ujazdowski Castle & A-I-R Laboratory Web Design: Berber Studio, Warsaw, Poland.

Resnik, M. (2018) *Lifelong Kindergarten: Cultivating Creativity through Projects, Passion, Peers, and Play.* The MIT Press, Cambridge, Massachusetts, USA.

Robinson, K. and Aronica, L. (2016) *Creative Schools: The Grassroots Revolution that's Transforming Education.* Penguin Books, New York, New York, USA.

Ruskin, J. (1870) Lecture III. The relation of art to morals. *Lectures on Art, Delivered before the University of Oxford in Hilary term, 1870.* Project Gutenberg Ebook #19164, release date: September 3, 2006. Available at: www.gutenberg.org/ebooks/19164 (accessed 2 September 2020).

Part III: Designing and managing creative tourism products

This central part of the book is divided into three sections focusing on different dimensions of designing, launching, adjusting, and propelling a creative tourism initiative forward. The first section, 'The cycle of product design, implementation, and iterative adjustments' (chapters 12 to 18), focuses on conceiving, designing, testing, and launching a creative tourism initiative. The second section, 'Marketing – reaching and engaging creative travellers' (chapters 19 and 20), examines the multifaceted aspects related to marketing, promotion and communications, commercialization, and market readiness, and is a practical compliment to the chapters in Part I on identifying and understandings different types of creative travellers. The third section, 'Managing and innovating creative tourism products' (chapters 21 to 23), looks at the longer trajectory of managing and evolving creative tourism initiatives and portfolios of creative tourism experiences over time.

The cycle of product design, implementation, and iterative adjustments

To begin, the essential question is: How to design a creative tourism product? This section brings together different experiences and perspectives on this process, offering advice and insights on important steps and factors to consider. These approaches have been used to inspire, inform, and guide the design and development of new creative tourism products in Thailand, Portugal, Slovenia, and Colombia. In Part IV, a 'companion' chapter by Larissa Almeida presents a fourth perspective on the creative tourism development process, focusing on the development of creative tourism initiatives embedded within communities as a local development and empowerment strategy.

In 'From theory to practice: The three principles of community-based design' (Chapter 12), Jutamas (Jan) Wisansing elaborates on the insights gained from an experimental learning journey, a creative tourism pilot project initiated by the Designated Areas for Sustainable Tourism Administration (DASTA) that was developed in villages in six designated sustainable development areas in Thailand. She explains the project's ten core principles, which were organized into three main skill-enhancing schemes: the 3Ss – story and storytelling, senses, and sophistication – and how these provided guidance and touchstones in the design and development of creative tourism initiatives. The creative tourism development programme was closely aligned with sustainable community-based tourism that embraces a decentralized, bottom-up development approach that aims to empower community members, preserve local heritage and revive endangered cultural practices, and provide opportunities for younger generations to get involved in cultural innovation.

In Chapter 13, 'User-centred design for creative tourism prototyping: The Maribor experience', Fiona Eva Bakas, Tiago Vinagre de Castro, and Ana Osredkar discuss how principles of user-centred design and emotional mapping can be used to help tourism and culture practitioners design prototypes of creative tourism experiences. They report on how a creative tourism prototyping lab held in Maribor, Slovenia, guided participants through a multi-phased user-centred tourism experience design and prototyping process. They also discuss how cultural mapping (applied in the CREATOUR® project) and emotional mapping (applied in Maribor) processes can guide people to think imaginatively about their place, what makes it so special to them, and how territorial connections can evoke emotions – all of which can serve as inspirational resources into the development of creative tourism products and embodied experiences. Reflecting on their experiences designing and managing the creative tourism prototyping lab, the authors provide advice for others who may wish to adopt this approach.

In 'Bogota through the 5 Senses: Building a creative tourism start-up' (Chapter 14), Diana Zuluaga and Diana Guerra present the start-up story of their company, 5Bogota: Travel with Locals, and explain the methodology they use to create and consolidate the experiences offered by 5Bogota. This methodology comprises eight stages, including context analysis, the study of the destination's characteristics, the establishment of the typology for the traveller and for the local host, the design of the tourism product, and the marketing strategy. In closing, the authors share the lessons they have learned through applying this methodology and identify key success factors for entrepreneurship in creative tourism.

In all these processes, identifying local cultural assets is an essential first step in designing place-specific and locally meaningful creative tourism (and other) initiatives that can create opportunities for bottom-up, community-led development. In Chapter 15, 'Deep mapping as a cultural mapping process and a creative tourism driver: Two examples', Kathleen Scherf offers insights on developing a cultural mapping or deep mapping process to capture the physical, intangible, and organizational resources that can provide a platform for the development of creative tourism activities. She presents two examples, the FoodLoops map she developed in Kamloops, British Columbia, Canada, and the Literary Atlas, which plots English-language novels in Wales, a compendium of information that is conveyed through a digital platform of literary geography. In presenting the examples, Scherf extracts a series of points of advice for others embarking on similar projects.

The next two chapters provide first-hand accounts of the launch and development of two creative tourism projects developed within the CREATOUR® project in Portugal.

In 'The creative tourist and the creating artist: The base for rural development' (Chapter 16), Catharina Sligting tells the story of her farm, Domínio Vale do Mondego, explores how arts-based creative tourism can contribute to engender positive attitudes of city dwellers towards rural and natural areas, and reflects on the impact of arts festivals and other activities on the local area. She explains how her arts-based entrepreneurial approach to the development of the rural art festival, Estival da Estela, provides an engaging environment for resident-participants in which there is no 'backstage', enables professional artists to imaginatively create new works of art inspired by local elements, and fosters collaborative approaches to arts making. She also discusses the challenges and emergent approaches to engaging local residents in activities largely attracting international visitors. Based on her extensive experiences over 10 years, she provides advice for others who may wish to create an arts-based creative tourism business in a rural locale.

In Chapter 17, 'COOLWOOL – Creative weekend at the city of Covilhã, a co-designed programme', Rita Salvado of the Museu de Lanifícios da Universidade da Beira Interior, and Guida Rolo of New Hand Lab tell us the story of the story of COOLWOOL, an inter-organizational collaboration to develop creative tourism activities based on their town's deep historical and contemporary connections with wool fabric producing and processing. Initially focusing on the organization of weekend programmes with creative workshop activities that would run all year long, they complemented this with smaller introductory workshops of various types. In the chapter, the authors reflect upon the challenges of designing and implementing a project focused on wool making and industrial cultural heritage in a city known for its seasonal focus on winter sports and snow. The chapter also addresses the importance of partnerships in designing and promoting creative tourism projects, mentioning articulation with the touristic *Wool Route – TRANSLANA*. As the COOLWOOL project is focused on networking, the city, and its industrial heritage, the involvement of local governmental bodies is identified as a determining factor to enhance its profile and its further evolution and development.

Passing from a wool-making industrial heritage to seaside landscapes, in Chapter 18, 'Co-creating subtle attractions', Britta Timm Knudsen and Jan Ifversen discuss how architectural landscape interventions can be an approach for inspiring new perspectives and providing new experiences in a seaside rural town in Denmark. Focusing on co-creation practices in tourism, the aim of their brief chapter is to discuss the possibilities of reconceptualizing the main features and forces of attraction of places. Instead of simply conceiving destinations as passive locations waiting to be visited, these authors focus on the idea of knowledge exchanges with tourists as a means to highlight the attractive capabilities of places. Based on two architectural interventions that explore nature as the centrepiece, the authors present five recommendations for coastal tourism and formulate suggestions for creative tourism planning.

Marketing—reaching and engaging creative travellers

With the chapters in Part I as background, this section provides an in-depth look at the many dimensions and details involved in effectively marketing creative tourism projects – identifying and reaching intended participants, knowing about market forces and opportunities, and addressing the challenges of communicating efficiently in an economy of consumption and experience overflooded with multilayered information. In this section, we present the insights of two women entrepreneurs in tourism who share their expertise in marketing strategy and trends applied to creative tourism and the behaviours of people who travel to be creative.

In Chapter 19, 'Linking creative tourism products to markets: Target marketing, promotion, commercialization, and market readiness', Marie-Andrée Delisle offers an array of information and insights on the many facets of marketing, based on her 30 years of consulting experience in tourism. The chapter includes comprehensive advice on how to understand the tourism supply chain in the context of creative tourism products. Within a tourism industry lens, it presents guidance and advice on strategic marketing, outlining four different strategies to foster successful business opportunities through partnerships and networking. Interlocking with the idea of market segmentation, Marie-Andrée Delisle discusses the concept of market readiness and pricing of tourism products. As a tourism planning-focused chapter, Delisle highlights the issue of market intelligence in organizations as essential to creative tourism initiatives, since it is vital to understand the tourism market, its driving forces, and the competitive environment. The chapter also addresses the context of the COVID-19 pandemic in a reflexive manner, mentioning new reconfiguration opportunities and possibilities for creative tourism partnerships and growth.

Elena Paschinger's contribution, 'Creative tourism marketing and social media management' (Chapter 20), complements Delisle's chapter and provides additional insights on marketing and social media. This chapter focuses on the link between creative tourism development and social media, in its scope and the underlying nature of the experience. The author argues that as the travelling population becomes more informed, alternative interest groups have become rooted in the proliferation of diverse social media networks across the globe. Paschinger addresses how creative tourism is also grounded in the digital economy, which has sprung forth with the ubiquity of the internet. She points to a shift from the four Ps of classic marketing (product, price, place, and promotion) to four Cs: creating, curating, connecting, and culture. The author presents the case study of Creative Tourism Austria to illustrate contemporary tendencies and provides a list of suggestions for successful communication with followers on social media.

Managing and innovating creative tourism products

The third section in Part III looks at the longer-term realities of managing creative tourism products over time, commenting on both the development phases as well as ongoing operational realities, the need to continually keep initiatives 'fresh', and how to nurture new and innovative offers.

The Municipality of Loulé's 'Loulé Criativo' creative tourism initiative is internationally known. In Chapter 21, 'Loulé Criativo: Our creative tourism development story', Marília Lúcio of the Municipality of Loulé provides an inside look at its development and implementation. This initiative is particularly noteworthy for its positioning of creative tourism within the small city's broader cultural policy and planning framework which spans from heritage skills training to fostering contemporary art and design practice. Operating for 5 years so far, the chapter outlines the process of establishing and managing the Loulé Criativo initiative and its different activities.

The City of Santa Fe, New Mexico, USA, is known as an early leader in the creative tourism through its hosting of the UNESCO Creative Cities Network's international conference on creative tourism in 2008 and the development of its Creative Tourism Initiative, which operated from 2009 to 2015. In Chapter 22, 'Santa Fe, New Mexico's Creative Tourism Initiative', Sabrina V. Pratt provides an inside look at the introduction and evolution of this initiative, and its legacies today. She provides advice for others wishing to develop a similar initiative.

In Chapter 23, 'Creative tourism development and innovation in Kenya: An entrepreneurship perspective', Lénia Marques looks at the challenges and possibilities to maintain innovation among creative tourism entrepreneurs over time. The chapter considers entrepreneurial initiatives in the Kenyan-Maasai context, focusing on creative tourism initiatives in Kenya and the different types of micro-innovations that can occur at the level of small communities and individuals. Relating to product/service development, Marques provides examples of how creative tourism rooted in locally embedded traditions can enhance the sustainability of communities. Marques' study concludes that while product innovation stands out in creative tourism entrepreneurial initiatives, in order for it to be fostered in the long run and gain critical mass, there is a need for managerial and governmental innovations to become more structural, allowing for space for continuous innovation to occur.

— N.D. and S.A.

12 From theory to practice: The three principles of community-based design

Jutamas (Jan) Wisansing

Perfect Link Consulting Group, Thailand

One of the challenges in developing tourism in many local communities, particularly in developing countries, is the danger of commodification of culture. Existing models of cultural tourism often see culture as a relatively static product to be 'sold' to tourists. By embracing local identity and intangible cultural assets, and concentrating on local creative processes, creative tourism can emerge as a fundamental tool for combatting such negative impacts of traditional models of cultural tourism.

This chapter reflects on an experimental learning journey, a creative tourism pilot project initiated by the Designated Areas for Sustainable Tourism Administration (DASTA) in Thailand. The main objective of this learning journey with DASTA was to develop a Creative Tourism Brain Bank (CTBB) (see Fig. 12.1), working together in a creative tourism lab which aimed to explore the following questions:

1. What constitutes creative tourism, specific to the Thai context?
2. What makes creative tourism different from other forms of tourism?
3. How can we transform community cultural tourism/activities into creative tourism?

The experimental learning journey: Methods

The DASTA team identified 39 potential cultural activities from local cultural assets in six designated areas around Thailand. Then, based on an extensive review of the core concepts of what should constitute creative tourism, ten core principles of creative tourism were formulated. These core principles were used as guidelines or a set of criteria to pre-assess the 39 activities, searching for local artists with high potentiality and readiness to participate in a learning journey.

The 17 selected learning cases started with their shared aspirations of what creative tourism should be and how each local artisan could be part of the activity. The self-assessment acted as a baseline to set individualized development goals that could be achieved in 12 months. The mission was to pursue the transformation of cultural activities into creative tourism activities. This became the important quest of the learning project and for the Creative Tourism Brain Bank to co-create creative tourism cases and answer the key question: How can we transform community cultural tourism activities into creative tourism?

The examination of the working principles revealed a need to develop a capacity-building programme to enhance a set of needed skills. The ten core principles were segregated into three main skill-enhancing schemes: the 3Ss – story and storytelling, senses, and sophistication. Literatures related to applied theatre and design thinking processes were used as a milieu of ideation, reflection, and formulation of simple but practical principles of how to create creative tourism. The ten core principles were re-evaluated during the post-assessment period to ensure their practicality after the intervention process (see Fig. 12.2). This chapter describes and discusses each of the principles.

Corresponding author: jutamas@perfectlink.co.th

 Creative Tourism: Activating Cultural Resources and Engaging Creative Travellers (eds. N. Duxbury, S. Albino, and C. Pato de Carvalho)
DOI: 10.1079/9781789243536.0012

Fig. 12.1. DASTA Creative Tourism Brain Bank (CTBB). Photo courtesy of DASTA, in Richards, Wisansing, and Paschinger (2019), used with permission.

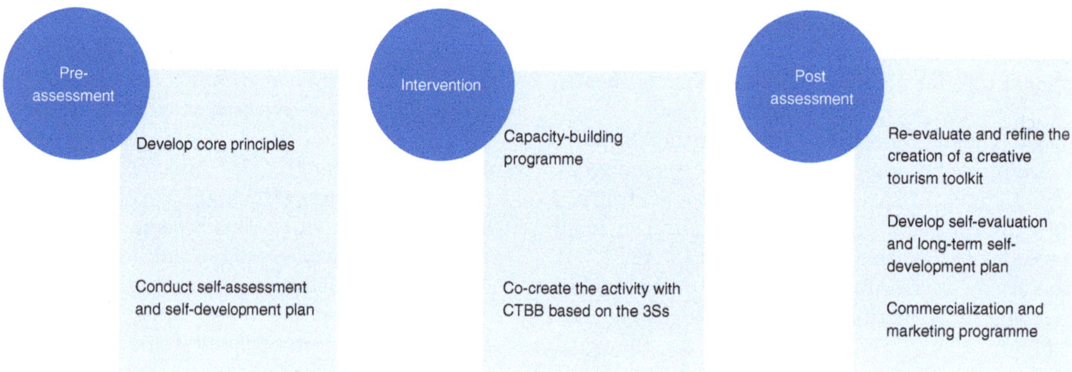

Fig. 12.2. Process of the experimental learning process. Image developed by author.

What constitutes creative tourism? The core principles

Creative tourism is a new term that appeared in the early 2000s coined by Richards and Raymond (2000). Richards (2005, 2011) emphasizes the significance of creativity in the planning and development of cultural tourism. In this context, creative tourism is a new way of redesigning the dull, non-interactive nature of sight-seeing cultural tourism. In that sense, Richards and Wilson (2007) conceive that the cultural landscape of the destination, including festivals, cultural routes, museums, and cultural heritage could be re-created to incorporate an interactive link and more engaging experiences between the visitors and the local population. Creative tourism arguably represents a notable departure from traditional models of cultural and heritage tourism, moving away from tangible heritage as the key asset towards creative and symbolic capital to overcome the dilemma in 'the massification of cultural tourism' (Richards, 2016, p. 31).

The UNESCO Creative Cities Network (2006) identifies creative tourism as 'travel directed toward an engaged and authentic experience, with participative learning in the arts, heritage, or special character of a place, and it provides a connection with those who reside in this place and create this living culture' (p. 3). Creative tourism has also been defined as a sub-segment of cultural tourism, focusing on interactive participation (Smith, 2009). Coinciding with other definitions, in this case creative tourism was invented due to a dissatisfaction with the 'sight-seeing' nature of mass cultural tourism in recent years (Smith, 2005). While Smith highlighted interactive learning, Thorne (2009) added that creative tourism should be a place-based approach of cultural tourism in the context of creative place interpretation.

The Creative Tourism Network agrees that cultural tourism includes creative tourism that provides a connection between visitors and residents through creative experiences (Rivas, 2011). Ultimately, creative tourism engages with active participation based on authentic and unique local experiences, coupled with artistic and cultural knowledge sharing. Authors and scholars who are interested in the topic concur that creative tourism emerged from cultural tourism and could either be a sub-segment or a spin-off from cultural tourism due to dissatisfaction with the lack of interaction between hosts, guests, and the rich authentic local landmarks or aesthetic places. This implies that developing creative tourism resources is not so much a question of building attractions or designing cultural itineraries. Instead, the basic questions that all creative destinations should ask themselves are: What can a creative tourist do here? What can they learn? What cultural processes can they be involved in? What skills and experiences do we have which others do not? This requires a new way of thinking about the culture of the destination.

In traditional cultural tourism development strategies, the development process begins with an inventory of tangible cultural resources. These are usually described in terms of artefacts, buildings, museums, and monuments. Creative tourism development begins from an inventory of local skills from all walks of life, something which requires a much more imaginative approach.

In Thailand, the development of 'creative tourism' has been specifically linked to the notion of adding value to community-based tourism, which is a mainstay of the Thai economy. Aware that conventional tourism models, particularly mass tourism, are subject to decreasing returns over time, Thailand has sought to identify new models that avoid the dangers of 'serial reproduction' (Richards and Wilson, 2006). This is based on the idea that diverse ranges of local culture and assets are unique to a specific place and that unique forms of creativity are attractive because they are linked to the pride of a locality. Creative tourism activities can therefore not only provide an authentic taste of Thai local culture for visitors but can also create strong senses of ownership which lead to the desire to protect the value of such assets.

The creative tourism programme headed by DASTA has been based on this notion, which is applicable to the six designated sustainable development areas. The selected activities are areas mostly taking place in villages situated in less popular cities (as opposed to main touristic places) around the country. Each one is linked to a craft or skill that demonstrates local creativity. The CTBB has been working with local leaders to engage residents to design creative tourism in these villages. The goals set forth in these initiatives are: to empower community members in leading sustainable tourism development, to preserve local heritage, and to revive endangered traditions while providing opportunities for younger generations to get involved in cultural innovation. The model employed in Thailand is a community-based approach because the central thrust of the programme is built 'by community for community' to achieve sustainability and to embrace local creativity. The features of creative tourism make it ideally suited to the development of sustainable community-based tourism. As DASTA has emphasized, creative tourism shares many characteristics with sustainable tourism. The characteristics of creative tourism according to DASTA are:

1. Visitors and hosts, each engaging the other.
2. Cross-cultural engagement/cultural experience.
3. Spirit of place/deep meaning/understanding of the specific cultural feature of the place.
4. Hands-on experiences.
5. Exchange of information/transformation and transformative experiences.
6. More participants than observers.
7. Tourists get the opportunities to develop their creative potentials and participate in the decision-making process (co-creating the tourism experience).
8. Authenticity in both process and product/genuine experience.
9. Memorable/I hear and I forget, I see and I remember, I do and I understand concept.
10. A tailor-made approach.

The ultimate goals of developing creative tourism are therefore to contribute to sustainable tourism development and increase community well-being.

A community-based approach: Prerequisites for creative tourism development

Sustainable development is increasingly accepted as a fundamental objective and guideline for public policy-making and tourism initiatives (Goodwin and Santilli, 2009). It encompasses a balanced integration of economic, environmental, and socio-cultural dimensions into development processes. Researchers in the field have made it clear that a 'better' sustainable tourism planning approach should constitute benefits for all aspects of community livelihoods, including sociological (e.g. promotion of community stability, family solidarity, and cultural identity), economic (e.g. employment, income), and environmental (e.g. conservation/preservation). It is argued that the community will benefit more from tourism development if the community members genuinely participate in making the decisions that affect their welfare and in implementing the desired actions/solutions. Therefore, a community-based approach is considered to be a better form of planning process because of its participatory orientation.

The main principle of the community-based approach is a quest for community inputs through their active participation in creative tourism development processes. This manifests a significant shift in tourism planning from being a centralized, top-down approach to a decentralized, bottom-up approach (Huras, 2015). Such an integrative and inclusive approach to planning puts an emphasis upon understanding the whole tourism system based on rigorous evaluative research involving extensive dialogues among all stakeholders involved. The goals for creative tourism development should therefore be derived from, and integrated into, the overall shared community visions and aspirations.

To achieve this ambitious goal, creativity is needed to stimulate an exchange of ideas and insights not only to discover the features of the local culture but also to expand the potential of that culture. Creative tourism must therefore be designed to facilitate exchanges between participants in order to increase their collective knowledge, skills, and understanding. As Richards (2011) points out, the catalyst that determines the intensity of participation and collaboration is creativity with environments, people, processes, or products as carriers. It manifests as either a background or an activity (Richards, 2011). Creativity as a background covers lower intensity interaction and activities, such as buying (shop window) or seeing (itineraries). It requires awareness and observation from tourists to experience the holiday destination and culture. Creativity as an activity covers high-intensity interaction throughout participatory activities, such as tasting (experiences, open ateliers) or learning (workshops, courses). In this form, it fully involves tourists and residents in collaborative and creative processes of skill development, cultural production, consumption, and place branding. The pilot project presented here focuses on creativity as an activity and addresses the issue of how to design a high-intensity interactive experience.

What makes creative tourism different from other forms of tourism?

Creative tourism developments need to be:

- active rather than passive;
- based on learning processes;
- integrated into the social fabric;
- innovative and dynamic;
- sustainable; and
- based on local resources (interview with Greg Richards, 2017).

Based on these creative tourism characteristics, the visitor journey and the existing cultural processes were carefully scrutinized. The developmental model was conceived for the learning journey, directing our attention to the necessity in engagement and interactions among visitor creativity, community creativity, and marketing creativity. Creative processes designed to unleash creative thinking and sharing between visitor and local artisan must be well designed to optimize the exchange of inspiration through the flow of the learning activities and to link each element to the formulation of the three categories of principles, the 3Ss – story and storytelling, senses, and sophistication, as shown in Fig. 12.3.

The backbone of the principles starts with an evaluative research to develop rich content that could be used as the basis to design insightful story and storytelling, which should be conveyed with full senses to achieve emotional engagement. A properly presented story, legend, or local folklore can be uniquely designed by a creative local studio. The integration of these 3Ss evokes an enrichment of the totality of a memorable creative journey (Richards *et al.*, 2019).

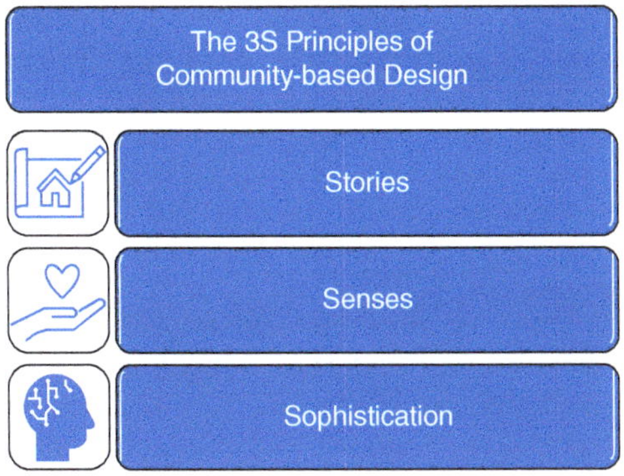

Fig. 12.3. The 3S principles of community-based design. Image adapted from Richards *et al.* (2019).

Each of these principles was further delineated by the Creative Tourism Brain Bank team as follows:

Story and storytelling

1. Evaluative research for rich content to create insightful 'stories' so that the creative tourism activity is designed from unique local cultural assets.
2. Storytellers or local artists could design an elevating story which could captivate the audience and create inspiration for visitors to learn and create their DIY masterpiece.

Senses: Conceptualize to create full 'senses'

3. Creative activity is designed to have participative and fun learning experiences with a depth of understanding from local experts (gurus or artists).
4. Creative activity is designed to have active participation and opportunities for host and guest to share experiences and cultural knowledge.
5. Creative activity is designed to have five senses: hear, touch, taste, see, and smell.
6. Place or creative space is designed to stimulate the expression of creativity.
7. Flow of the activities is well designed to stimulate the expression of creativity.

Sophistication: Beyond basic

8. Creative activity offers visitors the opportunity to develop their creative potential and could lead to the innovation of new ideas.
9. Creative tourism designers should identify the branding message and create marketing campaign.
10. Creative tourism designers should communicate the uniqueness of the activity and implement continuous improvement.

Story and storytelling

Creative tourism development begins with an inventory of local skills from all walks of life, something which requires an imaginative approach. The inventory provides an information bank of potential unique local content. However, it is important to recognize that this content is not a story. Creative tourism attractions and products must be developed from local contents which refer to a set of recognizable tangible, intellectual, spiritual, and emotional characteristics of that society. These characteristics encompass art and architecture, historical and cultural heritage, culinary heritage, literature, music, creative industries, and live cultures with their lifestyle, value systems, convictions, and traditions – each comprising both tangible and intangible assets. A story is the way we organize this voluminous content into a structured framework on which contents can be related. Our research found that the story is a fundamental building block for successful creative tourism, which can link the three vital elements for the design of the creative tourism activity: (i) storytelling; (ii) sense of place; and (iii) sophistication in how the unique local heritage is presented. Once these three elements were combined succinctly, the experience could then be better enhanced.

Story, storyteller, and storytelling skills play a significant part in engaging guests, creating interaction, and sharing memorable experiences. Storytelling and interpretation were identified as a core principle of creating creative tourism, considered as a process of looking for links between themes, resources, and spaces to find the best way to tell the story of the locality. However, it was clear from the pilot project that developing an elevating story that could captivate visitors was not an easy task. The challenging issue was to prioritize local contents into a creative and distinctive brand essence.

Storytelling must be well designed, based on this distinctiveness, and conveyed in a way that connects people, creates communities, and evokes emotions. Nowadays, people recognize the power of story and storytelling because a galvanizing story can help revive local culture and add value to local art, craft, and products. Story and storytelling have been used as a marketing tool as well as an important means of communication. From the pilot project, storytelling in creative tourism enabled the development of local pride through the meaning-making process in which both host and guest could have an emotional connection.

In our case, we applied the metaphor of a stage play in which the total visitor experience equals the combined effect of every element of the play. Experiential creative activities are analogous to theatrical performances; they include a physical environment (stage), local artists and creative travellers (actors and audiences), workshop processes (script), and back-office support (backstage crew). These experience elements and characteristics are often perceived and reflected as a theatre. The Creative Tourism Brain Bank therefore applied an adaptation of applied theatre to arrange an essential creative tourism system; each element is designed as a creative traveller journey, an experience that can be engineered or orchestrated to influence a traveller's overall assessment of an experience favourably. Unique local content (see Fig. 12.4) is conveyed through a story design. Script creation was used as a tool to guide the story design and storytelling process (see Table 12.1).

Senses

A growing body of evidence indicates that fostering cultural and creative activity is an essential strategy in building quality of place, maximizing talent, enhancing sustainability, and defining competitiveness for any destination. From large metropolitan areas to smaller towns, creative placemaking is becoming a crucial means of mobilizing creative entrepreneurship; supporting cultural diversity; attracting new residents, talented workers, and tourists; and leveraging investment (Richards and Duif, 2018). Consequently, policymakers are increasingly interested in supporting cultural infrastructure and facilities that support local talent, enhance diversity, and foster collaboration.

Drawing from this contemporary interest in creative milieu, for the fundamental placemaking of this pilot project, we first focused on the development of small-scale local studios within a destination. Local artists living in a destination could provide an essential stepping-stone for proactive strategies to become a creative tourism destination. Local artists could create essential ingredients for placemaking involving non-market processes and an effort to improve the quality of the lives of all those who use the place. An attractive external image should be a by-product of placemaking, not

Fig. 12.4. Lantern making: local content. Image courtesy of Community Innovation Agency (2019), used with permission.

Table 12.1. Example of how story and senses are composed to present sophistication in a lantern-making workshop

Scene/act 1 (sense of place and placemaking)	Location (the actual location)	Audio/script (compose storytelling)	Props (senses)
Life reflection	Cosy garden upon arrival	Theme of the activity will be presented. The lantern is historically recognized as a merit making ceremony in Buddhism. It involved hard work and concentration.	Local herbal drink is served. A small piece of bamboo will be circulated for guests to feel how delicate it is.

the goal. If a place is made more livable for those who are already there, it should also become attractive to others and could later become recognized as a creative destination or creative city.

Senses in this case are therefore focusing on the concept of servicescape as the environment of a local studio in which the creative activity is assembled and in which artists and visitors interact, combined with tangible commodities that facilitate performance or communication of the creative activity engagement. The aim of the core principles under this category is to conceptualize local studios to create full 'senses' ready to facilitate the participative and fun learning experiences with a depth of understanding from local experts (gurus or artists) and to stimulate the expression of creativity. There are two main aspects included in the creation of the essential five senses: designing creative space and the flow of the creative activity.

Creativity as a tourist activity is developed in the form of courses or interactive workshops, where participants are fully engaged in the process. The outcomes of this process are high-intensity interactions, the identity formation of tourists and residents, and tangible (e.g. pots, masks, or bags) and intangible (e.g. experiences, skills, and knowledge) tourism products. The active involvement of tourists in local activities blurs the distinction between tourists and locals. Tourists start taking an active part in the development of local narratives, rituals, and the image of the place (Lavrinec, 2010, 2014). In this model, the 'tourist routes' as well as routine everyday routes of the locals become an object of mutual reconfiguration. This can turn the locals into curious explorers of their cities and reserves a possibility for tourists to gain some local knowledge.

This mutual transformation through the co-operation of the locals and tourists is an important aspect of creative tourism. It has an impact on the revitalization of urban and rural localities and the place branding process. Creative space comes into prominence as a space for attracting artists to a locale (Markusen, 2006). Moreover, creative space that utilizes culture as tool for cultural regeneration enhances everyday cultural practice and accommodates social dynamics of the community (Evans, 2005). According to Santagata (2002), creative spaces that foster inspiration constitute links between the local community and its artists. Creative spaces can be fundamental instruments for protecting cultural assets and can also give identification to in-between spaces that constitute interfaces between the private and public and feature many activities that foster creative reinterpretation (Hertzberger, 1991; Evans, 2005).

In this context, a creative space can be defined as an in-between space that encourages social relations between the indoor and outdoor spaces (Can, 2012). In this conceptualization, we see a creative space as a 'space of hybridity, which is established by an interaction between different groups or individuals in a shared spatial encounter' (Mommaas, 2009, p. 3). On the other hand, the space could also be created to represent a distinctive characteristic of local artists with various linkages, such as between community life and vibrant cultural agendas; between work and leisure, or between working space and private residential space in a sort of 24/7 lifestyle; between knowledge-oriented and creative-oriented new facilities; or between informal networks and easy access to cultural exchanges. In this regard, creative spaces that provide art activities and a flexible use of public space can be defined as a combination of cultural life and local creative ecosystem. Consequently, connectivity between creativity and culture in such spaces provides community participation in arts, crafts, and group activity. Ultimately, creative spaces that can attract creative individuals and local artists create distinctive urban identity in terms of sustainable cultural tourism. In our project, we found it was significant for local artists and other creative tourism providers to understand their existing creative space and how to design the best ambience and environment for the exchange and expression of creativities. Figure 12.5 shows an example of how a local artist mapped out their space and identified the flow of the activities. This helped identify areas to improve for best interaction and experiences.

Sophistication

Once the step-by-step flow of activities is clearly marked out and the creative space is designed, the storytelling plan can then be better integrated and envisaged. The flow map helps elicit how the local heritage and skills could be interpreted. The ultimate goal of the holistic design is to offer visitors the opportunity to develop their creative potential which could lead to an exchange and the *innovation of new ideas*. The creative process is generally recognized as a cognitive process that produces new ideas or transforms old ideas into updated concepts. Creative tourism is challenging in illuminating this process because of tourists' notorious behaviour known as lacking enough time to appreciate the learning process.

The co-created model is a conceptual foundation, allowing the creative tourism planner to think through what should be considered in the entire process. The level of sophistication can go far beyond basic with continuous improvements, helping create a tailor-made process. We have learned, however, that each activity is not for all kinds of travellers, with some falling into a 'creative spectator' category, appreciating staged experiences, while others, passionate creative travellers, are longing to actively curate their artistic experiences with the artists. As Richards and Wilson (2006) have noted, the whole concept of creating creative tourism is dependent on the tourist as a creative co-producer and consumer of their experiences as well as the creative abilities of the experience creators. Figure 12.6 is an example of the print advertisement campaign, targeting creative travellers.

Creative tourism designers should identify the branding message and create a marketing campaign. In this pilot project, a collection of creative tourism experiences in Thailand was launched under the campaign, 'Be local artist for a day'. A first set of 17 creative tourism experiences have been offered by participating hotels in the selected areas. The collection has been listed on a digital platform and has been offered in collaboration with a Thailand-based Krug Thai Credit card holder and licensed tour operator Community Innovation Agency. Figure 12.7 illustrates the collection brand logo.

Fig. 12.5. Example and illustration of the flow of activity, mapped by local community and artist. Image courtesy of DASTA, in Richards *et al.* (2019), used with permission.

Concluding remarks

This project and learning journey with DASTA to develop a Creative Tourism Brain Bank will continue to inspire more potential local artists and will announce a new collection of activities on an annual basis. The mission set forth in the initiatives will continue to empower community members in leading sustainable tourism development, to preserve local heritage, and to revive endangered ones while providing opportunities for younger generations to get involved in the cultural innovation. The aims and objectives of the project will further reinforce six key actions: (i) to engage communities in co-creating creative tourism best practices; (ii) to build community capability and capacity to maximize benefits from tourism; (iii) to create local pride and self-sufficiency; (iv) to provide a learning platform for the nation to understand tourism as a tool for community innovation and development; (v) to offer a development framework/model for change agencies to engage communities in community-based tourism and creative tourism; and (vi) to create new ways to work across the tourism value chain.

Fig. 12.6. Creative potential of visitor. Photo courtesy of Community Innovation Agency (2019), used with permission.

Fig. 12.7. The collection brand logo. Image courtesy of Community Innovation Agency (2019), used with permission.

The ten core principles were tested and proven to be practical. Compositions of each element are elaborated into a practical toolkit (Richards *et al.*, 2019) which integrates the creative process into the development of creative tourism with the aim of enhancing the quality of community-based tourism. Creative tourism, if designed and implemented well, could be a mechanism to create local pride and stimulate cultural preservation, to reinforce a unique local sense of place based on visual quality and functional efficiency and, ultimately, to provide a better platform to engage different generations – 'the old meets new'. The ten core principles can be used to re-evaluate a current situation, what is being offered, how people currently engage with the current offer, and how one might build on this engagement. They can also help guide a process to think about who is currently visiting (or who one wishes to have) – how many people, who they are, and why they come. During this re-evaluation stage, it is also important to revisit the space in which creative tourism activities are implemented and to transform it into a 'creative space' or a unique local studio. Let your creative journey begin.

References

Can, I. (2012) In-between space and social interaction: A case study of three neighbourhoods in Izmir. PhD thesis. University of Nottingham, UK. Available at: http://eprints.nottingham.ac.uk/12637/ (accessed 24 September 2020).

Community Innovation Agency (2019) Available at: www.facebook.com/communityinnovationagency/ (accessed 23 July 2020).

Evans, G. (2005) Measure for measure: Evaluating the evidence of culture's contribution to regeneration. *Urban Studies* 42(5–6), 959–983.

Goodwin, H. and Santilli, R. (2009) Community-based tourism: A success? ICRT Occasional Paper, no. 11. Available at: www.haroldgoodwin.info/uploads/CBTa SuccessPubpdf.pdf (accessed 23 July 2020).

Hertzberger, H. (1991) *Lessons for Students in Architecture* (Trans.: I. Rike). 101 Publishers, Rotterdam, The Netherlands.

Huras, C. (2015) "Community-based creative tourism" as a strategy for poverty reduction. Master's thesis. Taipei National University of the Arts, Taipei City, Taiwan.

Lavrinec, J. (2010) Miesto studijos: trys žvilgsnio perspektyvos (Urban studies: three trajectories of approach). *Filosofija Sociologija* 21(1), 54–63.

Lavrinec, J. (2014) Community art initiatives as a form of participatory research: The case of street mosaic workshop. *Creativity Studies* 7(1), 52–65. doi: 10.3846/20297475.2014.933365.

Markusen, A. (2006) Urban development and the politics of a creative class: Evidence from a study of artists. *Environment and Planning A* 38(10), 1921–1940.

Mommaas, J.T. (2009) City, culture and identity: The city as third space. Paper presented at the Cultural Policy and Management Conference, Istanbul Bilgi University, Turkey, November.

Richards, G. (2005) Textile tourists in the European periphery: New markets for disadvantaged areas? *Tourism Review International* 8(4), 323–338.

Richards, G. (2011) Creativity and tourism: The state of the art. *Annals of Tourism Research* 38(4), 1225–1253.

Richards, G. (2016) The challenge of creative tourism. *Ethnologies* 38(1–2), 31–45. doi: 10.7202/1041585ar.

Richards, G. and Duif, L. (2018) *Small Cities with Big Dreams: Creative Placemaking and Branding Strategies*. Routledge, London, UK.

Richards, G. and Raymond, C. (2000) Creative tourism. *ATLAS News* 23, 16–20.

Richards G. and Wilson, J. (2006) Developing creativity in tourist experiences: A solution to the serial reproduction of culture? *Tourism Management* 27(6), 1209–1223.

Richards, G. and Wilson, J. (2007) *Tourism, Creativity and Development.* Routledge, London, UK.

Richards, G., Wisansing, J., and Paschinger, E. (2019) *Creating Creative Tourism Toolkit*. DASTA, Bangkok, Thailand. Available at: https://perfectlink.co.th/wp-content/uploads/2019/01/Creating-Creative-Tourism-Toolkit_Version-Eng.pdf (accessed 23 July 2020).

Rivas, M. (2011) *From Creative Industries to the Creative Place: Refreshing the Local Development Agenda in Small and Medium-sized Towns*. Óbidos: URBACT Creative Clusters.

Santagata, W. (2002) Cultural districts, property rights and sustainable economic growth. *International Journal of Urban and Regional Research* 26(1), 9–23.

Smith, M. (2005) Tourism, culture and regeneration: Differentiation through creativity. In: Swarbooke, L, Smith, M. and Onderwater, L. (eds.) *Tourism, Creativity and Development: ATLAS Reflections.* Association for Tourism and Leisure Education, Arnhem, NL, pp. 23–38.

Smith, M. (2009) *Issues in Cultural Tourism Studies*. Routledge, London, UK.

Thorne, S. (2009) Cultural tourism: A place-based approach. Presentation at Creative City Network of Canada conference, Fredericton, NB, Canada.

UNESCO Creative Cities Network (2006) *Towards Sustainable Strategies for Creative Tourism: Discussion Report of the Planning Meeting for 2008 International Conference on Creative Tourism*. UNESCO, Paris, France.

13 User-centred design for creative tourism prototyping: The Maribor experience

FIONA EVA BAKAS,[a*^] TIAGO VINAGRE DE CASTRO,[b] AND ANA OSREDKAR[c]

[a]*Lusófona University, Lisbon, Portugal;* [b]*Centre for Social Studies, University of Coimbra, Portugal;* [c]*Servis 8, Association for Service Design Innovation, Education and Workshops, Ljubljana, Slovenia*

This chapter discusses how the principles of user-centred design and emotional mapping can be used to help tourism and culture practitioners design prototypes of creative tourism experiences. Creative tourism is a novel interpretation of cultural tourism that incorporates within the tourism experience the dimensions of active participation, creative self-expression, learning, and community engagement (Duxbury and Richards, 2019), underlined by an immersive connection to place. On one hand, to create successful creative tourism experiences, it is important that the tourism experience designers empathize with the end users/participants and identify with their needs and motivations. By adopting a user-centred design approach, this empathy can be achieved in practical ways and then included in the design of the final tourism experience. On the other hand, emotional mapping deepens engagement with places, fostered by a multisensorial immersive exploring exercise that challenges participants to link places to their own emotions and feelings.

Although the processes of user-centred design and prototyping in the development of products have been widely used within industries such as consumer goods and informatics (e.g. games), they have been less used to help tourism professionals develop tourism products (Jung *et al.*, 2014). However, prototypes can support stakeholders to articulate their latent needs and provide vital input for new service development, and thus can be very useful in the development of new creative tourism experiences. Furthermore, a recent paradigm shift from 'design *for* users' to 'design *with* users' has fuelled a move towards design approaches that view the user as central to the product design process, empathizing with the end users in order to identify with their feelings, motivations, and priorities and to integrate them into the end tourism product (Dell'Era and Landoni, 2014).

Emotional mapping can be used as a supporting technique in the discovery and definition stage of user-centred design by helping to identify what the existing resources are, what potential strategies to engage with the public could be, and the stakeholders that could be involved in a new creative tourism experience. Mapping and acknowledging cultural assets are powerful instruments in designing tourism experiences, as 'the identity of a place is narratively constructed by the meanings that people give to it on the basis of the actual physical and social place and its history' (Eräranta *et al.*, 2016, p. 26). Emotional mapping is related to cultural mapping and focuses on the georeferencing of emotions, experiences, and/or perceptions (Pánek and Benediktsson, 2017) to help connect spaces with

*Corresponding author: fiona.bakas@ulusofona.pt; creatour@ces.uc.pt

^This chapter was written while Fiona Bakas was a post-doctoral research fellow at the CREATOUR project at the Centre for Social Studies, University of Coimbra, Portugal

DOI: 10.1079/9781789243536.0013

emotions. Emotional mapping allows for collecting and displaying to others the significant cultural information, traditions, stories, values, and hopes that locate people in their place(s). Emotional mapping can be used to construct a visual narrative about a place's identity and as such can provide 'an integrated picture of the cultural character, significance, and workings of a place' (Pillai, 2013, p. 153). The method used in this chapter to create creative tourism prototypes uses emotional mapping to inspire participants to think about local emotion–space connections that might be included within the tourism experiences they are designing.

The authors of this chapter were involved in organizing and participating in a 'creative tourism prototyping lab' in Maribor, Slovenia, as part of the 2019 edition of the 'Festival of Walks'[1] organized by House! Society for People and Places (House!),[2] a local cultural NGO. In this creative tourism prototyping lab, participants took part in an intensive three-day lab that guided them through a user-centred tourism experience design process (Fig. 13.1). Participant observation notes, lived memories, reflections, and photos taken by the authors inform this chapter.

User-centred design and emotional mapping in creative tourism

Since creative tourism research is a dynamic but relatively young field, a need has been identified for a better understanding of the intricacies in developing creative tourism activities, especially within a small city context such as Maribor (Bakas *et al.*, 2018). Within creative tourism, some research has explored how specific activities can be integrated into creative tourism experiences, such as exploring how geocaching is used in tourism to design new applications and how memorable experiences are created (Pirita, 2012); however, limited attention has been paid to research on using user-centred design in creating creative tourism experiences.

A user-centred approach to designing creative tourism experiences

Tourism experience design is broadly defined as the practice of designing products, services, processes, events, and environments with a focus on the quality of the user experiences – a deliberate, careful creation of a total experience for customers. Some of the approaches to experience design in tourism include naturalistic inquiries and empathic design to target experience narratives, participatory design involving tourists in design stages, integrative design research that includes explorative research as an essential part of designing, and the orientation of concepts and theories from multiple disciplines to apply to tourism contexts (Tussyadiah, 2013).

A user-centred design approach is useful for designing and delivering tourism experiences in a

Fig. 13.1. Team discussing at the creative tourism prototyping lab, Maribor, Slovenia.

way that considers users as sources of innovation (McDonagh *et al.*, 2013). Through a user-centred design approach, firms can identify unique insights by asking users about their needs. Innovation can be achieved by increasing and improving collaboration with an ecosystem's stakeholders and engaging users in the co-creation process (Guimont and Lapointe, 2016).

As tourism is often considered a service industry, there is a desire among tourism researchers and businesses to treat tourism as an experience production system wherein values and meanings can be designed and evoked through various contextual elements within tourism-related services and experiences. It is argued that designing for a tourism experience is not just a matter of creating a tour package or staging a theme park, it is designing the experiences tourists will have with the tour and at the park, which requires a focus on the users' feedback on what makes a good tourism experience. Hence, in order to design a good tourism experience one needs to adopt a user-centred, collaborative, and holistic approach to designing customer and service experiences that are an integral part of the whole tourism experience. Service design is an approach that does just that.

Service design aims to create seamless and quality service experiences by balancing the needs of customers with the needs of the business (Stickdorn *et al.*, 2018). It helps organizations design services that are useful, usable, and desirable from a user perspective, while also effective and unique, from a provider perspective (Mager and Sung, 2011). Service design thinking also provides processes and methods to create organizational structures and to understand the culture required to deliver tourism experiences within complex ecosystems such as tourism destinations (Stickdorn and Frischhut, 2012) and therefore could prove useful to designing tourism experiences.

Other principles that service designers adopt when developing new service or experience solutions are: (i) collaboration of all relevant stakeholders; (ii) iteration of process cycles, starting with small cheap experiments that allow us to fail and learn; (iii) reality-grounded needs and prototyped ideas; and (iv) holistic addressing of stakeholder needs, throughout the service or experience and across the business (Stickdorn *et al.*, 2018). These principles come to life when we adopt a designer's mindset while engaging in a structured iterative design process and using selected design methods and tools that best fit the context and object of design at hand. An innovative way of inspiring participants to connect their activity to the various spatially related emotions of the landscape where they will hold the tourism activity they are planning is by introducing emotional mapping within the tourism experience design process.

Emotional mapping

Although historically, cartography was mainly focused on representing that which is objective, visible, or measurable, critical cartographers have advocated mapping a space as people experience it (Pearce, 2008). Emotional mapping, for example, enables the display of subjective, qualitative, and bottom-up spatial information about the environment on a map (Pánek and Benediktsson, 2017). Emotional mapping adjusts cartography to enfold intangible, emotional, social, and cultural aspects (Duxbury *et al.*, 2015). Through emotional mapping it is possible to re-imagine landscapes, integrate emotional and affective dimensions into maps, and map stories onto everyday places. Recognizing that the identity of a place is constructed by the meanings that people give to it makes emotional mapping a powerful tool in identifying items to include within a creative tourism experience.

Emotional mapping has many similarities to cultural mapping as it can help connect place and local traditions and make them a key element within creative tourism activities. The use of cultural mapping in the design of these activities can add depth and richness by unveiling local traditions through the eyes of locals, rather than according to tourism promotion boards. Within the CREATOUR® project,[3] cultural mapping was used as a methodology to make participants think about their places and what makes them so special to them. In order to initiate the process of creative tourism product development, CREATOUR® participants were asked to bring items that represented their locale in 12 ways: photographic, sweet, creative, fragrant, historical, relaxing, symbolic, traditional, thrilling, musical, authentic, and innovative. As a result of mapping intangibilities through tangible objects, CREATOUR® participants rediscovered their regions' identity, which they then used to inform the development of their creative tourism products (Cabeça *et al.*, 2019).

Emotions, spaces, and places are very much connected (Pánek and Benediktsson, 2017) with emotions providing 'a *sense of place* and identity to specific locales' (Longley and Duxbury, 2016, p. 2). In the context of territorial connections, geographers have described emotions as subjective, relational flows between places and people. Every location can evoke an emotion and places can be attractive, boring, dangerous, or scary, among other emotions (Davidson *et al.*, 2016). Using emotional mapping techniques can reveal the ways in which meanings and values may be grounded in embodied experiences. This information can then allow tourism experience designers to build-in opportunities for tourists to have these embodied experiences too, moving closer to attaining the desired 'authentic everyday experience' that many tourists now crave.

Maribor Creative Tourism Prototyping Lab

The Maribor Creative Tourism Prototyping Lab, which is the focus of this chapter, was tailor-made for the Festival of Walks. The main aim of holding such a lab was to gather local practitioners from tourism and cultural sectors to co-create prototypes of creative tourism product experiences and help foster local networking in this area.

Four main influences came together to inform and create the Maribor Creative Tourism Prototyping Lab. First, it was inspired by the 'IdeaLab on Creative and Cultural Tourism' that was held in Maribor 6 months earlier, attended by cultural and tourism actors from Maribor and abroad (Greece, Germany, Spain, Ukraine, and Romania) who strategized on how to foster creative tourism in Maribor. As part of the 'IdeaLab on Creative and Cultural Tourism', a cultural mapping exercise was conducted, where local and foreign participants walked around the city and mapped cultural and creative elements, people, and places that later informed their strategy-making session. The second influence relates to the feedback on this cultural mapping experience, which indicated the need for a further lab to trigger the development of creative tourism experiences in Maribor. The third influencing aspect was input related to some of the organizers' previous experiences with emotional mapping in different contexts, such as a community activation and diagnosis tool for urban regeneration projects. The emotional mapping methodology used in the Maribor Creative Tourism Prototyping Lab was developed by the Portuguese NGO 4iS – Platform for Social Innovation (4iS), a partner organization of House!. The inclusion of an emotional mapping exercise was perceived as allowing participants to gain a deeper understanding of the potential touristic and cultural resources of their city. The fourth influence was the cooperation with an expert in user-centred design thinking from Servis 8, a service design thinking organization from Ljubljana, Slovenia, in the 'IdeaLab on Creative and Cultural Tourism', a meeting that highlighted the potential for combining emotional mapping and user-centred design principles in order to help professionals to prototype new creative tourism experiences.

To test this hypothesis, CREATOUR® joined forces with House!, 4iS, and Servis 8 to hold the three-day Maribor Creative Tourism Prototyping Lab. Eleven participants, who were mainly tourism practitioners, cultural managers, and marketing professionals (and one academic researcher), divided into two groups and developed and tested prototypes of creative tourism experiences.

The Maribor Creative Tourism Prototyping Lab experience was organized in three phases: (i) immersion; (ii) discovery and definition; and (iii) development and delivery (Table 13.1).

Phase I: Immersion

Phase I began with a presentation of the concept of creative tourism and CREATOUR® best practices, with the aim of introducing creative tourism to participants by providing examples of creative tourism experiences from within CREATOUR®,

Table 13.1. Phases of the Maribor Creative Tourism Prototyping Lab

Phase I	Phase II	Phase III
Immersion	Discovery and definition	Development and delivery
• Creative tourism concept and CREATOUR® best practices • Emotional mapping	• Understanding your audience • Re-defining your challenge	• Idea generation • First concept sketch • Scenario building • Prototyping • Prototype testing • Concept definition • Business model generation

such as immersive walks through the city of Caldas da Rainha followed by creative workshops related to ceramics, painting, and design; workshops in Montemor-o-Novo where people learn how to make 'bird basket houses' and 'rag dolls' using traditional craft techniques and local natural raw materials like reeds, cloths, and straw; the 180 Creative Camp in Abrantes, a festival where arts and creativity combine with the local cultural context and turns into a unique co-creation experience; a 'creative weekend' in the former wool-factory city of Covilhã, which includes city walks and workshops in textiles and photography, in the context of textile industry heritage (see Chapter 17 in this volume); a set of contemporary and experimental workshops that are offered in a famous local artist's house in Aveiro, now turned boutique hotel and cultural centre; and Roman mosaic-inspired workshops set in Conímbriga, the largest Roman heritage archaeological site in Portugal.

Next, participants were invited to participate in an exercise of emotional mapping, which aimed to make participants more aware of their surroundings in terms of what could be used as a resource to make a creative tourism product. The mapping exercise consisted of a city walk to observe and document different places of Maribor of natural and cultural interest and the emotions and feelings regarding those places. Participants were divided into two teams and each team decided which path to take for a duration of one hour and a half. During the walks, teams took pictures, drew, wrote, and recorded (sound and video) the city. They recorded whatever they felt important, meaningful, and interesting, as well as any kind of feeling or emotion these places stimulated for them.

After this exercise, each team prepared a storytelling session based on their experience and documentation, and presented it to the rest of the participants. This encouraged everyone to start thinking about how the resources around them (for the local participants these were everyday surroundings, as they worked and lived in the city centre where the mapping took place) could be potentially used as resources within the development of a creative tourism product. This narrative exercise also stimulated very useful inter-group discussions regarding the resources discovered on the emotional mapping journey and triggered a reflection exercise on Maribor's potential as a creative tourism destination. This exercise was particularly interesting because of its deeper meanings since the emotional mapping allowed participants to identify which places in the city they liked most, as well as the places they did not like and why. Following this immersion into understanding of creative tourism and the local context, it was time to start designing new creative tourism product prototypes. The challenge that was set for the participants was: How can we develop a creative tourism offer with superior audience experience?

User-centred design process (phases II and III)

Phase II and phase III were structured around the 'double diamond' design process (Fig. 13.2), developed by Design Council UK (2020), which enables designers and non-designers alike to look at an issue more widely and deeply (divergent thinking) and then to take more focused action (convergent thinking). Looking at Fig. 13.2, stage 1, 'Discover', at the start of the double diamond, was kicked-off with the challenge: 'How can we develop a creative tourism offer with superior audience experience?' in which the participants were encouraged to discover different aspects of the cultural context and how visitors currently experience Maribor by conducting interviews with visitors to Maribor. In stage 2 of the diamond, 'Define', the participants narrowed their focus by identifying specific visitor profiles and their needs and defining a clear opportunity for creative tourism experience development. In stage 3 of the diamond, 'Develop', the participants developed a range of different ideas and concepts that could shape these creative tourism experiences. In the final, fourth stage, 'Deliver', participants built prototypes that were tested with real users and defined a business model that could support the delivery of the creative tourism experience.

Phase II: Discovery and definition

In phase II, participants learned why it is important to understand who their audience is and how this can inform what their new creative tourism offer might look like. Various design research methods, such as data collection, observation, interviews, and mobile ethnography, can help us identify user needs, behaviours, and motivations, thus providing insights for problem definition. The participants were given instructions to conduct interviews to find stories of past creative tourism

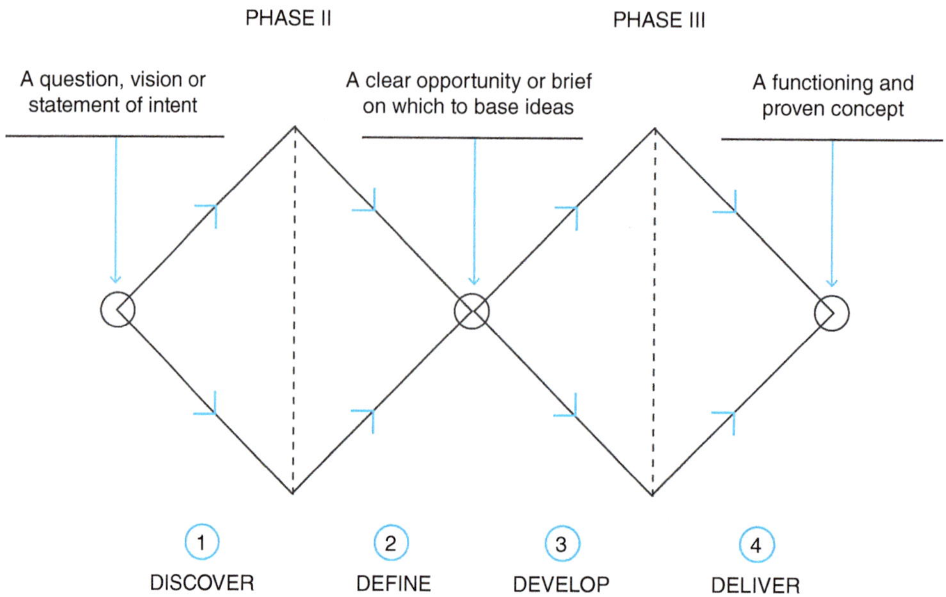

Fig. 13.2. User-centred design process (phases II and III), adapted by authors from the 'double diamond' design process (Design Council UK, 2020).

experiences by real people in the streets of Maribor. By doing so, the participants would be able to begin developing new offers based on real data instead of assumptions.

After participants conducted their research on what their audience could be, the two teams summarized their findings by visualizing and organizing visitor types into clusters on a research wall (Fig. 13.3). This enabled them to look for commonalities and differences between types of audiences, which were later visualized using a Persona template (Fig. 13.4). Persona is a design 'tool' that represents a particular group of people, such as a group of customers or users, a market segment, a subset of employees, or any other stakeholder group with shared needs and behavioural patterns. This profile is not a stereotype, but an archetype based on real research (Stickdorn *et al.*, 2018).

In order to further define actionable points of reference for the next steps in the design process, participants formulated several 'key insights' based on the data gathered for each of their personas. These key insights included user goals, needs, and problems. Lab participants then used these key insights and personas to formulate problem statements in the form of 'How might we…?' questions, which would serve as starting points for ideation in the next phase.

Phase III: Development and delivery

In phase III, participants started imagining new ways of serving their selected audience and showing how their creative tourism offer might work in practice. The first step was to generate ideas around selected problem statements. At this stage, the participants were invited to use the main creative tourism principles from CREATOUR®, which are active participation, creative self-expression, learning, and community engagement (Duxbury and Richards, 2019), underlined by an immersive connection to place, as guidance when generating ideas for their creative tourism experiences. The idea generation was done through a process of silent brainstorming, sharing with partners, and building on each other's ideas using 'idea sketches'. Participants then organized the 'idea sketches' generated into grids, prioritizing their ideas based on the potential impact (i.e. the extent to which it would address the user's need) and estimated effort needed (i.e. resources such as time, budget, and technological complexity). Idea sketches that were found as 'high impact' and 'low effort' were selected in order to be further developed into 'concept sketches'. These 'concept sketches' included giving the creative tourism activity idea a name, defining the target public, and describing how it works and what value it could bring to the audience.

Fig. 13.3. Creating visitor types. Photo by Fiona Bakas, used with permission.

The elements included in a concept sketch are outlined in Table 13.2.

They were instructed to start building elements of their concepts in order to validate that these elements could work and could be tested in real life. Again, the participants were reminded of the importance of addressing all four of CREATOUR®'s main creative tourism principles within their creative tourism experience prototype. Going back to their concept sketches, the two teams developed storyboards to visualize how their users might interact with their offer. Storyboards added more context to the concept sketches and helped the teams decide which elements of the creative tourism experience they needed to prototype in order to demonstrate its core value and to test for real user desirability and viability.

Participants were then given various prototyping methods for processes and experiences to choose from, depending on which elements or aspects of the creative tourism experience the team wanted to focus on. By building early versions of their ideas, they were able to explore a specific element of how their creative tourism experience could best fit their user's needs and expectations and learn through trial and error. In order to determine whether their ideas would be useful, valuable, and feasible in the real world, the teams had to then build a new version of their prototype, one they could use for testing their assumptions. This new version of the prototype had to demonstrate how, in a certain context, one or two aspects of their product would work from a user's standpoint.

The next step involved teams getting feedback and new insights from visitors using the methods of prototype testing and usability testing to evaluate the solution fit and then to adapt accordingly. Teams tested their prototypes in public, aiming to find the best representatives of their user profiles, and recorded their findings. A synthesis of these

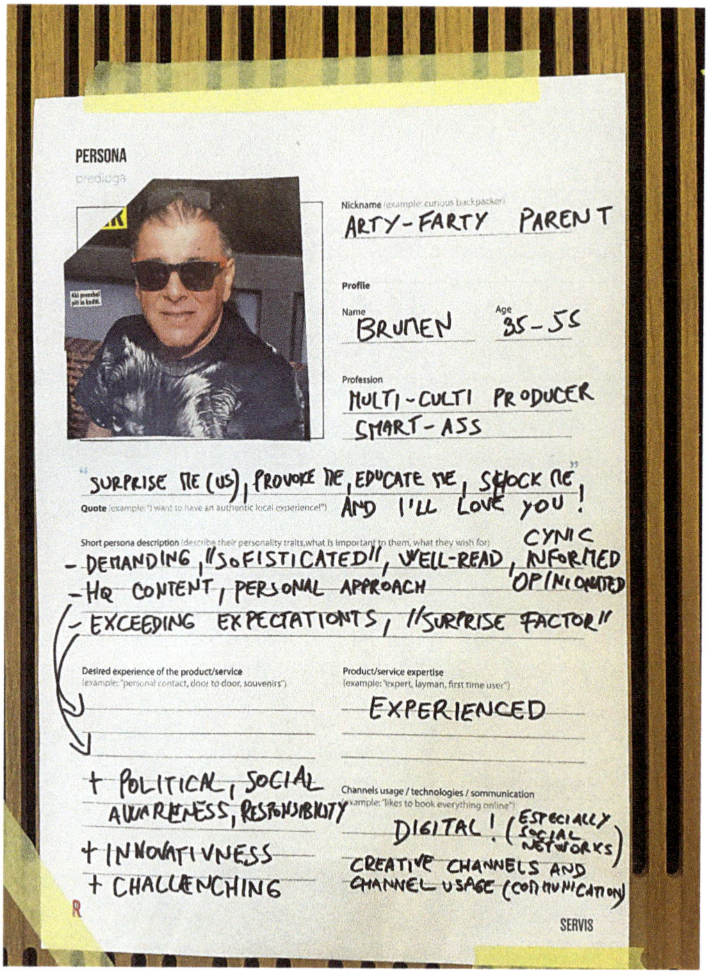

Fig. 13.4. Example of a persona. Photo by Fiona Bakas, used with permission.

findings helped them determine a set of action points on how to improve and fine-tune their creative tourism experience concepts. In the end, there was one more prototype that participants had to 'build', this time using an 'incognito' Business Model Canvas (Osterwalder, 2005) framework. This process encouraged teams to establish a common understanding of how their product would create, deliver, and generate value.

'Tune Yourself to Maribor'

Each of the two teams in the Maribor Creative Tourism Prototyping Lab created a prototype for one creative tourism experience each, namely: 'Tune Yourself to Maribor' and 'Treasure Hunt'. 'Tune Yourself to Maribor' was an experience allowing visitors to feel like a local for a day by participating in a musical tour of Maribor and also to be part of a choir where they sang a song. The 'Treasure Hunt' consisted of a tour around Maribor where visitors, after receiving a tour pack, walked around Maribor trying to find the items listed in the tour pack, thus taking in alternative parts of Maribor based on local people's perceptions of Maribor 'treasures'. Since one of the authors participated in the 'Tune Yourself to Maribor' experience prototyping process, and hence has lived experience of the process, this experience is chosen to explain how using a combination of emotional mapping and user-centred design can be used to develop a creative tourism prototype.

'Tune Yourself to Maribor' aimed to connect tourists to local people in order to experience local culture. By conducting short interviews with various visitors to Maribor, asking them what kind of things they liked doing on holiday in order to define a 'tourist persona', the team identified their target audience as 'Independent explorers looking for authentic local experiences'. One of the ideas that emerged from the brainstorming sessions was to form a partnership with a local association in order to engage the community, and a local choir was thought to be a suitable association to involve. Further brainstorming and user-centred design exercises led the team to decide that the activity's main aim would be to provide visitors with a music-related experience by singing in a local choir. During the concept sketch creation process, the team identified certain elements as key to the experience, always aligned with the four creative tourism elements. The team decided that the main goal of 'Tune Yourself to Maribor' should be to take tourists on a music-related tour around the city and enable them to experience local cultural activities together with locals.

Table 13.2. Elements to be included in a concept sketch

Concept sketch elements	
Element	Question to be answered
Name of product	What is the creative tourism offer called?
Short description of product	What does the creative tourism offer consist of?
Audience	Who is the target audience?
Value proposition	Why would people visit/use the creative tourism offer?
Active participation	How is active participation of the audience enabled?
Experience	What does the experience of the creative tourism offer look like?
Local community	How is the offer going to affect the local community?

In order to better think through the possible stages that this experience would have, the group used props such as LEGO (Fig. 13.5) to represent the experience and to show the other group how it worked. Finally, the usability testing took place as participants created a poster for the creative tourism experience and went out once more and conducted mini-interviews with potential users to obtain user opinions on aspects of the activity such as price, content, duration, and place of purchase.

After the whole process, the 'Tune Yourself to Maribor' creative tourism experience comprised six moments of interaction. Table 13.3 shows these moments (in chronological order) and how they correspond to the creative tourism elements of

Fig. 13.5. Using LEGO to represent part of the 'Tune Yourself to Maribor' experience. Photo courtesy of House! Society for people and places, used with permission.

Table 13.3. Creative tourism elements in Tune Yourself to Maribor

'Tune Yourself to Maribor'	Creative tourism element
Welcome moment	*Social and hosting element*
Visit to musical monument and some history about the city	*Connection to place + Learning*
Music museum short tour	*Connection to place + Learning*
Musical relaxation moment in the park	*Connection to place + Relaxing*
Singing in the choir of a local association	*Community engagement* is achieved by connecting the local choir association to tourists by hosting the tourist for a short period of time to coincide with the regular choir practice. *Active participation/ creative self-expression:* Since tourists are singing with a local choir, they are actively participating in an everyday event and they are learning something new, as they need to learn a new song
Café moment when participants can socialize before departing and are offered a picture and a recording from their choir experience	*Connection to place + added value of experience*

learning, community engagement, connection to place, active participation, and creative self-expression.

One of the more obvious ways in which the emotional mapping exercise influenced the design of this creative tourism experience was evident in the final prototyping stage of the 'Tune Yourself to Maribor' experience. One of the places a participant recorded on video with a narrative voice-over was Maribor central park, which was described as 'relaxing, quiet, serene' and a 'happy place to be' (participant quotes, March 29, 2019). Later on, in the usability-testing phase when the poster was created and tourists asked what elements they thought were essential within a creative tourism experience, many said that 'moments of relaxation' were essential to them within a tourism experience as their days were usually very hectic. Associating the park with relaxation from the emotional mapping exercise, participants (the experience designers) included a moment of relaxation in the creative tourism experience, which took the form of walking around the park and actively listening to the 'silent sounds' of nature, as a type of relaxing meditation.

Future steps in designing creative tourism prototyping labs

The approach and techniques used in the Maribor Creative Tourism Prototyping Lab offer insights into what may be useful ways of planning future creative tourism prototyping labs. This lab was designed in a 'design jam' format, an intensive three-day workshop where participants were learning new methods and applying them directly to problem solving in a co-creation process. This was also a safe space where participants could explore and experiment with new ideas through trial and error. The format proved to be very intense for the participants as there was a lot to take in within a short period of time. Some of the participants were also stepping in and out of the process because they were simultaneously running the Festival of Walks, which resulted in some disruption of team dynamics and the need to update team members on progress made in their absence.

In designing a future Creative Tourism Prototyping Lab, several important elements need to be addressed. First, it is important that all participants are able to fully commit themselves to the entirety of the Lab, to ensure that teams work at their full capacity and information flow is good. There are various other formats that the Lab could have taken, such as a Design Sprint. An investigation into the viability of the prototypes created in the Maribor Creative Tourism Prototyping Lab would indicate what formats might work better in the future, and what additional pieces might have been useful to incorporate.

Lab participants reported that the opportunity to apply a user-centred design approach in combination with CREATOUR® creative tourism principles and emotional mapping was a useful framework for their work in developing new creative tourism experiences. However, more research on the application of this framework and its parts should be conducted to more accurately estimate the impact of the lab itself. Questions such as 'Were there any attempts to further develop either of the prototypes, and how?'; 'Are participants using any of the methods in their other projects?'; 'What are their main challenges in designing creative tourism products today?'; and 'How is the community of practitioners thriving after the lab?' should be investigated in order to estimate how this process adds value to the work of tourism professionals and cultural managers.

Finally, it is important to find ways of supporting a participant Community of Practice based on the one loosely created within the Maribor Creative Tourism Prototyping Lab by organizing regular training, incubation, and network events. The combination of different types of expertise and backgrounds from both the cultural and tourism fields is paramount for the development of better and stronger creative tourism experiences. This fosters collaboration and cooperation between these sectors instead of viewing them as competitors or silos. One of the main aims of the Lab was to bring together tourism and cultural professionals, and making efforts to sustain participants' continued collaboration and cooperation would go beyond the temporary moment of togetherness created within the Lab to nurture more meaningful connections between the tourism and cultural field.

Notes

1. The Festival of Walks was held between March 26–31, 2019, featuring more than 30 free walks to discover the city's history, stories, and legends that make Maribor so 'mariborian' through its streets, underground trenches, markets, bars, and landmarks. For further information, visit the Festival of Walks website: www.rajzefiber.si/eng/blog/festival-of-walks-2019 (accessed 10 May 2021).

2. In Slovenian: Hiša! Društvo za ljudi in prostore.

3. CREATOUR® is a 44-month (2016-2020) research and application project focusing on creative tourism that involves five research centres working with 40 participating organizations, located in small cities and rural areas across Portugal in the Norte, Centro, Alentejo, and Algarve regions (for further information, visit www.creatour.pt [accessed 10 May 2021]).

Acknowledgements

This chapter was developed within the CREATOUR® project (no. 16437), which is funded by the Portuguese Foundation for Science and Technology (FCT/MEC) through national funds and co-funded by FEDER through the Joint Activities Programme of COMPETE 2020 and the Regional Operational Programmes of Lisbon and Algarve.

References

Bakas, F.E., Duxbury, N., and Vinagre de Castro, T. (2018) Creative tourism: Catalysing artisan entrepreneur networks in rural Portugal. *International Journal of Entrepreneurial Behavior and Research* 25, 731–752. doi: 10.1108/IJEBR-03-2018-0177.

Cabeça, S., Gonçalves, A., Marques, J., and Tavares, M. (2019) Mapping intangibilities in creative tourism territories through tangible objects: A methodological approach for developing creative tourism offers. *Tourism and Management Studies* 15, 42–49. doi: 10.18089/tms.2019.15SI05.

Davidson, J., Bondi, L., and Smith, M. (2016) *Emotional Geographies*. Routledge, Oxford, UK.

Dell'Era, C. and Landoni, P. (2014) Living lab: A methodology between user-centred design and participatory design. *Creativity and Innovation Management* 23(2), 137–154. doi: 10.1111/caim.12061.

Design Council UK (2020) What is the framework for innovation? Design Council's evolved Double Diamond. Available at: www.designcouncil.org.uk/news-opinion/what-framework-innovation-design-councils-evolved-double-diamond (accessed 4 May 2020).

Duxbury, N. and Richards, G. (2019) Towards a research agenda for creative tourism: Developments, diversity, and dynamics. In: Duxbury, N. and Richards, G. (eds.) *A Research Agenda for Creative Tourism*. Edward Elgar Publishing, Cheltenham, UK, pp. 1–14.

Duxbury, N., Garrett-Petts, W.F., and MacLennan, D. (2015) Cultural mapping as cultural inquiry: Introduction to an emerging field of practice. In: Duxbury, N., Garrett-Petts, W.F. and MacLennan, D. (eds.) *Cultural Mapping as Cultural Inquiry*. Routledge, New York, New York, USA, pp. 1–44. doi: 10.4324/9781315743066.

Eräranta, K., Leino, T., Seppälä, T., Viña, S., and Timonen, E. (2016) Mapping the pig tale journey: A multidisciplinary design framework for cultural mapping in an old abattoir. *City, Culture and Society* 7(1), 25–33. doi: 10.1016/j.ccs.2015.07.006.

Guimont, D. and Lapointe, D. (2016) Empowering local tourism providers to innovate through a living lab process: Does scale matter? *Technology Innovation Management Review* 6(11), 18–25.

Jung, H.S., Kim, K.H., and Lee, C.H. (2014) Influences of perceived product innovation upon usage behavior for MMORPG: Product capability, technology capability, and user centered design. *Journal of Business Research* 67(10), 2171–2178. doi: 10.1016/j.jbusres.2014.04.027.

Longley, A. and Duxbury, N. (2016) Introduction: Mapping cultural intangibles. *City, Culture and Society* 7(1), 1–7. doi: 10.1016/j.ccs.2015.12.006.

Mager, B. and Sung, T.-J.D. (2011) Special issue editorial: Designing for services. *International Journal of Design* 5(2), 1–3.

McDonagh, D., Thomas, J., Khuri, L., Sears, S.H., Peña-Mora, F., and McDonagh, D. (2013) Empathic design: Research strategies. *The Australasian Medical Journal* 6(1), 1–6.

Osterwalder, A. (2005) What is a business model? Available at: https://web.archive.org/web/20061213141941/http://business-model-design.blogspot.com/2005/11/what-is-business-model.html (accessed 4 May 2020).

Pánek, J. and Benediktsson, K. (2017) Emotional mapping and its participatory potential: Opinions about cycling conditions in Reykjavík, Iceland. *Cities* 61, 65–73. doi: 10.1016/j.cities.2016.11.005.

Pearce, M.W. (2008) Framing the days: Place and narrative in cartography. *Cartography and Geographic Information Science* 35(1), 17–32.

Pillai, J. (2013) *Cultural Mapping: A Guide to Understanding Place, Community, and Continuity*. Strategic Information and Research Development Centre, Petaling Jaya, Malaysia.

Pirita, I. (2012) Geocachers: The creative tourism experience. *Journal of Hospitality and Tourism Technology*, 3(3), 152–175. doi: 10.1108/17579881211264468.

Stickdorn, M. and Frischhut, B. (eds.) (2012) *Service Design and Tourism: Case Studies of Applied Research Projects on Mobile Ethnography for Tourism Destinations*. BoD–Books on Demand, Norderstedt, Germany.

Stickdorn, M., Hormess, M., Lawrence, A., and Schneider, J. (2018) *This is Service Design Doing*. O´Reilly Media, Boston, Massachusetts, USA.

Tussyadiah, I.P. (2013) Toward a theoretical foundation for experience design in tourism. *Journal of Travel Research* 53(5), 543–564. doi: 10.1177/0047287513513172.

14 Bogota through the 5 Senses: Building a creative tourism start-up

DIANA ZULUAGA* AND DIANA GUERRA

5Bogota, Colombia

The tourism industry plays a predominant role in the global economy. In the developing world, where informality is a daily occurrence, this industry presents an encouraging opportunity for entrepreneurs with low entry hurdles. It is in this context that the concept behind creative tourism takes centre stage for entrepreneurs due to its potential to become a distinguishing approach with important competitive advantages for a start-up in the tourism sector. Notably, while this type of tourism is to be self-sustainable, it must also have a positive effect on the parties in the value chain.

5Bogota: Travel with Locals was established in 2013 with the goal of promoting creative tourism in Colombia. This start-up was conceived as a marketplace connecting local hosts with foreigners wishing to explore the city through different eyes, while fostering local sustainable development in the places they visit.

Our offer currently includes ten set tourism experiences in Bogota and three in Medellin. One of the most popular experiences is the Farmer Market and Cooking Lesson (Fig. 14.1), where tourists learn about local recipes and chat one-on-one with their host, while getting to know their host's home and neighbourhood. The success of this particular experience has inspired us to fill our portfolio with offers developed around gastronomy, understanding it is the most genuine and powerful way to connect people from different origins.

The team of local hosts currently consists of ten individuals who chose to participate in the creation of these innovative experiences in the country despite not being tourism professionals. This has allowed for the redirection of financial profits from tourism towards other, non-tourism actors in the country.

Our experience with Colombian start-up 5Bogota (Bogota through the 5 Senses) underlines the importance of a methodical process integrating the creation of unique tourism experiences with the marketing strategies necessary for the development of a profitable business model, thus ensuring the social impact sought by the concept of creative tourism.

This brief text outlines the methodology used to create and consolidate the experiences offered by 5Bogota. The methodology comprises eight stages, including context analysis, the study of the destination's characteristics, the establishment of the typology for the traveller and for the local host, the design of the tourism product, and the marketing strategy. In closing, a series of conclusions are presented to share the lessons learned through the application of this methodology at 5Bogota, and to identify the main key success factors for entrepreneurship in creative tourism.

Methodology

1. Analyse the context

Every city in the world has a unique touristic portfolio: its local gastronomy, its tourist guides, its attractions, and its clichés. All of these constitute traditional tourism. This is why anyone visiting Paris wants to see the Eiffel Tower or visitors to Buenos Aires want to attend tango shows. This raises the first question: Does traditional tourism truly reflect the identity of a destination?

*Corresponding author: dianazuluagag@gmail.com

 Creative Tourism: Activating Cultural Resources and Engaging Creative Travellers (eds. N. Duxbury, S. Albino, and C. Pato de Carvalho)
DOI: 10.1079/9781789243536.0014

Fig. 14.1. 5Bogota local host and chef Jairo Amaya in the cooking lesson experience, preparing 'ajiaco', a typical Bogotanian soup. Photo by Catalina Torres, used with permission.

2. Identify the collective imaginary about the destination

The collective imaginary about a destination is what makes a traveller go to the destination promoted. In Colombia, for instance, some examples are coffee or the Caribbean beaches. While these elements are considered to be part of the culture, they are not by themselves enough to develop creative and innovative tourism products.

To contextualize and inform your development process, it is necessary to identify how a destination is promoted both locally and internationally. To this end, it is advisable to review the dissemination policy for the destination, that is, the concept under which it is advertised within the national or regional branding strategies. Another way to identify the imaginary at play is to review articles in magazines, blogs, or journals, all of which will reveal its main, most recognizable, attributes. Additionally, it is also recommended to rely on locals, asking them to describe how they think foreigners perceive the place. Determining the destination's 'public' attributes, which in many cases are the motivation for travellers to choose a place, is precisely the insight used to conceive creative tourism experiences.

The upshot of this stage will be a list of the most relevant and recognizable attributes for the destination in regard to other cities or countries. The list can then be used to create experiences based on the concepts discovered in this stage. For 5Bogota, and considering Colombia as the destination, we identified that the country is recognized for its good coffee, its biodiversity, its diverse fruits, its way of dancing, and its local kindness. Having identified these insights, the creation of experiences that incorporate in some way the generally recognized attributes in question can begin.

3. Recognize the identity of the destination

When seeking to establish the identity of a destination, many questions require answers: Who are the people who live there? What makes them unique? What distinguishes this from other destinations? What rituals and customs make up its history? What are the people like? What holidays do the locals celebrate throughout the year?

Travellers will likely visit the destination due to the appeal of the established attractions, but once *in situ* they must be able to find a number of possibilities to make them fall in love with the culture, visit the destination again, and/or recommend the destination to relatives and friends.

The answers you obtained regarding the identity of the destination will yield a new list of concepts that the locals feel identify them and through which they have a sense of belonging. To ensure the usefulness of the results from this stage, we suggest conducting a comparative analysis using the outcome from the previous step and the results produced in this stage. This analysis will unveil the correspondence between what foreigners recognize about a destination and that which the locals enshrine as their own.

Such a comparison may expose discrepancies between the sources of information: insights recognized by foreigners may be derided by locals, for example, drug trafficking and Colombia. This is a good example of a feature which in spite of its global recognition and its capacity to trigger curiosity in a certain kind of traveller, it does not represent a society or its members from any ethical standpoint.

4. Identify the creative traveller

The application of concepts surrounding creative tourism breaks open the paradigm between the traditional conception of a tourist and the new kind of traveller, understanding the latter as a person who approaches the act of travelling as a transformational experience in their life. This kind of traveller is searching for exceptional experiences that will bring about change and allow them to widen their perspective on their own values and beliefs. Hence, cultural exchange and learning become the main motivators in this context.

A number of questions must be solved in order to determine the quantitative and qualitative variables that define the creative traveller in your locale: Which cities have non-stop flights to this destination? From which cities or countries do most travellers arrive? How old are these travellers, usually? Why do they come to this destination? (Reasons may include business, leisure, culture, or health, among others.)

Once the target group is reduced through the consideration of demographic variables, the underlying reason that moves travellers to choose an experience rather than simply pick a destination must be elucidated. This could be either a feeling of social commitment, a personal transformation, the confrontation of extreme situations (mourning, absences, losses, etc.), or a mere reflection.

On the other hand, it is relevant to describe some of the characteristics of the traditional traveller in order to establish the clear differences between the creative and the traditional traveller. The traditional kind of tourist seeks out the most famous places and activities in a destination, is comfortable sharing their trip with large or medium-sized groups, and often engages the services of travel agencies and/or traditional tour guides. This type of traveller feels safer with package trips, preferably with full board. Moreover, they usually plan their trips ahead of time to coincide with high seasons, as their availability to travel is limited to vacation periods. What is more, this traveller is not interested in activities that require their direct participation, such as workshops or theme tours. Instead, they prefer to be spectators of what listings or personal recommendations label as *must see*, which leads them to visit popular or traditional sites. Finally, the traditional traveller would not be willing to deviate from a classic itinerary to visit sites off the beaten path, and would always stay in a hotel, never considering options such as Airbnb.

5. Foster a network of local hosts

Nowadays there is competition among destinations, and not just neighbouring or nearby territories. In today's globalized world, Bogota competes with Berlin, Spain with Indonesia, and Tuscany with Tokyo. This is relevant because there is a battle among destinations: it is possible to find a good hotel, a tour bus, or a good dinner in any city in the world, without having to move away from home.

This sets a stage where the secret formula is more commonplace than previously thought. It is there, daily, in the routine activities, in the cultural practices, in the inhabitants of a marketplace, in the butcher, in the architect, and in the artist, but also in the passers-by. All this paves the way for the following premise: 'The intangible heritage essence is synonymous with the ordinary'.

From this starting point, the next step in the methodology can be addressed: fostering a network of hosts with the ability to showcase the innermost and most representative aspects of a community, which is accomplished through exploring the destination and scouting not only the places, but also the people who could belong to the network.

An effective host selection process features some essential characteristics pursuant to the following definition: A local host is a person who commands certain aspects of their own culture and wishes to share them (see Fig. 14.2). They are interested in taking part in a cultural exchange with travellers, are open to new experiences, are willing to speak or learn other languages, and are ready to commit their time availability to set schedules.

With these premises in mind, the possible hosts are identified and invited to belong to the network, which is presented as an opportunity for additional

Fig. 14.2. 5Bogota local host and urban artist Marcel Marentes in the stencil experience, customizing the travel t-shirt. Photo by Catalina Torres, used with permission.

income in their spare time, taking on activities they enjoy and know by heart. It is worth noting that sometimes these activities match the hosts' day jobs or are a hobby.

6. Co-create the experiences

Co-creation represents the creative process whereby the prototype for each experience is drafted and created through a participative exercise. This stage requires the involvement of the hosts for each experience, along with an interdisciplinary team of collaborators (marketers, customer service representatives, tourism product designers, media planners, and community managers, among others) and someone who role-plays as the client, meaning they would share the attributes of the defined target.

For this stage, it is essential to foster an introspection that allows participants to be aware of their belonging and attachment to their own culture. To this end, it is crucial to posit all the questions drafted in stage 3 (Recognize the identity of the destination) as a starting point for what we call the 'Experience Lab'.

This part of the process is best completed through storytelling. Each host writes a story featuring what travellers will feel when participating in their experience. For this, hosts must imagine who the traveller is, what is their name, where do they come from, what moved them to travel, what are they looking for, and so forth. With these elements, hosts then build a story that covers all different stages, including a preface, an elaboration, and a conclusion. This story constitutes the main tool to design the experiences.

These stages are composed of moments that represent the different specific instants when a milestone is marked in the experience. Since moments set the pace in the experiences, all products conceived under the concept of creative tourism must be carefully planned in order to make them appealing and marketable, with the aim of replicating the experiences for each client. In other words, experiences are memorable not by happenstance but because of the co-creation process.

Throughout the process, a moderator is active and produces a document containing contributions from all participants. The document will help the design team when designing the experiences. This team will create the experiences considering two deciding factors:

1. The experiences must match the expectations of the travellers.
2. The experiences must also involve the creative tourism tenets, meaning that both the traveller and the host become learners, that the experience is both participative and representative of the destination's culture, and that it has positive impacts on local development, thus strengthening the social fabric.

7. Integrate the product with the host

Host training is an essential step in the process. The creative team will walk through all the moments in the experience, emphasizing that no moment is expendable, allowing some latitude for the host to assimilate the moment and add their personal signature. This means that while each experience has a set structure, it is up to the host to make it distinctive by adding details reflecting their own personality and life experience.

Since tourism experiences are unique and unrepeatable, they provide no second chances to impress customers. Therefore, service quality is paramount for new products. The final stretch of this stage must involve a mystery shopper whose task will be to provide a critical assessment of the experience and of the host in order to implement final adjustments prior to launch. This assessment measures timing, the host's ability to express themselves, their command of the foreign language, their adherence to a schedule, their execution of each established moment, and their resourcefulness to react in the face of unexpected situations.

8. Syndication and marketing

Once the tourism portfolio is ready, a marketing plan is necessary in order to present it to clients. It is important to mention that the concept of clients includes travellers, intermediaries who can link travellers to the company, issuing and receiving travel agencies, and hotels, among others.

Concerning the marketing plan and despite the many approaches by many experts on the matter, the entrepreneur should always choose the option with which they feel most comfortable or that suits them best. Therefore, the goal of this section is to provide tourism-specific tips that have proven to be useful during the development of the marketing plan for 5Bogota.

The following are some essential aspects to our corporate growth:

- *The internet is mandatory, not optional.* In order to implement the marketing plan, it is imperative to have a friendly website with attractive and simple content featured in the language clients speak; an efficient payment system that generates trust when making a purchase; and a social media strategy emphasizing the highlights of the experiences where the traveller is front and centre, ideally sharing these publications in their own social networks.
- *Specialized platforms, publications, and influencers in the tourism industry.* This refers to free access platforms that require the creation of a profile, such as TripAdvisor, where users contribute to the credibility of our portfolio of experiences. Secondly, publications such as Lonely Planet and Bratt – world-famous and recognized tourism guides with up-to-date editions for every destination in the world – can further our brand. While there is no sponsoring mode that guarantees featuring in these publications, good service and the recognition of our portfolio may lead our brand to be included in their pages. Lastly, specialized journalists, tourism bloggers, or generally anyone with credibility in this industry who post their opinions on their social networks should be engaged. Influencers can be reached through invitations to special events or to enjoy an experience. All of the above leads to the formation of a community of followers, which positions the brand organically on search engines like Google, among others.
- *International networks.* For 5Bogota, this means the Creative Tourism Network, which aggregates initiatives in this category in addition to calling for submissions of destinations or project ideas based on the creative tourism concept. There are many such networks in the world, which is convenient for tourism small- and medium-sized enterprises with a limited marketing budget.
- *Reviews by previous clients.* Tourism never gives second chances, which has become the main attraction for our clients who come to us in large part owing to the reviews of previous clients who have already lived 5Bogota experiences. Providing unique and impeccable service is essential for success since clients will feel the same obligation to report both bad and good episodes.

Advice to entrepreneurs

Entrepreneurs should bear in mind that the process necessary to implement a business plan and bring it to fruition requires three essential conditions: patience, steadiness, and a minimum financial support, considering that a start-up may take between three and five years to properly consolidate in most economies.

At 5Bogota, we learned that it is necessary to have an advertising and promotion budget, which will allow access to key communication tools such as a friendly web page, and search engine optimization and search engine marketing campaigns. As we did not have this budget, the process took much longer than planned.

Another lesson, and therefore another important recommendation, is to avoid high fixed costs. This will allow for flexibility in making decisions and for easy adaptation to the changing market.

Entrepreneurs should also clearly know that their products are intended for a set target market that, in most cases, will not match their own needs or preferences. In other words, product and service design should be based on what prospective clients might enjoy. A key aspect for 5Bogota was understanding the expectations of a travelling foreigner seeing as our local point of view prevented us from discerning the type of experience our target audience would find most attractive.

Conclusions and lessons learned

1. The experience creation process requires the participation of the hosts. Hosts must be allowed to actively contribute their knowledge and expertise during the co-creation stage.

2. Being a good host requires more than simply mastering a topic. Many people with vast knowledge of a topic do not have the personality required to be good hosts. Indeed, the most important attributes of a host are personality traits, which cannot be taught.

3. Not all tourists are open to creative tourism. Creative tourism is aimed at a specific market niche that seeks personalized and peculiar experience and is consequently willing to spend considerably more than the traditional tourist. All this constitutes a reliable customer filter as creative tourism and mass tourism have two distinctly different target markets.

4. Website, quick response, and excellent service. These three elements are crucial for the development of a creative tourism start-up. The website becomes your main tool, and as such it needs to be friendly, easy to use, and representative of your brand concepts. An efficient website is the element that allows for quick and effective communication with customers. This particular aspect has proven to be decisive for 5Bogota: much of our success has come due to being first to respond to customers, even outside normal business hours.

Time zones are an important consideration as start-ups may function in different zones than those of their prospective customers, who will in turn always need pertinent, personalized information. It is essential to remember that creative tourism strives to shape a community, and to that end customers must be given special treatment.

In closing, start-ups must be mindful that tourism does not give second chances. An unsatisfied customer is a person whose travel has been ruined, and that is why service must be impeccable from initial contact to after-sale service. As stated above, the success of creative tourism is people dealing with people, and that means customers will belong to our community and they will want to keep in touch after their travel experiences. Memorable experiences must be our absolute priority.

15 Deep mapping as a cultural mapping process and a creative tourism driver: Two examples

Kathleen Scherf

Thompson Rivers University, Canada

The importance of identifying your assets through collaborative planning

Tourism practitioners and local agencies wishing to develop a sustainable tourism sector are benefitting from tourist interest in local, distinct, place-based experiences and embedded cultural heritage, both tangible and intangible. The ability to provide visitors a sense of place created by locals with endogenous resources opens up the possibility of a tourism that leverages the unique cultural characteristics of *any* place, whatever the size or expressed parameters – as long as there are creative tourists who want to engage with it. This situation presents tremendous and exciting development opportunities for smaller communities.

Before such opportunities can be offered, however, it is necessary for communities to identify their assets. Creative tourism activities depend on locals to identify cultural assets that exist in their communities, and to tell the stories associated with them in ways that engage community stakeholders to realize that such unique-to-place resources can be identified, designed, produced, marketed, and mobilized as creative tourism products. Successful and sustainable creative tourism products are ones that are genuinely embedded in the community and its people, are willingly shared with visitors by locals, and are extensions of place-based skills and knowledge that exist regardless of the tourism product. We might term that situation an authentic, place- and culture-based experience for both tourists and residents.

The process of community-led development takes many forms, but it is usually an inclusive and iterative stakeholder conversation, which can be generated for any number of reasons, such as to address community questions, problems, issues, or opportunities. Usually, an end point for this process is to form a community plan. The process embeds residents in the discussion, commits them to the ensuing plan, and encourages them to express their belief and support to others.

There are many possible benefits arising from a collaborative approach to development and planning. For smaller communities especially, community-led economic development that involves cultural mapping can help forge a strong relationship between sustainable development and creative tourism. Identifying creative skills and aspects of lived experience, defining unique aspects of traditional local heritage, tracing creative networks, developing policies that integrate the needs of stakeholders, and connecting residents and visitors – all are strategies that can tightly bind creative tourism and sustainable development. As Elizabete Tomaz (2021) points out,

> The symbolic, socio-cultural, relational, and territorial elements of a place are activated in local development strategies to attract new residents, investors, or tourists, but also to strengthen the sense of belonging, identity, and trust of local communities and stakeholders. (p. 62)

The movement for community-led development became international during the 1990s. In Canada, for example, the Canadian Community Economic

Corresponding author: kath@tru.ca

 Creative Tourism: Activating Cultural Resources and Engaging Creative Travellers (eds. N. Duxbury, S. Albino, and C. Pato de Carvalho)
DOI: 10.1079/9781789243536.0015

Development Network was founded in 1999, and offers a very useful toolbox online for those interested in community-led development (https://ccednet-rcdec.ca/en/toolbox; accessed 11 May 2021). Another good resource, again from Canada, is from the Simon Fraser University's Community Economic Development Program. Its website offers information based on five principles for community-led development (SFU-CED, 2020):

1. Livelihood focused: ability of the economy to enrich livelihoods.
2. Diverse and inclusive: the economy equitably deliver benefits to all.
3. Sustainable: recognize complex systems for environmental, economic, social, cultural development.
4. Place-based: development grows from local assets while serving local people.
5. Community controlled: grass-roots, bottom-up, inclusive planning.

Community-led development is an essential framework for cultural mapping, an inclusive process that can be used to identify a community's unique characteristics: people, traditions, and organizations; networks and clusters of interest and expertise; skills, knowledge, and resources; and special locations, histories, and stories. For smaller communities especially, community-led economic development that involves cultural mapping can help forge a strong relationship between sustainable development and creative tourism. If you are interested in leading a cultural mapping process in your destination, the Creative City Network of Canada's publication *Cultural Mapping Handbook* (Stewart, 2007) offers a step-by-step approach that has been helpful for many communities and is openly available online (see References).

Mapping culture, creating opportunity, thinking tourism

The United Nations' World Tourism Organization (WTO) reports that in 2018 the international tourism industry sustained its ninth consecutive year of growth, with over 1.4 billion arrivals generating over US$1.7 trillion. In 2019, the top consumer travel trend was 'Travel "to change": live like a local, quest for authenticity and transformation' (WTO, 2019, p. 5). In its 2020 Barometer, the WTO reports that international arrivals in the first 6 months of 2020 were down 65% from June 2019, representing about 440 million fewer arrivals – a loss of US$460 billion (WTO, 2020, p. 1). The plunge in international arrivals highlights the importance of developing for locals; it also provides an opportunity to develop 'microtourism', or locally focused, place-based experiences for small numbers of people at a time.

Given the well-documented horrors of over tourism, smaller communities especially stand to benefit from 'under tourism', or smaller tourism that can facilitate the tourism trend to live like a local. In their 2018 book, *Small Cities with Big Dreams: Creative Placemaking and Branding Strategies*, Richards and Duif (2018) describe how community-led economic development that involves cultural mapping can help forge a strong relationship between sustainable development and creative tourism.

Identifying creative skills and aspects of lived experience, defining unique aspects of traditional local heritage, tracing creative networks, developing policies that integrate the needs of stakeholders, and connecting residents and visitors – are all strategies that can tightly bind creative tourism and sustainable cultural development.

Cultural mapping and deep mapping

A deep map uses the advantages of digital technology to display an online geographical map of a community, locating cultural resources via hotspots that, when clicked, deploy narrative devices such as images, text, video, audio, and links to tell a place-based story. Very place-specific elements of this story can be plotted in a geographical location. The process of building this digital story map, this *deep map*, can be part of a collaborative cultural mapping process that can uncover and articulate local specificities and, in itself, serve as a creative tourism asset.

As examples, I have two deep map examples to share with you. Please explore them both.

FoodLoops map

https://foodloops.trubox.ca

In early 2014, the City of Kamloops (population ca 90,000) in British Columbia, Canada, engaged in a community-led planning process to develop a path forward for local food sustainability, resulting in 2015's *Food and Urban Agriculture Plan* (FUAP). On the basis of this plan, the map-makers of FoodLoops (two semesters of students in a senior communications seminar) sought to create a deep map based on the FUAP (Fig. 15.1). We wanted to enliven the FUAP and use it to help share the passion of those people who care deeply about food sustainability in Kamloops. We found this passion and the activities that arose from it to be a cultural asset. The Thompson Okanagan Tourism Association (TOTA), which includes the City of Kamloops, has

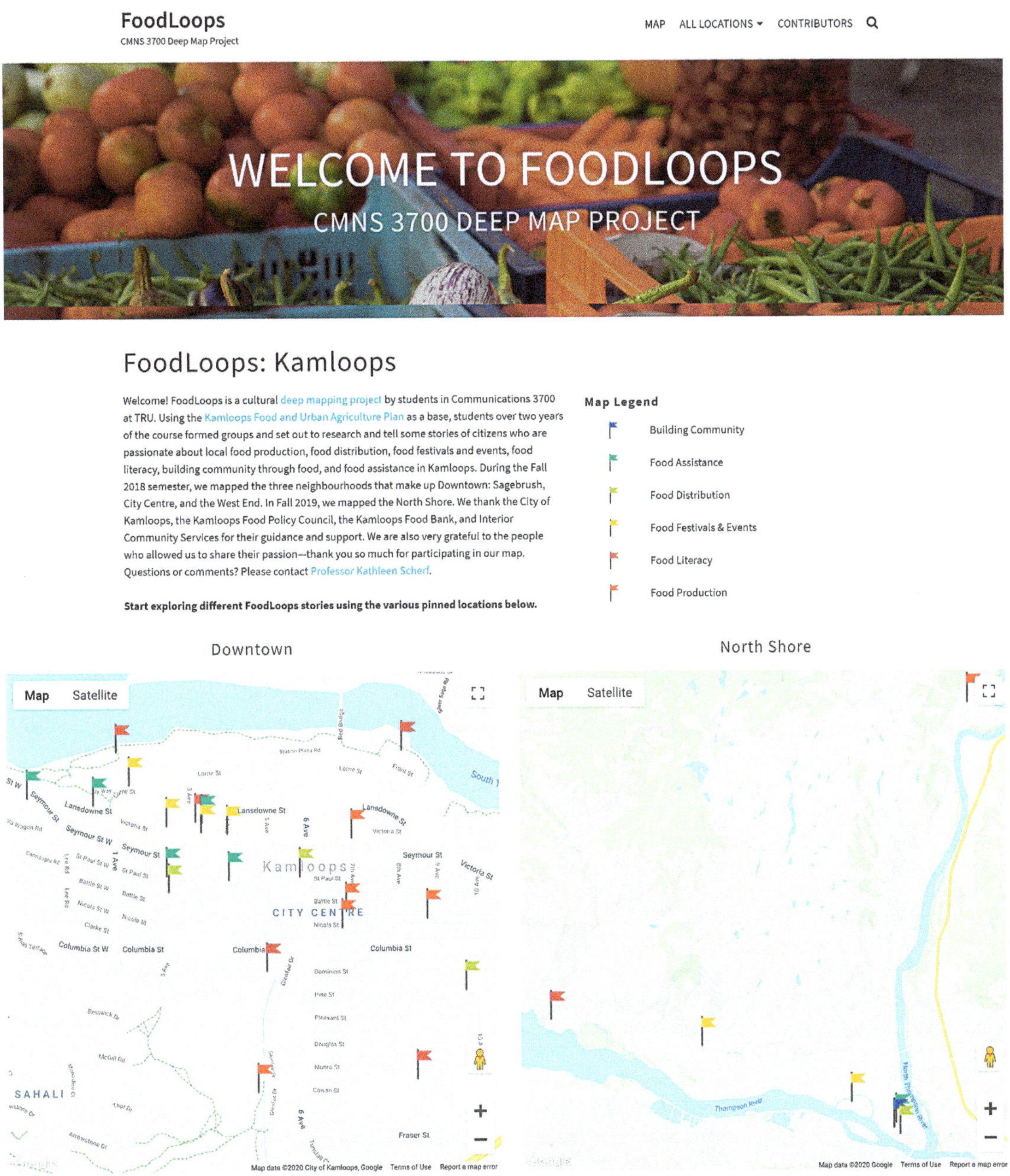

Fig. 15.1. FoodLoops website homepage, Kamloops, Canada. Image by author.

established three key themes for its activities: diversity and inclusivity; outdoor adventure; and food and drink (TOTA website, n.d.). The latter dovetails well with the FUAP, making this a great example of sustainable development goals crossing over into tourism development opportunities. We were very aware that many of the FoodLoops map stakeholders were interested in representing and presenting local flavours for visitors.

Building a deep map follows more or less the same process as cultural mapping. For FoodLoops, we definitely benefitted from the prior collaborative work that had been done on the FUAP. We already knew who comprised the food sustainability community in Kamloops, thanks to the FUAP's conscientious documentation of its wide consultation. This network of interested organizations also provides a web of possible agents for developing community-based and creative tourism initiatives; there is already an Ale Trail tour and many food festivals in Kamloops so an array of related activities could be developed.

A striking feature of a deep map is its power to capture iterative, collaborative work – its visual appeal and easy dissemination encourages community involvement. The first thing we decided as map-makers was that we would meet the community where it stood. We did not disrespect the work already done by imagining we were starting from scratch. We based our map on the FUAP, and asked some of its authors to present in classes. They appreciated our recognition and spoke for us as we began to research the food sustainability community in Kamloops. Their network became our network – this gained us immeasurable access to food assets and food stories. We were able to partner with local government and a number of food organizations in Kamloops, such as food banks, food assistance programmes, ranching and farming families, provincial organizations, educational agencies, urban farmers, urban bee-keepers, urban hen-keepers, community garden organizers, and so forth to hear their stories, share our map ideas, and get a sense of what the sustainable food community looks like in Kamloops. The partnership with the City of Kamloops was especially beneficial. The map's connection to provincial and regional tourism strategies is also useful. The *Thompson and Nicola Valleys Destination Development Strategy*'s fourth theme is 'The Visitor Experience', and the first goal in this theme is to develop niche sectors, such as 'enriching local flavours' (Destination British Columbia, 2020, p. 30) and growing the agritourism industry. The FoodLoops map provides these local agencies with a cultural planning object as well as a marketing tool.

> **Take-away #1:** As you begin to think about your deep map or cultural mapping process, take the time to do the research to identify work that has already been done and networks or clusters that already exist. Meet with them to explain your vision. Get them onside. Speak at gatherings in those interest groups. See if you can find champions.

Next, we began discussions on the themes emerging from our initial research. We wanted to represent the interests of the community, but we also had to find a way to select what information went into the map, and what did not. Also, with so much information, we felt we needed themes so that the end users of the map would be able to navigate a large amount of information. Again, the literal mapping aspect of the deep map is a useful way to organize information for the user. The need to select information and to represent a certain culture within Kamloops forced us to refine our topic – you need to know exactly what the map is meant to do before you decide what goes in it. We had these discussions as a group; with 30 students in each of two semesters and a large number of initial FUAP consultations, it was a very large iterative undertaking. Finally, in order to make a map that would tell the stories of people and organizations passionate about sustainable food in Kamloops, with a view to reinforcing that local food culture as well as expressing it to interested readers (and potential visitors) from away, we came up with the six themes you see on the deep map: Building Community, Food Production, Food Distribution, Food Assistance, Food Festivals and Events, and Food Literacy.

> **Take-away #2:** Part of your deep-mapping process is to be true to the purpose of the map and to its potential users. The sheer magnitude of information, plus the richness of media formats available for the hotspots, added to the number of possible voices and audiences, will make it easy for you to become distracted. Make sure you and your contributors have agreed to the reason for, purpose of, and users of your deep map. Reminders of these principles are a good part of the iterative process.

Map-making will soon get very real! Early on, map-makers must decide who can post on the map. The most fundamental decision here is this: Is the map open for users to upload content? This is a complex issue. Of course, the idea of 'open' is alluring, especially if creative tourism is part of why you are deep mapping; after all, the most important relationship in creative tourism is that between visitor and resident. However, an open deep map also poses some dangers. If you have agency or industry partners, they may be wary of sharing their logos if they might be associated with negative or offensive content. If you want to go with an open map, you may need a moderator. For FoodLoops, we decided against an open deep map, but with so many map-makers at work, we felt confident that we would present a wide array of voices. An innovative solution to this problem can be found in the second map to be discussed, the Literary Atlas of Wales project. Most of the map is not open, but one of the options on the landing page is 'Miplot', which allows users to locate and upload their own 'microfiction', on a map of Wales.

> **Take-away #3:** Ensure that map-makers and stakeholders agree on how open the map will be. Will users be allowed to upload? If so, will you require a moderator? And if the deep map is not to be open for users to upload, who is authorized to upload? Who makes the final editorial decisions? Is this a paid or volunteer position, or is this work added to someone's extant duties? How will the internal process of assembling the map be collaborative? Will there be a committee deciding on what gets posted? What is the content submission and approval process?

This discussion about an open map and who posts on the map presumes that you have a way to build a deep map. Unless you have some higher-level digital design skills, it will be difficult to make an attractive and functional deep map. Both FoodLoops and the Literary Atlas benefitted from being based at universities, where information technology departments provide not only institutional services, but are also committed to research and pedagogy. In each case, after the editorial process, one technical director was responsible for the final upload, appearance, and publishing.

> **Take-away #4:** Technical equipment, expertise, and time will be needed. If an agency or institution is spearheading the map, a commitment to an information technology service should be articulated and approved. If practitioners are leading the initiative, it is very useful to partner with an educational institution or a community organization – if such organizations are not already on board, is it possible that the deep map could express their interests such that they would contribute the information technology needs?

> **Take-away #5:** About a week before the deep map is actually published online, share a draft of it with all your stakeholders. Think about holding an event, a soft launch, at which you review the consultation process, acknowledge partnerships, thank contributors, review the intent of the map, and briefly describe the process you followed in building the map. If there are sections, explain those. Then take them through sample sections to show them how it works and the range of content. Ask for comments and promote discussion. Take notes! You will get some great feedback; do not hesitate to revise, but remember to stay true to the map's purpose.

How is FoodLoops related to creative tourism? It does promote some activities such as cooking classes and farm experiences but, to date, it is more of a cultural mapping process with the local community as its intended audience. However, using the FoodLoops deep map itself as a cultural asset, it would now be possible for creative tourism developers to use it to identify opportunities to provide creative tourism experiences. A better example of a deep map as a creative tourism product and driver is the Literary Atlas project.

Literary Atlas: Plotting English-language novels in Wales

www.literaryatlas.wales/en/

There is more than one map on this site. Really, it is a compendium of information that is conveyed to users in an amazing digital expression of literary geography. A massive undertaking, the Literary Atlas takes 12 Anglophone novels set in Wales, plots them on maps, and provides rich content around them. On the landing page (Fig. 15.2), scrolling down will show readers these options: (i) an introduction to the 12 novels, mapped and with a formatted presentation structure with lots of loops and links to zoom maps; (ii) a map of the 571 Anglophone novels published in Wales between 1800 and 2018; (iii) a section on how the novels were chosen, how to use the site, and how to connect with the map-makers; (iv) an 'Explore the Novels' section which repeats the content of the first section, but is organized by book covers rather than by geographical location; (v) the authors, with lots of ancillary links; (vi) artworks – one was commissioned by the project as a response to each novel – this is particularly striking rich content, which includes the art and detailed notes about it; (vii) how to follow and use the site on all devices;

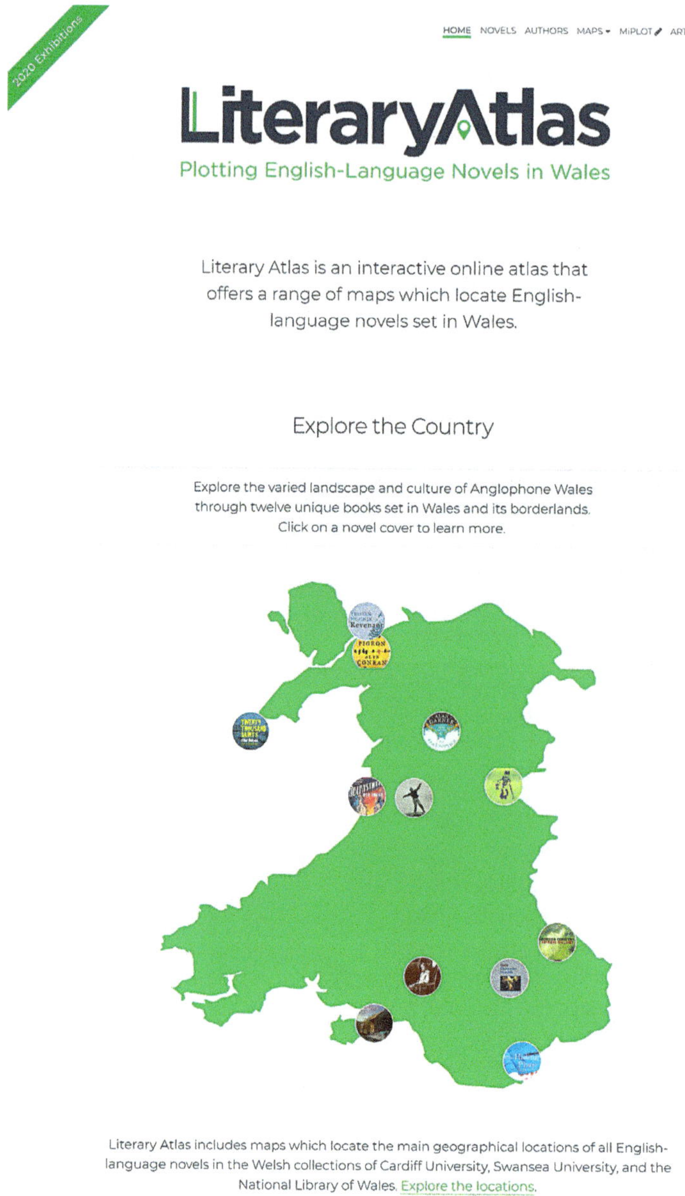

Fig. 15.2. Literary Atlas website homepage, Wales. Image by author.

and (viii) geographical 'plot points' for each of the 12 novels, so you can connect with a geographical reference in the novels near you. Regrettably, there is only one entry in Canada – Baffin Island, which is referred to in the novel *Mr. Vogel*, by Lloyd Jones.

> **Take-away #6:** A deep map is an opportunity to communicate a lot of information about a place.

> If engagingly presented, the map will interest potential tourists – in this case literary tourists – to explore place-based experiences for themselves.

As Kieron Smith (Smith *et al.*, 2021), one of the Literary Atlas map-makers, says,

> Creative literary tourism that utilises such maps can provide the basis for throwing into relief those

> 'intangible' cultural resources off the beaten track, while at the same time being sensitive to the diverse histories and stories of those places. Visiting through the lens of literature can open up touristic experiences that contribute to the 'ongoing composition' of places, offering new perspectives and co-creating sustainable new experiences and senses of place for today and the future.

The Literary Atlas is a website that must be maintained. The map-makers, for example, must keep up the count of Anglophone novels published in Wales. Also, its organizers offer field trips to the 12 literary locations as well as discussion groups – in themselves, two creative tourism products.

> **Take-way #7:** To make full use of the deep map as a tourism product, you will need to keep it updated and fresh. Do not underestimate the amount of time, information technology service, and human effort this will take.

As a collaborative cultural mapping process, deep mapping can encourage networks and partnerships in designing and offering cultural and creative tourism activities; as a product, deep maps can provide the basis for marketing your destination by providing a lively and engaging sense of place. Even if you are just seeking to bridge the pandemic and better times, deep maps can provide a site for virtual tourism. The use of digital media as a tool to express the complex process of cultural mapping and place creation, and to engage visitors in an authentic sense of the distinctive and unique features of the 'place' in a mediated setting, is timely in an age of digitally enabled tourism.

References

Destination British Columbia (2020) *Thompson and Nicola Valleys Destination Development Strategy*. Destination BC, Vancouver, British Columbia, Canada.

Richards, G., and Duif, L. (2018) *Small Cities with Big Dreams: Creative Placemaking and Branding Strategies*. Routledge, London, UK.

Simon Fraser University – Community Economic Development Program (SFU-CED) (2020) What is CED? Available at: https://www.sfu.ca/ced/About.html (accessed 11 May 2021).

Smith, K., Anderson, J. and Morgan, J. (2021) Literary Atlas: A digital resource for creative tourism in Wales. In: Scherf, K. (ed.) *Creative Tourism in Smaller Communities: Place, Culture, and Local Representation*. University of Calgary Press, Calgary, Canada, pp. 209–228.

Stewart, S. (2007) *Cultural Mapping Toolkit*. Creative City Network of Canada and 2010 Legacies Now, Vancouver, Canada. Available at: www.creativecity.ca/database/files/library/cultural_planning_toolkit.pdf (accessed 24 September 2020).

Thompson Okanagan Tourism Association (TOTA) (n.d). Thompson Okanagan. Available at: www.totabc.org/thompsonokanagan (accessed 10 May 2021).

Tomaz, E. (2021) The interplay between culture, creativity and tourism in sustainable development of smaller urban centers. In: Scherf, K. (ed.) *Creative Tourism in Smaller Communities: Place, Culture, and Local Representation*. University of Calgary Press, Calgary, Canada, pp. 61–78.

World Tourism Organization (WTO) (2019) *International Tourism Highlights, 2019 Edition*. WTO, Madrid, Spain.

World Tourism Organization (WTO) (2020) *Barometer*, 15 September 2020. WTO, Madrid, Spain.

16 The creative tourist and the creating artist: The base for rural development

CATHARINA SLIGTING

Associação Domínio Vale do Mondego, Portugal

In 2007 we discovered a hidden valley in the North Eastern part of the Serra da Estrela Nature park in Portugal. The road ended at O Cabo do Mundo, as the last village Vila Soeira on the meandering road along the mountain river Mondego was called, 'the end of the world'. It was June, it rained, the hill tops were clouded in misty fogs reminiscent of Ireland, the favourite country of my childhood. Even more so because of the goat-like sheep grazing among the rocky boulders of the harsh granite mountain ridges. The whole valley was green because of the evergreen leaves of the olive trees, and the cherries were as big as plums. We passed only five small hamlets composed of tiny houses handmade with granite stones. Some elderly people were about, taking care of their vegetable plots or gossiping in the café. There was something magical about the place. And we decided to stay.

As true romantics and idealists, being both ambitious and self-confident, we settled on our own farm, a *Quinta*, which we called 'Domínio Vale do Mondego', meaning the estate of the Mondego Valley. From the start we knew our goal: to create a healthy biodynamic farm with respect for nature and biodiversity, producing healthy food – and to become a meeting place for interesting people (meaning: people who are interested in culture, heritage, ecology, and art). As long-standing practitioners of culture and agricultural management, over 12 years we developed a programme of art-and-nature workshops, summer festivals, philosophical convivia, art residencies, and field studies. Our farm is certified organic and biodynamic, and we have shared many wonderful meals with many people from all over the world. If this sounds like paradise, let me enlighten you: it is a hard world out there, so be sure before to throw all your comforts of urban life overboard!

Living in the city is very practical, efficient, and comfortable in comparison to a true rural life. By a true rural life, I mean being dependent – socially, culturally, and economically – on the amenities, facilities, and products that living and working in remote, sometimes isolated, conditions provide. For a week or two during holidays, people love the outdoors, the 'simple' life in the middle of nature. Back to basics. We all crave silence, space, hummingbirds, and butterflies. But in reality, city dwellers turn their backs towards the rural area. And in doing so, we have become alienated from nature, from agriculture, from our own food production.

In our re-location, we have embraced and shared the experiences of other urban-to-rural migrants. We have come to deeply understand the transition we have taken, like Henk Keijzer, long-time director of outdoor art festivals:

> When we started making works in rural areas, we shared a somewhat romantic idea of what was going on there: beautiful nature, harmonious life, a fairy tale. Getting to know the people, listening to their stories, learning about what is life in rural areas and farming nowadays changed our attitude and opened us to active listening.

Throughout the past decade, we have come to understand the relevance of being embedded in the local community, to be integrated in the social and economic fabric of the region. We realized our

Corresponding author: info@dominiovaledomondego.com

 Creative Tourism: Activating Cultural Resources and Engaging Creative Travellers (eds. N. Duxbury, S. Albino, and C. Pato de Carvalho)
DOI: 10.1079/9781789243536.0016

activities did have an impact on the existing community. By moving to the valley and working full-time on its development, we created a new reality. The important question here is: Was the change for the good, for the benefit of the locals, for the natural environment? Have our activities contributed to rural development? What exactly contributed to its benefit, and if we could enhance this, what element, what activity, what programme would improve and increase its value?

This brief essay explores how arts-based creative tourism contributes to positive attitudes towards rural and natural areas, both by participating creative tourists and artists as well as by the local community. It shows the necessity of a sense of place, an awareness, a consciousness of the valorization of art as a creative force to break the negative downward spiral of negligence, outmigration, isolation, and dependency of small rural communities in rural and natural areas.

We are based in Faia, Guarda, located in the Serra de Estrela National park in central Portugal, a location where, for many centuries, nature and culture have blended in a very harmonious way. With the rugged mountains directly around the valley, which provides a beautiful view of the green valley and the small rural villages, the medieval atmosphere, and an authentic population that still works the land as their forebears did, the sense of place is immediately felt and is inspirational. No tourists, no industry, no highways. It is a lively although secluded world, characterized by ancient walking trails, Roman roads, Romanic bridges, mountain rivers, rocks, women washing at the communal washbasin, the local *festas*, the romantic ruins, and, above all, space and silence. The *Genius Loci* – the uniqueness of the place – inspired us to settle here.

In line with our ambitions and dreams, we developed activities centred on eco-agrotourism, cultural heritage, nature conservation, biodynamic farming, artists-in-residence, an annual theatre and arts festival, land art projects, and slow food. The artist in us is at the heart of our entrepreneurship. We worked along the lines of Scherdin and Zanders' (2011) definition: 'Art entrepreneurship is about the discovery and pursuit of new art ideas, using a multitude of artistic expressions and organizational forms as vehicles by which to express and convey these ideas to the public' (p. 3).

Perhaps our best and most rewarding example of arts entrepreneurship has been our rural art festival, *Estival da Estela*, which we have organized for the last 10 years at Domínio Vale do Mondego. The event is deeply rooted in an 'arts entrepreneurship' sensibility. This means that the artistic mission is at the core of the planning, strategies, and other activities through which the event is realized.

In short, the week-long 'gathering' of international participants, volunteers, and artists – in total, approximately 150 people of all ages participate – establishes a very special vibrant energy, shared by festival participants and locals. At the core is the artist – musician, dancer, performer, poet, or painter – who creates a new work of art on the basis of what already exists, be it a traditional song, folklore music, oral history, cultural traditions, rural architecture, or the natural environment. Working with these local ingredients, the artist creates something new: he or she adds to the public domain, and by doing so a connection is developed between the rural and the urban. The festival participants join the artists in their creative work: the opera singer assembles a choir of 30 participants, rehearses several times, and the whole choir is on stage together with the professional singer one night. The land art artist creates a huge wooden sculpture, for which she needs the co-operation of five people during the week, who all contribute to the final artwork (see Fig. 16.1). The felt artist demonstrates how to felt from rough woollen fleece, and 3 days later the participants of this workshop present their own creative work in a general exposition in the atelier. The poet assists the participants of the poetry workshop in writing their own lines, which are displayed a few days later during a poetry slam in the amphitheatre. In this fashion, Estival offers a myriad of creative workshops every day of the week, catering for everyone's wishes and needs.

One of the unique features of the small-scale festival is the absence of the backstage. There is no physical division between the paying participant and the performing artist: they all stay at least a week and share the same 'space', they sit at the same the dinner table, they brush their teeth in the open-air campsite bathroom, and they dance together at night on the river beach to the tunes of the DJ. The poet might participate in the workshop of the painter, the musician contributes spontaneously to the work in progress of the dancers (a co-operation that could continue long after the festival ended); and they all – those who are young at heart and healthy – take part in the demanding 4-hour 'river hike' through the Mondego mountain river.

Fig. 16.1. Land art project with Marjorie Sloof during the rural art festival, *Estival da Estela*, 2018. Photo by Lisette Zeegers, used with permission.

The experience of art works created together adds to the public memory. The festival specifically requests artists to engage not only with the participants of the festival, but also with the local community, to share their passion through workshops and presentations, in situ performances and public interventions (Fig. 16.2), communally built installations, and street parades. Our intention for the festival

Fig. 16.2. Dance performance in Faia, during the rural art festival, *Estival da Estela*, 2018. Photo by Lisette Zeegers, used with permission.

has been to create 'the conditions that facilitate community-embedded arts-based development in rural area – sparking a creative fire' (Balfour *et al.*, 2018, p. 229). But what are these conditions, and what have we learned from our experiences of the last 10 years? And to what extent can these experiences be copied by us as well as other creative tourism operators?

For a start, to genuinely create an arts-based creative tourism business, one has to engage with professional artists. By this I mean artists who have professional training and live from their art and are true to their passion: this could mean the outcome is unpredictable, but definitely unique and artistic. The artist needs freedom and space, and the organizer has to respect that. Although entrepreneurs do not favour unpredictable outcomes, it is exactly this principle that is essential to produce a unique experience. It all comes down to trust. Whereas you can duplicate your method, the content of your activity may be different from year to year. This is great if you want to attract repeat visitors – they will be surprised by a completely new programme every year. That is exactly what we did. Estival has been frequented by various families and individuals for many consecutive years.

Yet on the entrepreneurial side one must reckon that 'the main purpose of the entrepreneurs in the arts is not profit, but rather the realization of cultural values, and business is only a tool used in the end: ... this means that the cultural entrepreneur, in creating these values, should not in any way follow the logic of business' (Klamer, 2011, pp. 154, 141).

Although most participants of the festival travelled from afar and used their full holiday budget to partake in the festival week, it was not always easy to attract our neighbours in the next village to visit us. We learned by trial and error that asking an entry price did not work, nor lowering the price, nor giving it for free. What we offered was unknown to the residents of the local villages, who had no past experience with (international) theatre on location. We had to overcome the cultural difference by offering something which was known: a Fado concert. From the moment we presented Fado, which was liked by the foreigners as much as by the locals, it became an established part of the programme.

The biggest success was to bring the whole festival to the village, instead of inviting the villagers to the festival location. Three main elements were decisive for the success. Firstly, with this move, suddenly the organization became a joint effort with representatives of the local community. We needed their support to prepare the location, the access, the facilities, and the communication. It also facilitated the involvement of the whole village, being so small there was no chance of escaping it. Secondly, to integrate two different cultures (Dutch and Portuguese), we found that food provides an excellent means to connect. With the support of the municipality, we were able to invite all the villagers – and, as it turned out, all their family members in the region – to a sumptuous street dinner of roasted presunto (which was less imposing than a whole pig on a spit), *caldo verde* soup made by the local women, and salad from our kitchen for the vegetarians (not surprisingly, only present among the international tourists). Thirdly, and most importantly, the success was determined by the incredibly versatile creative contribution of the artists: it was through their efforts of blending modern dance with folklore music, the local church choir with a professional electronics musician, a parade with 100 handmade lanterns ending in a village dance on the square ... just to mention a few examples that created a unique festival atmosphere that transcended cultural differences and connected the rural with the urban, the young with the old, and the artist with the audience. For some, it was a life-changing experience and, for others, the manifestation of a hopeful future. And definitely for all, an unforgettable evening.

So back to the fundamental questions at the beginning of this essay: Did our involvement have a sustainable impact on the social, economic, and ecological circumstances of the local community? What elements, if any, contributed specifically to that development? And to add to this, did we succeed in establishing an organization that attracts creative travellers on an annual basis? Is there something to learn here for other creative tourism entrepreneurs?

The answer to these questions is 'yes', and the reason is because we believe we created a fundament, literally a foundation, which has served as a permanent and solid base for input from local and international resources as well as for output of creative and artistic programmes and projects. This permanent base has proven to be essential in the composite of existing local cultural resources and innovative creations by (international) artists. We have succeeded in establishing a cultural 'hub' or incubator with a firm connection to the local while being, at the same time, open to engage with an open mind, to share the act of creation, and to co-create – this combination has resulted in new connections, in the development of new values, and in processes of rediscovering the rural. As Andrew (2017) notes,

> Rural areas offer a unique set of challenges for arts organizations seeking to serve those communities. Due to the geographic isolation and expanse of rural settings, access to transportation, jobs, healthcare, and general public services are a challenge for residents of rural communities. Arts organizations looking to serve rural communities must address challenges unique to rural audiences. Through collaboration with organically congregational centers, arts organizations can capitalize on the particular resources that are available in rural communities and mitigate barriers to art participation in rural areas. Arts organizations that create partnerships with these community gathering spaces are able to provide more arts programming to rural residents while insuring institutional longevity through the building of public value. (p. 1)

In hindsight, we may conclude that both the creative tourists as well as the creating artists have been indispensable for the success of ten years organizing the Estival festival. The next step is to professionalize and to partner with local cultural organizations, to engage in networking activities on creative tourism, and to collaborate with international artist residence organizations. The Rural Art Festival organization has the ambition to develop into a genuine Rural Art Hub for the benefit of creative tourism, art entrepreneurship, and sustainable local development. To us, its value is apparent and inevitable.

Acknowledgements

This chapter was developed within the CREATOUR® project (no. 16437), which is funded by the Portuguese Foundation for Science and Technology (FCT/MEC) through national funds and co-funded by FEDER through the Joint Activities Programme of COMPETE 2020 and the Regional Operational Programmes of Lisbon and Algarve.

References

Andrew, B. (2017) Places, perceptions, and public value: Building rural arts participation through community partnership. Master's thesis, Goucher College, Baltimore, Maryland, USA. Available at: https://mdsoar.org/handle/11603/4346 (accessed 20 April 2020).

Balfour, B., Fortunato, M., and Alter, T.R. (2018) The creative fire: An interactional framework for rural arts-based development. *Journal of Rural Studies* 63, 229–239. doi: 10.1016/J.JRURSTUD.2016.11.002.

Klamer, A. (2011) Cultural entrepreneurship. *The Review of Austrian Economic* 24(2), 141–156.

Scherdin, M. and Zander, I. (eds.) (2011) *Art Entrepreneurship*. Edward Elgar Publishing, Cheltenham, UK.

17 COOLWOOL – Creative weekend at Covilhã, a co-designed programme

RITA SALVADO,[a*] AND GUIDA ROLO,[b*]

[a]*Museu de Lanifícios da Universidade da Beira Interior, Portugal;*
[b]*New Hand Lab, Portugal*

COOLWOOL – Creative weekend at Covilhã is a creative tourism programme to discover the wool industrial heritage of the Portuguese industrial city of Covilhã. It proposes a singular experience of immersion into the city factory and into the factory ambience. The project is co-organized by the New Hand Lab (www.newhandlab.com) and the Wool Museum of the University of Beira Interior (www.museu.ubi.pt) and was developed as part of the CREATOUR® project (www.creatour.pt; accessed 10 May 2021).

Covilhã is a small academic city in the interior centre of Portugal, located in the southwest slope of the highest Portuguese continental mountain – the Serra da Estrela mountain. The city has 51,797 inhabitants, but only 23,040 as its active, working-age population (www.cm-covilha.pt; accessed 10 May 2021), while the University of Beira Interior, located at Covilhã, has currently more than 8200 students. The place and local spirit of the city are strongly marked by a long tradition in the wool industry, which reached, during the twentieth century, such brightness that Covilhã was once considered to be the 'Portuguese Manchester' (Pinheiro, 1998, 2009; Pinheiro and dos Santos Silva, 2012; Folgado, 2009). Nowadays, the wool industry is still an important economic activity. At Covilhã, the wool industry is thus a living heritage.

The first co-organizer, New Hand Lab, is a project that promotes creativity, innovation, and entrepreneurship through the implementation of ideas, products, and initiatives. The factory António Estrela / Júlio Afonso, now home to the New Hand Lab, was built on the site of a factory from the seventeenth century and was a working wool mill until the early 21st century. Distinguished over this time by its creativity, in 1904 the factory received an award at the Universal Exhibition in St Louis, USA, and in 1976 received the prestigious 'Brilliant Pen' award from Men's Fashion Writers International. Today, it collects and preserves collections composed of fabrics, yarns, technical files, colour palettes, blending palettes, and fabric samples from its operations between 1950 and 2002. The New Hand Lab holds a number of creative laboratories in photography, multimedia, fashion, product design, plastic arts, painting, and music. Since its foundation in 2013, it has developed a unique set of interdisciplinary projects and products that can be purchased in its shop.

The second co-organizer, Wool Museum, is a science and technology museum at the University of Beira Interior. Its primary venue is located at the Royal Textile Factory, a state manufacture founded by the Marquis of Pombal in 1764, with an 18th-century dye-house that has been classified as a National Cultural Heritage since 1982. Beyond the diverse sites of the Wool Museum, other faculties and services of the University are installed in old industrial buildings that have been renovated and given back to the city for a vibrant academic campus and local use. Located within the academic campus, the Wool Museum is by vocation a cultural hub among the academy and the citizens. Being a member of the Portuguese Network of Museums since 2002, the Wool Museum has the mission of preserving the heritage related to the wool industry in the territory of Serra da Estrela, with Covilhã as its historical centre. To promote the collective valuing of this heritage, the Wool

Corresponding authors: lrbss@ubi.pt; guidarolo@gmail.com

 Creative Tourism: Activating Cultural Resources and Engaging Creative Travellers (eds. N. Duxbury, S. Albino, and C. Pato de Carvalho)
DOI: 10.1079/9781789243536.0017

Museum proposes the cultural and touristic *Wool Route – TRANSLANA*, with diverse itineraries to explore and better understand the countless marks linking the territory to wool production and wool industry (Pinheiro, 2009).

The partnership of the two institutions, a public one, the *Wool Museum,* and a private one, *New Hand Lab*, while assembling different approaches to heritage conservation share the same commitment to value it. The co-operation enlarges the cultural mapping of the industrial heritage of the city, allowing an immersion into diverse ambiences and historical periods, from an 18th-century dyehouse to a 20th weaving factory. The creative tourism programme reveals a collective memory of the industrial heritage by uncovering the city's DNA, as well as the unique ambience of the working space of Júlio Afonso, an award-winning creative woven designer.

This partnership, although challenging from the beginning, aimed to co-create a new touristic product to fill the gaps found in the region's touristic offer, in three ways. Firstly, it focused on addressing the seasonality inherent to the touristic product that drives most tourists to the region – snow and ski – looking for a sustainable tourism industry that works all year round. In particular, the private institution involved in this project suffers from this inconsistency in touristic demand, facing sharp declines in the remaining 8 months of the year (April to November). Secondly, it was inspired by the enormous opportunities for a touristic offer linked to the place's cultural identity, appealing to tourists seeking experiences and knowledge and enlarging the *Wool Route – TRANSLANA*. Lastly, it aimed to address the lack of a creative tourism product that, through its gradual growth, could motivate local networking and the growth of parallel activities, such as small businesses, small creative industries, and hand crafts, thus positively intervening in the community and in the well-being of inhabitants.

The COOLWOOL product

The creative tourism product was initially proposed as a weekend programme, *COOLWOOL – Creative weekend at Covilhã,* which included creative workshops that were also offered separately as *COOLWOOL – Creative Workshops*.

COOLWOOL – Creative weekend at Covilhã

COOLWOOL – Creative weekend at Covilhã is a city break programme that invites participants to discover the wool industrial heritage of the city. It aims to offer creative and relaxed tourism activities to discover the local culture, through being introduced to crafting techniques and by sensing the wool heritage. The programme is conceived to reach a very specific group of tourists, experienced people who have accumulated both a taste for creative experiences as well as an enthusiasm for textiles and wool culture. One of the promotional flyers is presented in Fig. 17.1.

Covilhã is nationally well known for tourism in wintertime, to experience the snow and skiing and, in general, to access the natural area of Serra da Estrela. This programme, offered all year round, aims to offer alternatives to winter sports, challenging visitors to discover the wool culture. The aim in the future is to enlarge the audience, bringing to Covilhã more visitors interested in industrial wool heritage. It is thus a programme for curious people who like new experiences, to be challenged, and to know the places they visit through their history and identity. The full-weekend programme, from Saturday 14h30 to Sunday 17h00, includes:

- Guided tours to both cultural institutions, inviting participants to learn more about the wool industry, its history, craft memories, and multidimensional impacts on the city of Covilhã, and offering experiences of immersion in the industrial atmospheres, throughout the 18th to the 20th centuries.
- Creative workshops in which creativity combines with knowledge and handcrafting techniques, where participants explore materials and learn techniques in hands-on workshops and feel the spirit of a textile factory.
- A walk through the city, following the itineraries of the *Wool Route – TRANSLANA*, to get a sense of the identity of the place and uncover the cultural DNA of *Covilhã City Factory*, through its toponymy and urban structure.
- Opportunities to taste the typical local gastronomy and discover traces of the wool industry in customs and traditions.
- A screening of the 1921 silent movie about the city, which integrates the narrative of the *Wool Museum*, accompanied live by musicians playing an original soundtrack while enjoying dinner in

Fig. 17.1. Promotional flyer for *COOLWOOL – Creative weekend at Covilhã,* held in October 2019. Photo by João Pedro Silva, graphic design by Guida Rolo. Image courtesy of COOLWOOL, used with permission.

the factory atmosphere of the *New Hand Lab*. Accommodation and dining in the city, offering a genuine and quality experience (for further details, see Salvado and Rolo, 2020).

COOLWOOL – Creative Workshops

The *COOLWOOL – Creative Workshops* are two-hour hands-on activities running more frequently or by request. Three offers were tested:

- *Illuminated Textiles / Wool Museum.* This workshop combines electric conductive materials with wool and other conventional textile materials to create interactive products as well as lighted cloth. Using simple handcrafting techniques, participants create electrical circuits and textile sensors, forecasting the future of wearable technology. It is an activity that promotes a new approach to textiles and handcrafts, good for curious people who like to make things work. Through this workshop, students and researchers from the University of Beira Interior co-operate with the Museum to introduce an emerging topic, wearable technology, which is creating new opportunities for the textile industry and bringing new values to fashion.
- *Punch Needle / New Hand Lab.* The embroidery technique known as punch needle is excellent for getting started in textile crafts as it allows beginners to reproduce patterns with ease. In this workshop, participants explore New Hand Lab's collection of fabric designs, creating contemporary patterns based on historic ones from the wool industry (see Fig. 17.2). Participants take home the work produced in the workshop, and also the needle and the frame to carry on creating original art works afterwards. The workshop is conducted by Ana Gonçalo, who also organizes textile workshops on weaving, sandals of rope, macramé, and manual felting, among others.
- *Heritage in Images / New Hand Lab.* Documentary photography plays a decisive role in the preservation and dissemination of cultural heritage. In this workshop, the participants have the opportunity to work with the materials available in the factory, but in another way, through the analytical look and the recording of memories through photography. João Pedro Silva, resident photographer of the New Hand Lab, discusses different creative possibilities in the field of documentary photography, creating, together with the participants, a small documentary of the Factory António Estrela / Júlio Afonso. This activity can be designed for parents and children.

Fig. 17.2. Participants in the COOLWOOL punch needle workshop at New Hand Lab. Photo by João Pedro Silva, courtesy of COOLWOOL, used with permission.

Reflections on the creative tourism offers

COOLWOOL is a creative tourism programme that approaches creativity through offering a unique experience of immersion into the textile industrial heritage of Covilhã. The COOLWOOL weekend programme was tested once, with a small group of participants, who were very pleased with it. Participants were looking to meet other people who share the same interests. Answering this motivation demands a dynamic programme of creative tourism activities and more flexibility in terms of dates, duration, prices, and cultural content. COOLWOOL results from a co-design process, developed during the working sessions of the CREATOUR® project, that made both organizers reflect on diverse approaches to culturally value the wool heritage and about the role of creative tourism in it. Despite the operational improvements that are still required, the COOLWOOL product seems to be a good seed for networking based on creative tourism in Covilhã.

In parallel with the offer of creative tourism for the weekend, short-term activities were designed as part of the same programme. These activities were designed with the tourists from neighbouring counties in mind, in order to offer a 'first contact' with the local cultural identity and a learning experience. The creative workshops looked for a combination of novelty and usefulness, aiming to offer a relaxed learning experience. However, creation requires dedication to experiment and construct the product. The two-hour hands-on activities have to be tightly monitored and creativity is surely limited by this. Thus, the workshops simply aim to unveil a theme and excite further learning. A longer-term serial programme may be an option for residents in the region, but other formats are also needed to deepen skill development and creative production for creative travellers from further away.

Reflecting on the project experiences to date, although both co-organizers had the same desire to develop a touristic approach to value the local memory of the wool industry, trying to take into account the segment's diagnosed flaws while meeting their individual objectives and missions proved to be a real challenge. Well known to each other in general, these institutions have always followed different paths. New Hand Lab is an informal project born in 2013 within the community and the Wool Museum was founded in 1989 by the University of Beira Interior; both have the aim of meeting the needs of conservation and valuation of the heritage associated with the wool industry in the city of Covilhã. The starting point to co-organizing COOLWOOL was identical. Each institution submitted their own project idea to the CREATOUR® open call and, based on the analysis of the two separate projects, the CREATOUR® team considered that there was an interest in joining the projects, which worked on the same theme – wool – and had similar objectives.

The challenge was accepted with enthusiasm by both parties and steps were immediately taken to unite the two proposals. As in the initial proposal, a weekend programme was foreseen and it was considered that it made sense to divide this time equally, with the organization of the Saturday activities under the responsibility of New Hand Lab and the organization of the Sunday activities under the responsibility of the Wool Museum. Later on, however, we came to the conclusion that it could have been more innovative, for the final product and for the tourists, if the two institutions had achieved a true harmony in the organization of the activity. An integrated programme with the joint participation of the two entities could have provided all the elements involved with greater learning and growth experiences and a better use of the resources of each of the entities, naturally resulting in a more consistent touristic offer. It was likewise realized that if efforts had been made to unify the touristic product, other partners would have been more easily integrated in the same spirit of unity.

Although the co-designed solution that was found seemed simpler and more effective, we concluded that it weakened the touristic product, as the physical distance between the partners led to an extreme difficulty in finding a common ground to support a coherent programme. After carrying out the activities and analysing their implementation, it has become clear to the co-organizers that in order to continue this experience, it is necessary to have a new approach. The design of this touristic proposal would have to involve all the elements of the institutions linked to the implementation of the project. In this way, they would have to get to know each other better and share their different experiences and views on the conservation and enhancement of this heritage. An articulation with the touristic *Wool Route – TRANSLANA* and its shared experiences and ideas would certainly enrich the proposal. It seems also to be necessary to more deeply identify the strengths and weaknesses of each co-organizer in order to find a solution of real complementarity. On this basis of sharing knowledge, trust, and generosity between the two institutions, it would be possible to involve other local partners who share this spirit in this creative project.

The path of COOLWOOL began with a partnership that is expected to grow as there is still a wealth of possibilities to explore by enlarging the cultural map to incorporate other cultural, leisure, and tourism facilities and organizations. An important objective has been to involve local residents and collaborators in an integrated way, engaging them into the project itself and rising awareness of the potential role of COOLWOOL project on the city's growth and development. The partnership established through this initiative is generating and enlarging the forms of community involvement, engaging a high number of local artists, craftsmen, college students, and researchers through both institutions, and enriching the COOLWOOL programme.

Although COOLWOOL is still at a very early stage, it has already created some synergies, offering activities that may mobilize more tourism to the city, increasing the profits around it and contributing to the economic development of the region. As the project is focused on networking and on the city and its industrial heritage, the involvement of the city council appears to be a determining factor to its further evolution and development, to increase the visibility of the project, to articulate it with local policies and strategies for tourism, and to enhance its impacts in other local growth and development strategies. We look forward to the challenges and rewards of fostering these collaborations.

Acknowledgement

This chapter was developed within the CREATOUR® project (no. 16437), which is funded by the Portuguese Foundation for Science and Technology (FCT/MEC) through national funds and co-funded by FEDER through the Joint Activities Programme of COMPETE 2020 and the Regional Operational Programmes of Lisbon and Algarve.

References

Folgado, D. (2009) Covilhã, a cidade que também foi fábrica. In: *Monumentos, Cidades, Património,*

Reabilitação. Instituto da Habitação e da Reabilitação Urbana, Lisbon, Portugal, no. 29, pp. 88–97.

Pinheiro, E.C. (1998) *Roteiro do Museu de Lanifícios da Universidade da Beira Interior: Núcleo da Tinturaria da Real Fábrica de Panos.* Universidade da Beira Interior, Covilhã, Portugal, pp. 15–35.

Pinheiro, E.C. (2009) *Rota da Lã TRANSLANA: Percursos e marcas de um território de fronteira: Beira Interior (Portugal) e Comarca Tajo-Salor-Almonte (Espanha), Vol.1 – Reconhecimento e valorização patrimonial,* pp. 365–384, pp. 415–460. *Vol. 2 – Inventários das vias agro-pecuárias e do património edificado associado à indústria de lanifícios,* pp. 526–717. Museu de Lanifícios da Universidade da Beira Interior, Covilhã, Portugal.

Pinheiro, E.C. and José dos Santos Silva, M. (2012) A Covilhã: uma paisagem cultural evolutiva. Algumas notas sobre a (re)construção das memórias industriais da cidade. *Ubimuseum – revista online do Museu de Lanifícios*, no. 1, pp. 1–21.

Salvado, R. and Rolo, G. (2020) COOLWOOL—Fim de semana criativo na Covilhã. In: Duxbury, N. and Silva, S. (eds.) *CREATOUR: Desenvolver Destinos de Turismo Criativo em Cidades de Pequena Dimensão e em Áreas Rurais.* Imprensa da Universidade de Coimbra, Coimbra, Portugal, pp. 204–220.

18 Co-creating subtle attractions

Britta Timm Knudsen* and Jan Ifversen

Aarhus University, Denmark

The west coast of Denmark with its rough North Sea, sandy beaches, and tall dunes is a very popular tourist destination, particularly for German tourists. From 2016 to 2019, a group of researchers from Aarhus University and the Aarhus School of Architecture worked together with the municipality of Ringkøbing-Skjern to rethink tourism in a coastal city. We had two main goals for rethinking tourism in the area: The first goal was to reconceptualize the attractive force of places. Instead of simply conceiving places as passively waiting to be visited, we wanted to highlight their force of attraction. Since the municipality brands itself as 'the realm of nature', we set out to investigate how 'nature' – always man-made to some extent – could be viewed as a force of its own. Our second goal was to step away from the conventional understanding of tourists as simply consumers of places and treat them not as strangers and guests to the places, but as potential hosts and designers of place experiences. Furthermore, we reversed traditional identity mechanisms by treating those tourists who returned year after year as part-time locals (Gravari-Barbas and Guinand, 2017). We chose to study the co-creation between 'nature', locals, and part-time locals through a set of interventions where we tested new scenes of engagement between different actors. A number of interventions were conducted at different locations in the municipality. Here we will present two interventions that through their different mediations of nature landscapes enhance and create experiential qualities.

In a broader sense, these interventions are framed as contributions to creative tourism. In the first place, they view tourism as activities based on participation and co-creation performed by tourists in the everyday life of destinations (Richards, 2011). With its emphasis on creative and reflexive engagements, creative tourism is developing in opposition to the conventional massification of the tourist experience and the serial (re)production of culture (Richards and Wilson, 2006; Richards, 2011). Our interventions do not, however, understand creativity as simply a partaking in existing local activities of traditional craft making, which is prominent in some conceptualizations of creative tourism (e.g. Tan *et al.*, 2012) but has widened over time (Duxbury and Richards, 2019). In our view, a traditional crafts-focused understanding runs the risk of essentializing and exoticizing traditions. While our interventions were not explicitly organized with the purpose of facilitating learning encounters nor transmitting knowledge from local to tourist (with the risk of simply reversing the power geometry of the tourist gaze), we certainly intended to see assemblages of human and non-human actors as processes during which creativity and improvisation entailed learning and experiential outcomes for all partners involved (Ingold and Hallem, 2007).

Tourists hosting experiences

The first intervention directly challenged existing conventions of the tourist gaze (Urry, 1990). What if tourists are not simply transforming places according to their leisure desires? And what if the tourist industry does not just respond to these gazes? One way to break with existing assumptions of how tourists act was to give them voice. Instead of seeing them as trapped in discourses, we intended to let the part-time locals distribute their knowledge of places and peoples to the rest of us. They were asked to produce a tourist guide. The idea of

*Corresponding author: norbtk@cc.au.dk

DOI: 10.1079/9781789243536.0018

letting locals function as tourist guides is well known and often used in concepts such as 'Meet the Locals' in metropoles around the world. It is not new either to view guided tours as a relational game played out jointly by tourists and professionals (Larsen and Meged, 2013). To let tourists directly take on the role of guides themselves is, however, quite new.

Although we chose to deviate from the more typical guidebook with its historical framing, its practical recommendations, and its lists of holiday activities to be undertaken, we still kept 'nature' as a central scene. Official tourist brochures from the area spoke of 'the grandiose nature' that 'should be experienced up-close' and recommended 'to sense the west coast'. We therefore asked the authors both to activate their knowledge as part-time locals and to narrate their 'experiences of locations "up-close"'. The preparation of the guidebook was based both on using cultural probes (Gaver *et al.*, 1999) – packages of material such as postcards, photos, and diaries to be produced by the authors – and on interviews. A criterion for selecting tourists was that they had visited the area several times over a number of years. Some potential authors accepted to take us on longer tours in the area while taking pictures and conversing with us; others accepted to draw up the guidebook in conversations on tourist sites. A total of 12 families participated in the intervention. Most of them were middle-aged families or elderly couples with quite a strong connection to the area. One of our part-time locals had visited every year for almost 50 years.

They were asked to personalize their tourist guides and to condense their feelings of particular places into a few metaphors and concepts. Their conceptual work gave valuable information on exactly how to pin down the attraction of the area. The prevalent metaphors used by the authors to condense their experiences were health and cleanliness. Health was directly linked to the force of nature; cleanliness worked as a metaphor for a local culture of order and responsibility towards others. We also asked our authors to translate their sensorial experiences of nature into words. To help them, we used so-called phenomenological interviewing techniques that specifically ask the interviewed persons to render their experiences in minute details (Bevan, 2014). They divided up their experience of 'nature' into dimensions that could be sensed such as wind, sea, sand, light, salt, and breaking waves. All senses were activated, from seeing and hearing to smelling and feeling. Their bodies became the interfaces in a communication with 'nature' and our aspiration was to let the inspiration these part-time locals get from nature come through in their narratives and, in this way, to inspire others. One of our authors, a woman in her 50s, invited us into her quite concrete bodily experience with this description: 'It is the hearing, it is the hearing, it is the roar; it is ... eh ... the sand, the sand is also different. The sand is simply, it is different! How shall I describe it? When you, for instance, are in the Caribbean, it immediately sags away. And here the sand – if I could not use my eyes – would have a kind of stability'.

Our hosts thus take us on a guided tour into their sensorial experience of nature or rather into the poetic rendering of these experiences. They offer rich communications with the different forces of nature. These are communications marked by reciprocity and respect. Through the authors, we are moved into a world gently led by the forces of nature. Instead of being reduced to a resource for entertainment, nature becomes a central actor in the narrative.

Based on the authorial work of our tourists in their capacity of part-time locals, sensing bodies, and interpreters of the concrete poetry of places 'up-close', we get access to a genuine co-creation of the tourist guide. Their work can then enter an archive of experiential knowledge produced, for instance, in digital forms that can include images and soundbites as well as commentaries from users. The results of the intervention was offered to the local tourist authorities as a supplement to already existing digital guides based on photos taken by tourists with small comments added to them. The idea is to develop a digital guidebook where the part-time locals are not only given a voice, but where new, artistic forms of experiencing places are rendered. In this sense, tourists are creating new experiences.

The attractive and infra-ordinary wind

Adding to the up-close poetic redescriptions of natural landscapes, we also wanted to create subtle and sustainable site-specific experiments addressed to both locals and tourists with the goal of challenging the divides and incommensurabilities between different approaches to the landscape. Our second intervention revolved around wind as a contested key natural resource in the area. Offshore wind turbine installations have been accompanied by increased protests in the area, highlighting that different gazes and uses of the landscape are at play in this locale. Wind is used for energy production purposes, but it is also just an infra-ordinary feature, an ordinary and trivial overlooked aspect of everyday life at certain places, as French avant-garde writer George Perec coins the concept (Perec, 1974). To look for the infra-ordinary and render it through meticulous descriptions is to examine how small, everyday routines and the descriptions of those routines interact, for example: 'Describe your street and yet another street and compare them. Pile up the content of your pockets and ask yourself about the provenance, use and future of the objects you find there. Ask about small spoons, and look at how people compose text messages, where and when' (Perec, 1989, pp. 12–13). Through the close-up descriptions, letting the insignificant and the unimportant 'speak', Perec hoped to let the disorder and the messiness, but also the accidental beauty, of the everyday appear. A recurrent observation from our interviews with wind-users in the area – windmill engineers, coastal security personnel, nature guides, windsurfers, cyclists, sailors, meteorologists; 22 people in total – was that wind in the area becomes noticeable only when it is *not* there. Otherwise, it is a constancy: contested as it is tantamount to wind turbines, but also trivial and/or annoying to some. As a former resident of the area formulated it, 'You normally get it [the wind] in your mug. It's always windy out here, dammit' (young woman). With our intervention, we intended to let people sense and immerse into the wind as a primary atmospheric actor and not only as an extractable natural resource.

We wanted to focus on wind as a site-specific phenomenon of everyday life in the area. To take notice of the infra-ordinary is to go counter to the extraordinary that the tourism industry normally markets, and it is also to go counter to the wind uses that just extract its force for water-based activities. To visualize the infra-ordinary wind, we collaborated with two London-based artists, Lise Autogena and Joshua Portway, who in their artistic work draw on computing technologies to deal with climatic changes in inflicted landscapes.[1] For our intervention, they created a prototype design, a data-visualization of wind in the form of interactive art, which we tested in the dunes around 50 metres from the sea. The small test area was close to the famous lighthouse, Lyngvig. It was around 8 square metres in size, symbolically enclosed by an open tent (see Fig. 18.1). The equipment consisted of a small table, a virtual reality-headset, two joysticks and a huge computer, plus cables. Using the headset, the augmented reality application allowed the user to visualize the physicality of the wind by making the flow of air visible as it swirls and eddies around the shapes of the dunes. The testing took place over a period of 10 days in August 2018. The intention was to superimpose a digital model of the wind conditions of the landscape and let people immerse themselves into the simulation while being in the real environment and be simultaneously exposed to sounds and smells from the surrounding landscape in a kind of augmented real experience (Guttentag, 2010). In our intervention, the heavy and bulky technology of the virtual reality glasses, serving as an enhancing mediator between landscapes and bodies, increases the experience of proximity to a visible wind that is then both *felt and seen*. The overall purpose of the art installation was to create an interactive experience that gave the user a strong sense of being a part of – and not master of – the wind forces of the place. The experience of virtually walking through the landscape and being in the middle of visible flows of air as they swirl around the shapes of the dunes was supposed to let our test persons – and in particular, the younger audiences – immerse themselves into the landscape through the use of virtual reality.

Around 150 people tried out the prototype during the test period. Our testing produced data in the form of first-hand testimonies and short talks with the audience after each experience. Some were interested in the technological side of the experiment (how to translate wind into visible movements via drone photogrammetry and computational fluid dynamic), some felt alienated by the digital component in the experience of nature, but most of the test persons, which included tourists, part-time locals, and locals, were thrilled by this adventure: 'It's magic. I feel I am in the middle of the creation' (a German tourist in his 50s).

We suggest that *infra-ordinary* phenomena are highly interesting for rethinking tourist experiences at smaller places because they offer a perspective on the everyday as an alternative to the typical veneration of the exotic and extraordinary expected by tourists whose gaze is partly disciplined and governed in quite standardized manners by the industry and the destinations themselves. We also suggest that interactive art installations can be a suitable form to experience deep connections with

Fig. 18.1. The spectacle of the wind-installation. Photo by Britta Timm Knudsen, used with permission.

the natural landscape if the immersive environment corresponds to the specific location and its audiences. To data-visualize the wind – through advanced technological devices – seemed the right choice in our area with heavy wind turbine technology. At the same time as making the wind – in itself an invisible natural force – visible in order to stay in this contemplating and aesthetically rich experience, it functions as an alternative to the abundant water- and wind-based offerings such as windsurfing, sailing, and fishing in the area.

Based on our two interventions, we have formulated five more general recommendations around coastal tourism. Firstly, co-creation should depart from an acknowledgement that tourists possess expertise on site-specific experiences. There cannot be any co-creation without a consideration of *knowledge exchange*. Secondly, experiences have to be liberated from the typical framings of the unique. Experiences of sites such as natural landscapes also derive from that which goes unnoticed; in our cases, the experience depends on a heightened awareness of the *created and co-created poetry of that which is simply there*. Thirdly, the unnoticed needs to be enhanced through *up-close concrete renderings of experiences* instead of narratives and stories of places and through *mediating designs and technologies*. Fourthly, experts in the form of researchers and artists can be part of successful experiments if their interventions – however provoking to some – build on acknowledged values in the concrete area. Data visualization instead of a more artistic interactive installation seemed the better choice for us. Fifthly, the place-enhancing experiences depend on *intense partaking in the processes* around the experiments. We are sure that the persons who have been involved in both the part-time local guide production and the persons testing the wind-intervention will never forget that experience, forever a part of the newly created memory of that particular spot.

Creative, playful, and reflexive engagement with places is certainly present in the two interventions that take nature in general and wind in particular as natural rough resources that are important drivers of the experience economy of the west coast of

Denmark. Thinking of our experiences from the perspective of creative tourism planning, we have formulated the following suggestions:

- To avoid the risk of traditionalism, secure local diversity through the inclusion of all the different groups with stakes in the sites.
- Look for overseen, unnoticed, infra-ordinary resources – also non-human resources – to provide the inspiration for creative tourism initiatives.
- Co-create low-key, everyday place-enchanting, and sustainably up-scaled experiences and processes (e.g. interactive installations, workshops, intimate concerts, terroir initiatives, etc.).
- Format creative initiatives that many different people can participate in, both locals and tourists, young and old. In particular, do not forget the younger parts of the population – much of the tourism offers cater unreflexively to middle-aged, well-off groups or to families with children.
- Treat tourist knowledge as a resource in understanding the locality and avoid reducing tourists to simply consumers of experiences (the tourist gaze).
- Creative activities in nature could enhance and confirm nature connectedness on a whole range of levels for both residents and tourists, connecting creative tourism to criticism of instrumental uses of natural resources in general.

Acknowledgements

This research was made possible with a grant from the Danish Innovation Fund.

Note

1. For more about their art practice, see: www.autogena.org (accessed 11 May 2021).

References

Bevan, M.T. (2014) A method of phenomenological interviewing. *Qualitative Health Research* 24(1), 136–144.

Duxbury, N. and Richards, G. (2019) Towards a research agenda for creative tourism: Developments, diversity, and dynamics. In: Duxbury, N. and Richards, G. (eds.) *A Research Agenda for Creative Tourism*. Edward Elgar Publishing, Cheltenham, UK, pp. 1–14.

Gaver, B., Dunne, T., and Palcenti, E. (1999) Design: Cultural probes. *Interactions* 6(1), 21–29.

Gravari-Barbas, M. and Guinand, S. (eds.) (2017) *Tourism and Gentrification in Contemporary Metropolises: International Perspectives*. Routledge, London, UK, and New York, New York, USA.

Guttentag, D. (2010) Virtual reality: Applications and implications for tourism. *Tourism Management*, (31)5, 637–651.

Ingold, T. and Hallam, E. (2007) Creativity and cultural improvisation: An introduction. In: Ingold, T. and Hallam, E. (eds.) *Creativity and Cultural Improvisation*. Berg Publishers, Oxford, UK, pp. 1–27.

Larsen, J. and Meged, J.W. (2013) Tourists co-producing guided tours. *Scandinavian Journal of Hospitality and Tourism* 13(2), 88–102.

Perec, G. (1974) *Especes d'espaces*. Galilée, Paris, France.

Perec, G. (1989) *L'infra-ordinaire*. Seuil, Paris, France.

Richards, G. (2011) Creativity and tourism: The state of the art. *Annals of Tourism Research* 38(4), 1225–1253.

Richards, G. and Wilson, J. (2006) Developing creativity in tourist experiences: A solution to the serial reproduction of culture? *Tourism Management* 27(6), 1209–1223.

Tan, S.K., Kung, S.F., and Luh, D.B. (2012) A model of 'creative experience' in creative tourism. *Annals of Tourism Research* 41, 153–174.

Urry, J. (1990) *The Tourist Gaze: Leisure and Travel in Contemporary Societies*. Sage, London, UK.

19 Linking creative tourism products to markets: Target marketing, promotion, commercialization, and market readiness

Marie-Andrée Delisle

Université du Québec à Montréal (UQAM), Canada

Marketing creative tourism initiatives is a challenge, as so many aspects are interrelated and linked to the tourism industry value chain. This chapter aims to provide a broad view of the tourism industry structures, the role that promotion and distribution networks play, and the requirements attached to them, in order to demystify the workings of the tourism industry for creative tourism developers. It aspires to help creative tourism entrepreneurs increase their knowledge about where and how they fit in, the scope of their target markets, and how they can expand their knowledge and participation in tourism partnerships and channels.

Tourism is an immense and complex industry. Economic aspects are at the forefront, accompanied by the social and cultural changes that are closely linked to it. Basically, it is a people-to-people industry, be it at the level of operational structures and activities or at the level of interactions between a local population and its visitors.

Tourism professionals are always looking for new products to promote and sell, but little research has been conducted about small tourism business entrepreneurs and marketing (Tsiotsou and Ratten, 2010; Agusdin, 2018; Pret and Cogan, 2018; Yachin, 2019; Remoaldo *et al.*, 2020). In order to understand how the tourism industry operates, creative tourism entrepreneurs need to become acquainted with the industry basics, as it is essential to establish strong ties between creative tourism suppliers and their tourism industry counterparts for promotional and commercialization purposes. Moreover, small businesses also require development and promotional support from local and regional government entities with an overview of the destination, which comprise an important part of the overall operating environment.

As a tourism consultancy practitioner and instructor for over 30 years, I have had the opportunity to engage in tourism marketing as part of my praxis and my doctoral research on creative tourism. As a travel writer for the tourism trade, I meet with buyers and sellers of different parts of the world, observing how this industry operates and unfolds. I have worked as a travel agent for leisure and business travel, as a world group tour co-ordinator, and have instructed first-time sellers about marketplaces and given workshops to industry professionals about media relations, packaging, marketing, and networking.

This chapter presents a tourism industry perspective about key aspects of market thinking and provides tips and strategies for building a strategic marketing plan for a creative tourism initiative. The visitor profile is examined, not so much as demographic or psychographic (i.e. not what they are, which is explored earlier in this book), but to provide a collection of tools to pinpoint where they are! Market segmentation, the product/market match, and product evolution and lifecycle are then explored. Building on this, four strategies to build

Corresponding author: info@madelis.com

 Creative Tourism: Activating Cultural Resources and Engaging Creative Travellers (eds. N. Duxbury, S. Albino, and C. Pato de Carvalho)
DOI: 10.1079/9781789243536.0019

business opportunities through partnerships and networks are outlined. Next, key elements of the structure and requirements of the major marketing elements of the tourism industry are discussed, including the issue of market readiness and pricing structures. Bringing these elements together, the chapter closes with guidance about market planning.

Visitors: where to look for them, and how

The tourism industry is an important market provider for creative products, services, experiences, and workshops since it continuously looks for novelty, trend seeking, product differentiation, and competitiveness. As more and more travellers now look for off-the-beaten-track experiences, slow-paced itineraries, and mingling with locals everywhere they go, creative tourism is ideally positioned. Creative tourism activities attract a niche market of people looking for authentic encounters with local populations, tied to a quest for active learning about other cultures and ways of life: people want to meet people! Tourism marketing strategies – combining communication, promotion, and distribution/commercialization – link the products of creative tourism entrepreneurs to visitors[1] motivated by the cultural and creative aspects of a destination.

Many scientific articles study and analyse creative tourism visitors' profiles (Raymond, 2007; Cohen, 2009; Frochot and Batat, 2012; Tan *et al.*, 2014; WTO, 2016, 2018; Matos, 2018; Remoaldo *et al.*, 2020). The international Creative Tourism Network's website (www.creativetourismnetwork.org; accessed 10 May 2021) describes different types of creative tourists, as do the chapters by Diana Zuluaga, Elena Paschinger, and Jaime Serra *et al.* in this volume.

But where exactly can we find these creative tourists in order to market to them? What do creative tourism entrepreneurs know about their market? What is their reach radius? Where are they coming from? Locally, regionally, nationally, internationally? Are they long-haul or short-haul visitors? When are they arriving, and for how long? What kinds of trips are they taking – group tours, special interest tours, individual travel, adventure tours?

Primary, secondary, and emerging markets

Tourism product/service providers have many questions and often have difficulties finding wayfinding guides. They often know little about the cultural background of their target markets. Where and to whom should they turn? To start, national and regional tourist offices can usually supply some information about the number of arrivals to your area from various countries. From this beginning, one should consider different types of primary, secondary, and emerging markets.

In many cases, a *primary market* would be located between 50 and 100 km away, which corresponds to day visitors or excursionists (see Fig. 19.1). They are usually linked to an interest drawn from a collective identity and/or traditional savoir-faire. They will also be easier to reach, at cheaper costs. A second natural primary market could also attract people with a common historical past, such as sharing a language. For example, France is a natural international market for the French-speaking province of Québec. Similarly, one would consider the natural market of Brazilian tourists interested in visiting Portugal. A third type of primary market may be tied to supplying a creative tourism experience to a receptive tour operator[2] on a regular basis. The tour operator would then be the first market to serve since they would bring a series of visitors on an annual basis.

A *secondary market* may target visitors coming from further away, regionally, nationally, or even internationally. Secondary markets also require continued promotional and commercialization efforts but to a lesser degree. As an example,

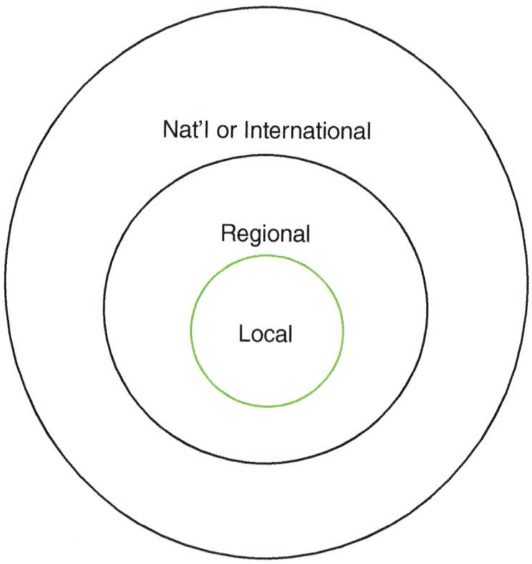

Fig. 19.1. The radius from the product to the visitor. Developed by author.

Mexicans mostly like to visit Canada outside of the high season. However, they represent lesser numbers on an annual basis than their French counterparts. Since these numbers are growing during lower seasons, it is worth maintaining that market share. Often times, budget limitations will dictate which markets are worth spending on, which usually means which are most profitable.

Emerging markets should not be neglected, with continuous promotional strategies of a smaller kind. The creative tourism concept would have to be an important motivating factor to put time and money on potential markets versus actual markets. Here again, consulting statistics and marketing orientations from one's national and regional tourist offices, also called destination management organizations (DMOs), would give an idea where efforts are best devoted.

Destinations undertake a market intelligence process to understand the tourism market and analyse its competitive environment. For example, Destination Canada offers an insight on its various target markets, with fact sheets on each market, be it primary, secondary, or emergent (see www.destinationcanada.com/en/markets; accessed 10 May 2021). This data is complemented by marketing recommendations for each of them: a clear illustration of the vital link between data collecting and marketing strategies.

Once the target markets are identified, it is easier and more cost effective to establish marketing strategies. So, the question remains: Which visitors should be targeted, and how?

Researching your local or primary market

Assembling an inventory list as a reference, creative tourism organizations can measure the extent of their knowledge about their business environment while scouting for possible collaborators and partnerships. Here are some useful examples to get you started:

- Inventory of local and/or national receptive tour operators coordinating ground arrangements for international tour operators, who are usually looking to expand their products and experiences to offer them.
- Inventory of small tourism and cultural businesses interested in expanding their markets by networking and partnering with organizations able to complement their tourism offer; for example, small hotels and inns, local cuisine restaurants, and cultural venues. Proposing demo workshops may be an attractive means to stir their interest in partnership building.
- Inventory of other activities/attractions/experiences around that already attract compatible market segments, for partnerships and packaging purposes, including concert halls, festivals, conventions' spouse programmes, community and social events, and so forth.
- Listing of festivals and events (big or small) and their themes, to create new projects such as running a creative workshop, testing a pilot project, or offering an open house day in sync with their theme, while benefiting from their promotional activities.
- Inventory of local and regional media (digital and paper), radio, television, etc.
- Inventory of tour operators offering month long winter stays in the sun, such as *Transat Holidays* packages that include flights, hotels, and transfers. Hoteliers may have an interest in promoting creative tourism activities to their long-stay clients.
- Inventory of surrounding resort hotels listed on e-commerce reservations systems such as booking.com, with packaging possibilities in mind.

These inventories cost very little except for time, while building relationship with new business acquaintances and prospects.

Market segmentation, targeting, the product/market fit, and evolution

Market segmentation

After having gone through various inventories and fact checking, creative tourism entrepreneurs need to know a significant amount of information about potential market segments since they will have a great impact on the creative tourism experience structure. Is the target market close (local, regional) or far (national, international) from the destination's main attractions? Is it potentially worth partnering with other tourism businesses for packaging purposes?

Market segmentation can be established based on demographics (age, gender, family status, income, education), psychographics (lifestyle, personality types) and geographic markets (e.g. UK, US, Brazil). It could be activity or purpose oriented. Interestingly, they could be segmented by groups of individuals with similar motivations, such as creative tourism travel seekers, with shared values and sustainable tourism behaviour.

Specific target markets derive from market segments. Target markets worth selecting are determined by two major factors: the most attractive and the most profitable in terms of growth and size. Locals, day visitors, expats, VFR (visiting friends and relatives), Third Age University groups, alumni (university graduates), convention and event participants, student groups (see Student and Youth Travel Association [SYTA] website: https://syta.org/conference2020/about/; accessed 10 May 2021), and educational travel groups (see Educational Travel Consortium website) are all specific market segments to address with specific strategies. Long-haul (i.e. long-distance flights) monthly winter vacationers, long-stay university students, or tourists taking part in a larger tour, circuit, or itinerary via a receptive tour operator are also specific market segments to address differently. In order to prioritize what to look for, each market segment must be precisely identified and evaluated in order to validate the most effective marketing strategy to reach them: one size does not fit all!

The product/market match

Each market that will be targeted will imply that some aspects of your creative tourism experience be personalized to the target market profile. In other words, the same offer could be used for any market so long as there is some adaptation to suit each segment differently. Figure 19.2 illustrates various matches based on the same product or experience.

Here, the market profile becomes the variable that helps modify (and diversify) the actual creative tourism experience by adapting to it. The content and the duration of the experience, the season, the origin and language spoken by the visitors, and the extent of the target market (i.e. short haul, long haul) will dictate the promotional and distribution networks and partners to choose from. Will the experience be sold all by itself, will it be packaged with their products and services, or will it be addressed to a specific market such as conventioneers? Will the location vary? Bringing these dimensions together helps personalize the strategies while selecting which promotional and distribution channels will be most suitable for each targeted market.

Research and development extensions and combinations

A creative tourism workshop should always lead to follow-ups in terms of research and product development (R&D), to keep innovating from the basic content and strategies. R&D is needed to explore new trends and to dig into cultural roots in order to design original experiences that reflect the identity of the place and expand the sustainability of the creative tourism activity (Stipanović and Rudan, 2014). Finding new ways to link other or new components spikes the interest of tourism partners and brings an added value to the visitor. For example, the product development can be extended to include team-building experiences (e.g. Eyexplore, n.d.), a corporate workshop (e.g. Atelier MJB, n.d.), or a happy-hour workshop including wine tasting (e.g. Artspace, n.d.), among other possibilities. These combinations of activities can be inspired and informed by the market information research conducted earlier.

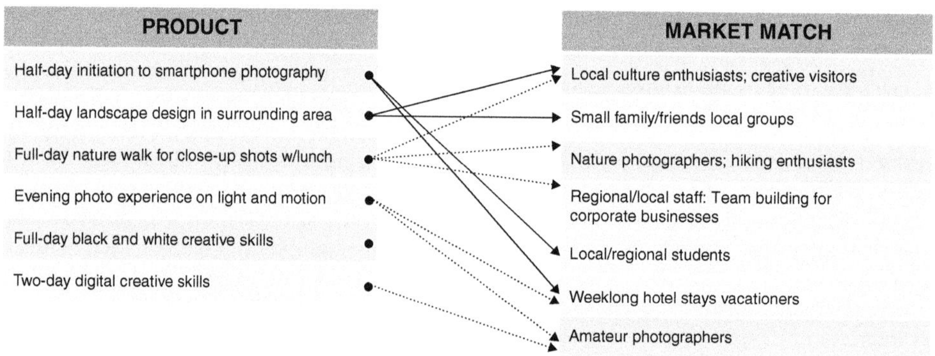

Fig. 19.2. Product–market match example. Adapted from Patterson (2016).

Partnering and packaging may also be useful for the creative tourism initiative to be part of a larger package, where the extent of the experience is spread over a few days. For example, in Crete, a 5-day cooking seminar is linked with botanical hikes, organic agriculture, culture, and cuisine (Cooking in Crete, 2019). It is offered three times a year at specific summer dates. It gathers enthusiasts and special interest groups on biodiversity while the destination is economically benefiting from longer stays.

A 'co-opetition' exercise (co-operation among 'competitors') can enable different actors to collectively bring out the distinctiveness of a place. This is where packaging ideas can enhance the offer. For example, a Vietnamese activity links local market, boat fishing, and cooking into a half-day experience (Cookly, n.d.). Creative Iceland, through Viator tour operator, cross-posts its half-day knitting experience with nature walks and authentic local food, to complete a full-day experience (Creative Iceland, n.d.). Scanning offers developed by other agents internationally can offer insights and ideas (see Box 19.1).

Box 19.1. What about the competition?

Creative tourism experiences and workshops are bound to have competitors and local providers seldom identify the variables that can work for or against them. By looking at members of the Creative Tourism Network and analysing their offer, and by consulting various websites such as Viator, Evaneos, Airbnb Experiences, TourScanner, TUI Destination Experiences, TourRadar, Get Your Guide, Tours by Locals, Go Learn To, and Craft and Travel (websites are listed in the References list), creative tourism providers can start comparing elements such as content, price range, booking conditions, marketing tools and partners, user friendly information, and so forth.

Even though these sites mostly concentrate on attractions and activities for mass tourism, creative tourism providers can note if those offers represent the type of market segments they are looking for or not. The wording, the call to action, and the use of photos and videos are interesting to analyse both as sources of information about the intended market and as selling points.

Competing with other creative tourism initiatives is part of the tourism business and a reality check is needed, not only to see how others fare but how an entrepreneur can distinguish their own initiative as distinctive instead of copying what others do (Richards and Wilson, 2006). The competitive advantage is linked with the sustainability of the initiative. As the initiative grows in a unique context reflecting the local culture's distinctiveness, the creative tourism entrepreneur will enter a phase of consolidation where creativity and imagination become precious tools for their product development.

The lifecycle evolution of tourism product/business/destination

The lifecycle of a creative tourism workshop or experience must also be considered in order to situate it on a time frame. Butler's (1980) research on the lifestyle evolution of tourism products remains true to this day. I find that this model helps locate where a product, a destination, or a tourism business stands on its lifecycle evolution. As Fig. 19.3 illustrates, the growing curve of the start-up of a product or activity with few visitors and its gradual growth in developing its content, promotion, and sales while building a consumer base. As seen here, the time that must be devoted to reach a successful stage is appreciable. The next stage brings the business into a stagnation phase, where neither product or visitor show increase or improvement, while the challenge to constantly improve the experience may be exhausting/wearing. The last stage reflects the entrepreneur facing the decision to revitalize, remodel, or check out when its profit and consumer base decline, which announces the end of its lifecycle. Sometimes, the product has had a great life, but it is time to re-evaluate, adapt to new trends, or exit the market.

From the perspective of target markets, the lifecycle of a product, destination, or business sparks two questions: What is the degree of attractivity of the broader location/destination? Is there a critical mass of interesting and different things to see and do, and for how much time to be spent? If the most important market segment is situated close to the location, then its lifecycle will be short. If the main market segment is farther away, the lifecycle will be much longer. For instance, a major museum in a large city usually has a permanent exhibition that will be replaced every 5 years or so, but it will have many temporary exhibits in the meantime. The permanent exhibition will be well suited for international tourists while the temporary one will be essential to attract local and regional visitors again and again, so as to repeat the museum experience.

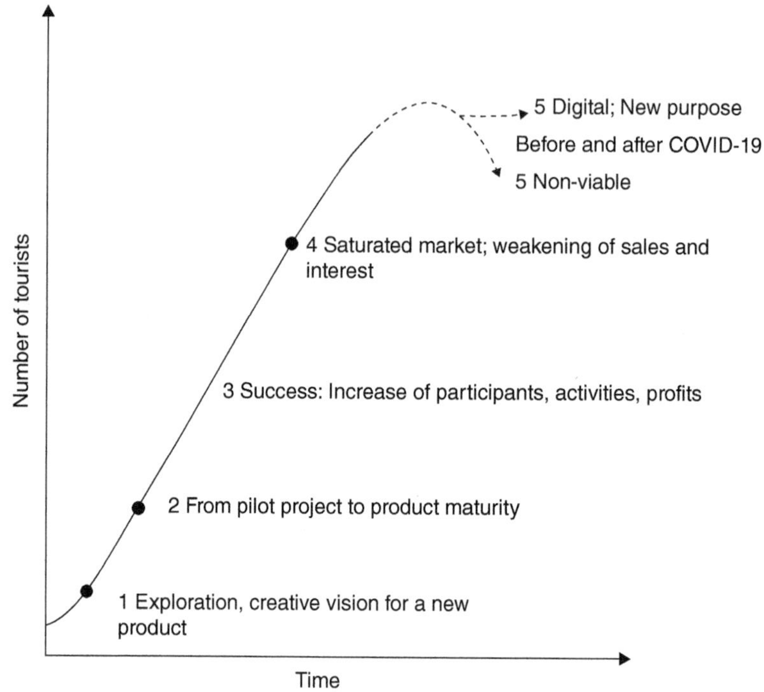

Fig. 19.3. Tourism lifecycle evolution. Adapted from Butler (1980).

Business opportunities through partnerships and networks: Four strategies

The most important partnerships to work on at the local and regional level are cultural and tourism organizations and businesses. A new creative tourism experience or workshop must be known and promoted and, usually, the first visitors to participate in creative workshops are local and regional people. Consequently, if the most important market is local and regional, one should look at specific strategies. The following four strategies are presented in a chronological approach but should work in parallel.

First strategy: Communication

This strategy targets the local and regional media (radio, television, social media networks, influencers, local papers, etc.) who are always looking for news. Consider memberships in the tourist office, chamber of commerce, or visitors' bureau. These organizations should be visited in person with photos, videos, and a flyer or fact sheet[3] about the content of the experience or the workshop. In certain cities, local guides and greeters[4] are great emissaries of local creative tourism initiatives. Often, when visiting their own tourism board or office, providers find out that their product is unknown or partially understood: a golden opportunity to familiarize these offices with the entrepreneur and its product, and to bring them some flyers. As we say, it is not the mountain to come to the cowboy, it is the cowboy's job to travel to the mountain!

Second strategy: Sales intermediates

This strategy concentrates on connecting with receptive tour operators as intermediates between creative tourism initiatives based on small-group encounters and the international tour operators with which they do business. As creative tourism is an alternative to mass cultural tourism, local and sustainable cultural experiences are also being requested by international buyers, although their needs and motivations may differ.

Small tourism businesses usually cannot afford the cost of international marketplaces and many are unable to meet the demand from buyers (frequency, capacity, volume, stable business, successful marketing strategy, etc.). However, they still need to cover both angles of promotion and distribution to attract consumers. Conducting their own operation with limited human resources, lacking market knowledge and time to devote to marketing planning and strategies, small tourism businesses often choose to work closely with their regional and national receptive tour operator to bridge the gap.

By the same token, they can benefit from their advice and expertise, learning about the industry's requirements, about what buyers are looking for, what trends are coming up, and what tools to provide, like videos and photos. These partnerships are based on trust and reliability, reflecting an iterative process. Keeping their receptive tour operator informed on a regular basis is an indisputable form of relationship building. Face-to-face appointments are much more efficient than emails. However, the supplier/provider must be able to demonstrate the relevance of the experience he or she wishes to sell and be knowledgeable about the pricing structure of the industry (see Box 19.2).

Third strategy: Packaging

The third strategy focuses on other tourism products with packaging purposes in mind. Densifying the experience by offering a critical mass of things to see and do at a destination redoubles the marketing efforts in promoting and selling the experience while sharing costs. Local and regional hotels, inns, restaurants, museums, art galleries, artists' workshops, and tour guides – as well as small- and medium-sized enterprises (SMEs) offering walking tours, nature walks bike tours, and so forth – represent many opportunities to share promotional costs and marketing campaigns when packaging complementary products. The same applies to promotional stakeholders such as DMOs, tourism boards, tourist offices, visitors' bureaus, social networks, and media (see Table 19.1). These types of organizations are always eager to promote new offers showing the dynamism and extent of their tourism businesses' creative experiences.

One should remember that remoteness could become an issue when the creative tourism experience is not included in a package. It is often the responsibility of the artisan's sense of entrepreneurship to identify and contact other tourism partners. However, it may be in reverse, where a hotel could be looking to host a workshop in collaboration with an artisan.

Some destinations organize marketplaces where local and regional tourism stakeholders, partners, sellers, and buyers get together on a one-day exchange of information, primarily to get to know each other and to find potential matches. Kiosks, roundtables, mini-presentations, and other means facilitate exchanges among the participants about potential co-creation, co-production, and co-promotion of creative tourism activities. On a greater scale with regards to larger target markets, many tourism industry destinations hold annual professional marketplaces for buyers and sellers (see the 'Marketplaces' section). However, due to COVID-19 sanitary constraints in 2020, many of the international marketplaces, such as WTM (London), have been adapting the appointments by using virtual methods (for example, Zoom) and some, such as Top Resa (France), have rescheduled theirs for 2021. The virtual meetings are run as real appointments.

Fourth strategy: Local people

Local people and businesses are key ambassadors when it comes to spreading the word. As we know, word-of-mouth remains the best and most effective marketing tool. Consequently, as a fourth strategy, these long-time partnerships must be nurtured. One should not believe that everyone knows about one's creative tourism product, experience, and/or workshop.

Pilot projects should be tested with local stakeholders (e.g. receptive tour operator, municipality, chamber of commerce, tourist office, cultural association, artisans, local supporters, media, hoteliers, etc.). Open-day activities during events and festivals, and familiarization experiences for tour operators and media, are convincing examples for various exogenous resources to experiment with creative experiences or workshops. Buying a car from a brochure or a website does not compare to test driving it!

Some of those strategies, such as an inventory of resources and small-scale networking, may seem insufficient. However, they reflect a step-by-step approach, with the product, its promotion, and its distribution remaining a work-in-progress. Results come in the middle- and long-run, while the creative tourism initiative keeps improving in approaches and strategies.

Table 19.1. Tourism industry promotional networks. (Source: Developed by author.)

Tourism industry distribution/commercialization networks		
SUPPLIER *Transportation, accommodation, food, attractions, activities, experiences, festivals, etc.*		
ANDOR		
Minor actors	**Major actors**	
Local, regional markets	National market	International market
Local/municipal tourism office	Regional tourism offices	Regional tourism offices
Regional tourism offices	Regional tourism offices	National tourism office
	Sectoral associations: businesses, accommodation, restaurants, convention halls, culture, art, attractions, etc.	
	Festivals and events	
	News media / social media	
	Regional and national partnerships Municipalities, ministries: communications, culture, heritage, environment, transport, commerce, parks, etc.	

The tourism industry structure and requirements

Distribution networks

A creative tourism provider can decide to reach the potential consumer directly, which implies some costs, or decide to use an intermediary to sell its product to a larger market. In this case, there are fees involved, either by paying a commission or by selling at a net price to these mediating organizations. Travel trade stakeholders such as international tour operators and receptive tour operators can play an important part in the distribution and commercialization of tourism products. Understanding how these networks are structured helps differentiate direct contacts with consumers (B2C) versus the use of intermediate sellers (B2B) to reach the selected target markets. Table 19.2 presents various levels of sales intermediates pertaining to leisure travel.

Tour operators produce detailed catalogues to present their itineraries, tours, products and services, and selling price. Those comprehensive brochures are then distributed electronically or in paper formats to a large range of travel agencies, reaching bigger volumes of travellers. A brochure will take many months of preparation and detailed products and services coordination. That is why international tour operators will need a market-ready product, often 2 years before it can appear in a brochure. Regional and national receptive tour operators will not publish brochures for tourists, because their work is about presenting tour programmes to special interest groups or specialized international tour operators (B2B). They will work with net prices collected from their tourism providers. Aggregators such as Viator booking software for suppliers (merged with TripAdvisor) offer small providers or sellers to enlist their products with them. Their website demonstrates how to get started with them.

In the case of national and international marketplaces, DMOs and buyers such as national and international tour operators work hand-in-hand. Basically, buyers will look for new products and reinforce business ties with their actual tourism providers. DMOs will inform buyers about the destination's highlights and novelties while communicating leads to their respective sellers such as hotels, tour companies, major attractions, transportation, and other experiences and services.

These networks work in similar ways throughout the world, operating roughly in the same manner. They represent a significant leverage effect for tourism products and services businesses.

Pricing structure

Pricing a product is often a trial-and-error exercise. An analysis of similar products will give some information about the public price range for an

Table 19.2. Distribution/commercialization tourism networks. (Source: Developed by author.)

Sales intermediates	Direct contact
Travel trade professional buyers: • Tour operators (international and national) • Receptive/incoming agencyground operators (regional and local) • Attend national and international marketplaces such as FITUR, ITB, WTM, etc.	Website (production and management; transactional) Targeted emails; newsletters, blogs Own social media (Facebook, Instagram, YouTube) Ads (Facebook, Google, etc.)
Online booking sites, ex. Online Travel Agencies (OTAs), Booking.com, Viator, etc.	Brochures, posters and other promotional material design and production, photos, and videos
Intermediaries between tour operators and consumers: retail travel agencies, outgoing travel agencies	Participation in consumer shows and follow-ups
Social media networks, influencers, start-ups (negotiable contracts), hotel concierges, etc.	Knowledgeable marketing and digital skills human resources
Cost	**Cost**
Intermediaries take up a percentage of the sale as a commission that may represent between 10 and 25% of the selling price. Travel agencies may get up to 10%, while hotel concierges get cash. Commissions are paid when sales are completed.	Above costs are variable. Direct contacts are more complex and more time consuming since sales are dealt on individual basis, while sales intermediaries bring volumes.
CONSUMER	

activity. One must also consider what the consumer is willing to pay, depending on the context, the location (place, environment, ambiance), the perceived value, the degree and level of involvement, and the uniqueness of the product, what is included, and what is not. The basic pricing structure is always presented on a per person basis, which is different from the hotel industry where prices are usually published on a room per night basis.

The main issue is about covering costs. The *total fixed costs* are the operating costs, regardless of having visitors or not. The amount of yearly costs must be divided among the number of creative tourism activities during a year. This will include, for example, administration costs, rent, equipment, mail, phone, electronic equipment, rental, insurance, marketing, memberships, etc. For SMEs, partnerships and in-kind services from municipalities, cooperating organizations, and associations can help alleviate those costs. The question here is the reasonable extent of fixed costs to cover when a creative tourism entrepreneur is at a start-up stage.

The *total variable costs* represent the costs to run a creative tourism workshop or experience, once there are enough participants. For example, one might consider preparation time, supplies, production, equipment, and any human resources such as a co-ordinator or co-facilitator. The question here is how many participants are needed to cover a minimum of costs and what would be the minimum/maximum ratio of participants to operate the experiment.

The fixed and variable costs represent the total costs of the operation, to which a profit will be added. Calculating the *profit margin* or mark-up is a currency amount or a percentage added after calculating the costs (fixed and variable) in order to establish the net price. The profit margin can be flexible depending on the volume of sales an intermediate will bring to the business.

A *net price* is what the creative tourism entrepreneur will get from selling the experience to a sales intermediate (e.g. tour operators, packaging partners, etc.). The expression 'net price' means that it is net of commissions or money retributions. To that net price, a commission to sales intermediates will be added, which could range from 10% to 25%. Travel agencies and hotel concierges work on a smaller commission basis.

The selling/retail published price should not change even if the sale is through direct contact with the consumer or through a sales intermediate. This will protect the relationship and trust between the tourism entrepreneur and the sales intermediates he or she is working with. The total selling or retail price represents a basic pricing structure for the tourism industry. A hypothetical example is presented on page 144.

Box 19.2. Pricing structure: Costs, profit margin, net price, commissions, and selling price – an example

Costs + Profit Margin = Net Price (wholesale) + Commission (10% to 25%) = Selling/Retail Price on a per person basis

For example:

Fixed costs	8 Euros
+ Variable costs	16 Euros
= Net costs	24 Euros
+ Profit margin	10 Euros (variable)
= NET PRICE	34 Euros (or wholesale price)
+ Commission 15%	6 Euros (variable from 10% to 25% of net price; negotiable as per the sales volume)
= SELLING PRICE	40 Euros per person (retail price, the advertised price)

Market readiness

For local and regional visitors, market readiness has few requirements. For example, a 'visitor-ready' business would have the necessary licences, permits, and insurance to operate legally, and offer a reliable product. When a creative tourism business does not have any track record, it must first work on a pilot in order to test it, price it, and market it on a small scale. Repeating the pilot with different types of participants and collecting their comments also brings a better understanding about operating the creative experience or workshop, preparing for a Plan B, revising costs, redesigning the content, and reviewing the communication skills of the artisan or facilitator. Hence, the 'visitor-ready' level would suffice at the beginning.

'Market readiness' is a major consideration for a buyer such as a receptive tour operator to engage in a product. When they meet with suppliers, they expect them to know about their tour operating business and to meet various requirements, sometimes even asking for the supplier's annual marketing plan. They need to ascertain themselves that what the seller is presenting is what their clients will get. In short, the requirements from such buyers are track records, trust in the supplier, reliability and consistency of a product, dependability, immediate email and phone responses to receptive tour operators' requests, Plan B, a calendar of activities months in advance, a pricing structure, liability insurance, and reservations, payments, and cancellation policies, to name a few.

'Export readiness' requirements for international visitors are even more demanding in regard to long-term engagement, marketing plans, price structures, and negotiations, with price lists set for a year or even two in advance. As mentioned previously, since the creative tourism businesses are small and represent an alternative to large groups of visitors, the international tour operators are likely not to be as suitable.

Marketplaces

Any marketplace represents an opportunity to exchange ideas, to network, and to build relationships. Interestingly, the face-to-face encounters are the most important factor in establishing contacts since it is an industry of people more than anything else. Getting to know a person helps create a feeling of confidence, another key word in doing business in a professional environment. Understandably, a tour operator would not put his or her clients in the hands of a provider or a product they do not totally trust, since they could be sued and lose their reputation if the 'deliverable' is not as promised.

Usually, national and international marketplaces (also called international trade fairs) such as the Bolsa Turismo Lisboa (BTL) and Rendez-vous en France, or international travel trade shows such as ITB – Internationale Tourismus-Börse in Berlin, FITUR in Madrid, WTM World Travel Market in London, and IFTM – International and French Tourism Market in Paris are 2 to 4 days long, so as to give enough business time for appointments between buyers and sellers – more than 5000 delegates from over 180 countries participated in the WTM event in 2019 (Newbound, 2019; the websites of these trade fairs are listed in Online Resources). Marketplaces also invite the media, namely, specialized tourism journalists and chroniclers, who write articles for a travel trade publication or magazine with a B2B perspective while covering the event (see TravMedia's International Media Marketplace website in the Online Resources). Since international marketplaces are quite costly, and are geared mostly to international markets, they gather representatives from larger tourism businesses along with national tourist offices and international sectoral associations (e.g. the Creative Tourism Network,[5] see Online Resources).

Having attended more than a dozen tourism marketplaces throughout the years in Japan, South Africa, Mexico, Australia, the United States, and Canada as an observer, a consultant, and a travel trade reporter, I often found that some of the sellers were not *market ready* enough to tackle international markets. While being eager to tell everything about their business during their appointments with the buyers,[6] the sellers tend to talk too much instead of listening in order to respond to the needs of the buyer in front of them. The most serious pitfall lies in the fact that many sellers forget to follow-up on the leads they collected and the business partners and buyers they met.

Educational tours

Educational tours are familiarization tours, called FAM tours. They are provided through the collaboration of tourism partners such as airlines, national tourist offices, hotels, and other tourism businesses involved in the proposed itinerary. Tour operators, travel planners, and travel agents are invited to experiment a destination. The tours usually last between 2 and 10 days, depending on the travelling distances involved.

The invited participants' costs are mostly covered. They visit attractions, experiment with various activities, check different types of accommodation, use various means of transportation, and taste the local cuisine. They bring back the information collected to their own business target markets. This is an efficient strategy to promote a creative tourism activity at a low cost while getting acquainted with professional buyers.

Press trips and press tours

These tours target specialized tourism writers to familiarize themselves with a destination (country or region or city). Travel writers, chroniclers, bloggers, and sometimes influencers are invited on a complimentary basis covering all or most costs. They usually write articles for magazines, electronic papers, and social networks for consumers.

In contrast, travel trade reporters usually participate in press tours offered before or after a marketplace. They write articles in travel trade papers and electronic magazines, providing detailed business information on the destination for key tour operators and travel agencies – audiences already interested or engaged in its promotion.

Tools and technology

Brochures, fact sheets, flyers, posters, business cards, websites, and newsletters are traditional tools still in use. In addition, photos and videos (clear of rights) showing the creative tourism workshop or experience in action is always a winner. Non-traditional/digital tool basics are email addresses, mobile websites for smart phones and tablets, along with photos, videos, map geolocation, social media (e.g. Facebook page, Instagram, YouTube, etc.), blogs, podcasts, and the use of #hashtags. Transactional websites also make it easier for the visitor to check the availability of the product, book it, and pay for it immediately.

E-marketing, or electronic marketing, has become the new norm and, as of now, promotion, distribution, and direct sales will all be done by digital means. It represents a serious challenge for SMEs and specifically for creative tourism entrepreneurs, since they do not necessarily have the knowledge, the expertise, or the financial and human resources to adopt them. Many will turn to professional firms specialized in e-marketing for help.

Market planning

Gathering information, namely creating inventories, collecting data, researching promotion and distribution networks, and identifying partnerships and market segments gives enough information to elaborate on a marketing plan. In this plan, five questions need to be asked: What is the offer? Why is it of interest? Who is it for? How to reach them (with which marketing tools and strategies)? and When?

The last question is tricky: the creative tourism entrepreneur needs to commit to specific dates months in advance, either on their website, via a tour operator, or through a receptive agency since they work on planning and programming from 6 months to 2 years in advance. An international traveller interested in creative workshops would need that information in advance in order to plan their trip around it. For instance, if I book my flights some weeks or months in advance, I may start researching information on creative tourism workshops at the same time. The benefit is twofold: the provider gets an idea of how many people are interested or ready to book the activity while it helps the 'creative' traveller organize their itinerary accordingly. Receptive tour operators will need confirmed dates and prices even a year in advance in order to present the activity to marketplace buyers who will use the information many months after.

Marketing strategies between direct and indirect sales should be balanced. Small moves should be part of monthly planning and respond to direct marketing strategies: continuous website updates, cross-referencing with partners, data collection and follow-ups, searching for product distribution partners such as receptive tour operators, and the use of electronic word-of-mouth (E-WOM) by visitors and participants as user-generated referrals in their own social media. A marketing mix between direct sales and indirect sales by intermediates 'should improve the product, pricing, distribution and promotion' (Stipanović and Rudan, 2014, p. 515).

Market planning will necessarily include a marketing mix of the traditional four Ps: product, price, promotion, and place (i.e., distribution). Added to this, a fifth P for local people is becoming more and more a part of the marketing mix, since creative tourism experiences and workshops are about local people, their artists and artisans, their culture, their place, and their role as ambassadors.

A marketing plan does not have to be complicated. It should cover a 1-year period quite precisely, and a 3-year scope on how it will develop. It does not need to be long. An eight-page document if followed with discipline, with simple actions such as inventories, partnership building, and follow-ups, would be a good start.

Product development should also follow-up on trends, sustainability, and community well-being to improve the initiative and adapt the content accordingly, as more visitors become aware and knowledgeable about the impacts of tourism on the viability of cultural, environmental, and economic community resources.

Conclusion

Creative tourism initiatives, apart from tapping into their local roots, are marketed like any other initial tourism product or service. The most challenging aspect remains the technological advances of e-marketing. Delegating this task to an expert allows creative entrepreneurs to focus on what they should do best, concentrating on partnership building and what is called relationship marketing, which involves fostering business relationships with 'consumers, producers, sellers, distributers, and other stakeholders' to create business through a variety of consumer-seller, producer-seller, seller-distributer, etc. relationships (Uğurlu, 2016, p. 509; see also Tsiotsou and Ratten, 2010; Delisle, 2020). Needless to say, a marketing co-ordinator hired by a municipality or a regional development organization, a non-governmental organization, a not-for-profit organization, or external sources such as a receptive tour operator can be a key element to help small creative tourism entrepreneurs promote and market their product.

Communicating and sharing market knowledge and successes with other creative tourism entrepreneurs facilitates capacity building. Connecting the dots between the experience or workshop and market segments is a long-term development process while the product improves (Delisle, 2020). Localized/regional, national, and international networks can be extremely helpful in building collective 'market intelligence', confidence, and strategy, as well as skills and capacity. Of note, the network CREATOUR® International, launched in June 2020, aims to further knowledge exchange and capacity-building among creative tourism researchers and small-scale practitioners internationally.[7]

Other types of organizations or associations are worth considering as well. As an example, the company Vacation with an Artist (VAWAA) is quite dynamic with short apprenticeship workshops, multiple-day and (recently introduced) online workshops – and it is constantly adding artists throughout the world (see interview with Geetika Agrawal, Chapter 10). It shows how such a platform works in aggregating similar experiences, and how such a platform would be suitable for the set-up of a co-operative for local and regional creative tourism experiences.

University research can also provide insights for practitioners. For example, the Portuguese CREATOUR® research-and-application project has conducted incredible research on creative tourism initiatives and shed a light on the role of mediators in managing the coordination and promotion of their pilot projects (Bakas *et al.*, 2019; see also Chapter 28 by Fiona Eva Bakas in this volume).

I cannot close this chapter without mentioning that some communities are working on their tourism projects apart from the tourism industry system. Although promoting and selling remain key elements to the survival of tourism initiatives, new collaborative tourism methods distance themselves from the industry's commercial approach. For example, *Les Oiseaux de passage,* a recently established co-operative, present in several French cities and rural areas, created an online platform bringing together local tourism businesses with different types of accommodation, activities, itineraries, and creative, artistic, and cultural propositions (see the Les Oiseaux de passage website in Online Resources section). Shared values of hospitality, co-operation, and humanity compose this new ecosystem of stakeholders and local partners under a common mode of operation, a kind of a community-based tourism. The Fairbnb co-op, started in 2016, also proposes a sustainable and more equitable alternative in terms of accommodation rentals in cities and rural areas (see the Fairbnb co-op website in the Online Resources section). The fees generated from bookings are split half-way between the co-operative and local community projects.

A final note regarding COVID-19. Although it is impossible to predict the future with certainty, this difficult pandemic period will eventually end. Meanwhile, we have been provided a moment to reflect on the meaning and purpose of travel. Confinements throughout the world have pushed marketing agendas into a renewed interest for local tourism, with a brighter and more sustainable outlook. Small groups and small-scale local tourism activities and experiences targeted towards markets of proximity represent a positive impulse for creative tourism endeavours. Connecting culture, arts, and significant places and landscapes with travellers through practices like creative tourism workshops offered by micro and small enterprises fits well into this scenario.

The creativity and innovation from artists and artisans embodying this movement as an alternative kind of tourism for the benefit of local communities is giving it meaning and purpose. Most recent research articles about the current period come to similar conclusions, foreseeing a shift towards a more equitable and viable tourism focused on host communities' well-being (Higgins-Desbiolles *et al.*, 2019; Romagosa, 2020; Tomassini and Cavagnaro, 2020; Everingham and Chassagne, 2020; Lew *et al.*, 2020; Haywood, 2020). As Lew *et al.* (2020) have remarked, this transition needs to be grounded in specificities at the local level: 'Social innovation needs to take place creatively at the community level to find the best balance between the benefits and excesses of tourism development' (p. 4). In this context, a spotlight is being directed towards smaller locally grounded enterprises:

> At this point, we should make special reference to the myriad of micro and small enterprises that are deeply rooted in the destination: those companies that offer ecotourism products or products based on the local natural and cultural heritage and do not contribute to overcrowding, offering both high-quality experiences for tourists and high added value to the destination. (Romagosa, 2020, p. 4)

Although international travel is postponed for a while, the tour operators are working on developing scenarios to conquer domestic markets with roughly the same marketing strategies as before, but most tools are of a digital and virtual nature, including marketplaces. Some even use AI for data collecting and most are showing more interest in local initiatives based on cultural and creative resources and networks, 'guided by an ethics of care, social and environmental justice' (Benjamin *et al.*, 2020, p. 479). At this time, creative tourism entrepreneurs should focus on building and strengthening partnerships in their locale and region (Carvalho, 2020), becoming more visible, more connected, and consequently more recognized as valuable and vital resources in their community.

Notes

1. A visitor (domestic, inbound or outbound) is classified as a tourist (or overnight visitor) if his/her trip includes an overnight stay, or as a same-day visitor (or excursionist) otherwise (United Nations – Department of Economic and Social Affairs, 2010, p. 21).

2. Receptive tour operators, also called incoming, inbound, or ground tour operators, run on a business-to-business (B2B) basis, coordinating international tour operators' demands with all ground operations such as products and services at destination. An example of a receptive/incoming tour operator is Tourlink in Italy (https://tourlink.it/en/; accessed 11 May 2021).

3. A fact sheet is a one-page description of the experience or workshop with main highlights, level of difficulty, info such as dates, duration, price, location, reservation and payment procedures, contact name and phone/email, etc., in English.

4. For more information about greeters, see Global Greeter Network (https://globalgreeternetwork.info/; accessed 11 May 2021).

5. In November 2019, the WTM International Trade Fair in London featured a session entitled 'Is Creative Tourism the New Multi-Tool of the Tourism Industry?' (WTM, 2019).

6. In the US and Canada, appointments between buyers and sellers are limited to 10 minutes.

7. For information about CREATOUR® International, visit www.creatourinternational.net (accessed 11 May 2021).

References

Agusdin, A. (2018) Contribution of creative tourism to entrepreneurship. In: Sotiriadis, M. (ed.) *The Emerald Handbook of Entrepreneurship in Tourism, Travel and Hospitality*. Emerald Publishing Limited, Bingley, UK, pp. 387–401. doi: 10.1108/978-1-78743-529-220181028.

Artspace (n.d.) Art happy hour. Available at: http://artspacenc.org/art-happy-hour/ (accessed 16 September 2020).

Atelier MJB (n.d.) Corporate workshops. Available at: https://ateliermjb.com/en/corporate-workshops (accessed 16 September 2020).

Bakas, F.E., Duxbury, N., and Castro, T. (2019) Creative tourism: Catalysing artisan entrepreneur networks in rural Portugal. *International Journal of Entrepreneurial Behavior & Research* 25(4), 731–752. doi: 10.1108/IJEBR-03-2018-0177.

Benjamin, S., Dillette, A., and Alderman, D.H. (2020) "We can't return to normal": Committing to tourism equity in the post-pandemic age. *Tourism Geographies* 22(3), 476–483. doi: 10.1080/14616688.2020.1759130.

Butler, R.W. (1980) The concept of a tourist area cycle of evolution: Implications for management of resources. *The Canadian Geographer/Le Géographe canadien* 24(1), 5–12. doi: 10.1111/j.1541-0064.1980.tb00970.x.

Carvalho, R. (2020). The need for creative actions in tourism in the Post-COVID 19 pandemic world. In: Figueira, L.M. and Oosterbeek, L. (eds.) *Turismo Mundial, Crise Sanitária Futuro: Visões Globais Partilhadas*. Instituto Politécnico de Tomar, Tomar, Portugal, pp. 177–190.

Cohen, S. (2009) The search for 'self' for lifestyle travellers. PhD dissertation. University of Otago, Dunedin, New Zealand. Available at: www.researchgate.net/publication/312631840_THE_SEARCH_FOR_'SELF'_FOR_LIFESTYLE_TRAVELLE (accessed 22 April 2020).

Cooking in Crete (2019) Crete's culinary sanctuaries: Seminars celebrating biodiversity (sample programme). Available at: https://cookingincrete.com/CCS-Biodiversity-Seminars-2019.html (accessed 16 September 2020).

Cookly (n.d.) Vietnamese local market, basket boat, fishing and cooking experience. Available at: https://cookly.me/do/1737-vietnamese-local-market-basket-boat-fishing-cooking-experience/in/hoi-an (accessed 16 September 2020).

Creative Iceland (n.d.) Knitting workshop Iceland. Available at: https://creativeiceland.is/arts-and-crafts-knitting-workshop-and-walking-tour-mosfellsbaer-iceland (accessed 16 September 2020).

Delisle, M.-A. (2020) Marketing creative tourism experiences. In: Gonçalves, A.R., Marques, J.F., Tavares, M., and Cabeça, S.M. (eds.) *Creative Tourism: The CREATOUR Recipe Book*. CINTURS, University of Algarve, Faro, Portugal, pp. 34–36.

Everingham, P. and Chassagne, N. (2020) Post COVID-19 ecological and social reset: Moving away from capitalist growth models towards tourism as Buen Vivir. *Tourism Geographies* 22(3), 555–566. doi: 10.1080/14616688.2020.1762119.

Eyexplore (n.d.) Team building photo workshop. Available at: www.eyexplore.com/team-building-photograph-workshop-japan (accessed 16 September 2020).

Frochot, I. and Batat, W. (2012) *Marketing and Designing the Tourist Experience*. Goodfellow Publishers, Woodeaton, Oxford, UK.

Haywood, K.M. (2020) A post-COVID future—tourism community re-imagined and enabled. *Tourism Geographies* 22(3), 599–609. doi: 10.1080/14616688.2020.1762120.

Higgins-Desbiolles, F., Carnicelli, S., Krolikowski, C., Wijesinghe, G., and Boluk, K. (2019) Degrowing tourism: Rethinking tourism. *Journal of Sustainable Tourism* 27(12), 1926–1944. doi: 10.1080/09669582.2019.1601732.

Lew, A.A., Cheer, J.M., Haywood, M., Brouder, P., and Salazar, N.B. (2020) Visions of travel and tourism after the global COVID-19 transformation of 2020. *Tourism Geographies* 22(3), 455–466. doi: 10.1080/14616688.2020.1770326.

Matos, O. (2018) Profile of creative tourist in north-western of Portugal: An exploratory approach to creative experiences. Presentation at International Conference Cultural Tourism – Theory and Practice: Research Status Quo, Practical Applications and Development Perspectives, Poznan, Poland. Abstract available at: www.researchgate.net/publication/328748925_Profile_of_creative_tourist_in_north-western_of_Portugal_an_exploratory_approach_to_creative_experiences (accessed 22 April 2020).

Newbound, J. (2019) WTM London 2019 sees increase in industry leaders. Available at: https://hub.wtm.com/press/wtm-london-2019-sees-increase-in-industry-leaders/ (accessed 22 April 2020).

Patterson, C. (2016) Match products with the markets identified in a marketing analysis. Excerpt from *Sustainable Tourism with Web Resource. Human Kinetics*. Available at: https://canada.humankinetics.com/blogs/excerpt/match-products-with-the-markets-identified-in-a-marketing-analysis (accessed 15 September 2020).

Pret, T. and Cogan, A. (2018) Artisan entrepreneurship: A systematic literature review and research agenda. *International Journal of Entrepreneurial Behaviour & Research* 25(4), 592–614.

Raymond, C. (2007) Creative Tourism New Zealand: The practical challenges of developing creative tourism. In: Richards, G. and Wilson, J. (eds.) *Tourism, Creativity and Development*. Routledge, London, UK, pp. 145–157.

Remoaldo, P., Serra, J., Marujo, N., Alves, J., Gonçalves, A., Cabeça, S. and Duxbury, N. (2020) Profiling the participants in creative tourism activities: Case studies from small and medium-sized cities and rural areas of Continental Portugal. *Tourism Management Perspectives* 36. doi: 10.1016/j.tmp.2020.100746.

Richards, G. and Wilson, J. (2006) Developing creativity in tourist experiences: A solution to the serial reproduction of culture? *Tourism Management* 27(6), 1209–1223.

Romagosa, F. (2020) The COVID-19 crisis: Opportunities for sustainable and proximity tourism. *Tourism Geographies* 22(3), 690–694. doi: 10.1080/14616688.2020.1763447.

Stipanović, C. and Rudan, E. (2014) Development concept and strategy for creative tourism in the Kvarner destination. In: *Tourism and Hospitality Industry 2014, Congress Proceedings: Trends in Tourism and Hospitality Industry* (pp. 507–517). Available at: www.researchgate.net/publication/323656862.

Tan, S.-K., Luh, D.-B., and Kung, S.-F. (2014) A taxonomy of creative tourists in creative tourism. *Tourism Management* 42, 248–259. doi: 10.1016/j.tourman.2013.11.008.

Tomassini, L. and Cavagnaro, E. (2020) The novel spaces and power-geometries in tourism and hospitality after 2020 will belong to the 'local'. *Tourism Geographies* 22(3), 713–719. doi: 10.1080/14616688.2020.1757747.

Tsiotsou, R. and Ratten, V. (2010) Future research directions in tourism marketing. *Marketing Intelligence & Planning* 28(4), 533–544.

Uğurlu, K. (2016) Relationship marketing in tourism sector. In: Avcikurt, C., Dinu, M., Hacioglu, N. *et al.* (eds.) *Global Issues and Trends in Tourism*. St. Kliment Ohridski University Press, Sofia, Bulgaria, pp. 509–517. Available at: www.academia.edu/30180968/Global_Issues_and_Trends_in_Tourism-.pdf?email_work_card=title (accessed 22 April 2020).

United Nations – Department of Economic and Social Affairs (2010) *International Recommendations for Tourism Statistics 2008*. United Nations, New York, New York, USA. Available at: https://unstats.un.org/unsd/publication/Seriesm/SeriesM_83rev1e.pdf (accessed 22 April 2020).

World Tourism Organization (WTO) (2016) *Tourism and Culture Partnership in Peru – Models for Collaboration between Tourism, Culture and Community*. WTO, Madrid, Spain. Available at: www.e-unwto.org/doi/pdf/10.18111/9789284417599 (accessed 20 March 2018).

World Tourism Organization (WTO) (2018) *Tourism and Culture Synergies*. WTO, Madrid, Spain. doi: 10.18111/9789284418978.

World Travel Market (WTM) (2019) Is creative tourism the new multi-tool of the tourism industry?, trade fair session held 5 November 2019. Available at: https://london.wtm.com/en/Sessions/80339/Is-Creative-Tourism-the-New-Multi-Tool-of-the-Tourism-Industry (accessed 16 September 2020).

Yachin, J.M. (2019) The entrepreneur–opportunity nexus: Discovering the forces that promote product innovations in rural micro-tourism firms. *Scandinavian Journal of Hospitality and Tourism* 19(1), 47–65. doi: 10.1080/15022250.2017.1383936.

Online resources

Airbnb Experiences. www.airbnb.com

Bolsa Turismo Lisboa (BTL). https://btl.fil.pt/?lang=en#

Craft and Travel. www.craftandtravel.com

Creative Tourism Network. www.creativetourismnetwork.org/

CREATOUR® International. www.creatourinternational.net

Destination Canada Markets. www.destinationcanada.com/en/markets (accessed 16 September 2020).

Educational Travel Consortium. https://educationaltravel.travel/about/

Evaneos Tailor-made Tours with Local Agents. www.evaneos.com

Fairbnb co-op. https://fairbnb.coop/

FITUR International Tourism Trade Fair. www.ifema.es/en/fitur

Get Your Guide. www.getyourguide.com

Global Greeter Network. https://globalgreeternetwork.info/

Go Learn To. www.golearnto.com

IFTM – International and French Tourism Market. www.iftm.fr/en-gb.html

ITB – Internationale Tourismus-Börse Berlin. www.itb-berlin.com/Exhibitors/OnlineOrders/

Les Oiseaux de passage. https://lesoiseauxdepassage.coop [in French]

Rendez-vous en France. www.rendezvousenfrance.pro/en
Student and Youth Travel Association (SYTA). https://syta.org/conference2020/about/
Tourlink Incoming Tour Operator & DMC. https://tourlink.it/en/ (accessed 16 September 2020).
TourRadar. www.tourradar.com
Tours by Locals. www.toursbylocals.com
TourScanner. https://tourscanner.com
TravMedia International Media Marketplace. https://travmedia.com/immuk/imm-showcase.html
TUI Destination Experiences. https://gotui.com/en/main
Vacation with an Artist (VAWAA). https://vawaa.com
Viator. www.viator.com
World Travel Market (WTM) – London. https://london.wtm.com/

20 Creative tourism marketing and social media management

Elena Paschinger

Creative Tourism Austria, Austria

As an emerging trend, creative travel is rooted in the rise and mass democratization of cultural travel. The Association for Tourism and Leisure Education (ATLAS) shows that the experiences most enjoyed by cultural travellers are those offering a real taste of the local culture (Richards, 2013). In the past, cultural travel has largely been based on cultural heritage, such as visits to museums or monuments. With the rise of the creative class, a socioeconomic class whose key driving force is economic development through education and entertainment, this focus is gradually shifting towards more active engagement. Furthermore, the more experienced travellers are, the more likely they are to add specific educational goals to their travels over mere rest and relaxation (Richards, 2013). Meanwhile, the wide-ranging and expanding use of social media in many spheres of the economy, and wider society, is contextualizing these travel dynamics. Determining how to most effectively use social media in order to attract and engage these travellers is of growing interest in the field of creative tourism today.

The link between creative tourism and social media

Understanding the demand for cultural and creative tourism has long focused on socio-demographic and motivational criteria, attributing to tourists a certain age, education, income, social status, willingness to learn, and so forth. However, with the advent of the so-called 'experience economy', there is a need to reconsider this traditional way of pigeon-holing tourists: as the travelling population becomes more and more informed, alternative interest groups emerge in favour of traditional categories such as nation states, class, or gender. Increasingly, these have to do with, and/or are rooted in the proliferation of social media networks across the globe (Poria *et al.*, 2004).

The growth of social media, as well as the ubiquitous availability of the internet, has succeeded in stimulating a more direct exchange between travellers and locals. Sites designed to connect the two abound, including Couchsurfing, AirBnB, EatWithALocal, and many more. They share and recycle knowledge, presenting geographically isolated areas with new means of accessibility that were not easily available in the past and enabling the development of new types of products and services (Paschinger, 2015).

The use of social media in the tourism and travel industry is growing constantly, and social media has become an important marketing instrument (Surugiu and Surugiu, 2015). Currently, most marketing strategies cover measures related to social media, and the mobile industry is continuously providing solutions to support the interaction between tourism and social media in every single stage of travel. In an Expedia/Egencia study of 2014 that examined how mobile devices impact travel, three out of four travellers declared that smartphones play a crucial role in their tourism experience (Expedia/Egencia, 2014), and over 5% of mobile apps available on app stores were dedicated to travel in 2015. Numerous mobile apps and information and communications technology frameworks have been developed to engage travellers when visiting a certain spot (i.e. a museum, a city, a region, etc.). Social media as a marketing tool helps businesses to save costs, build and

Corresponding author: reachme@creativelena.com

 Creative Tourism: Activating Cultural Resources and Engaging Creative Travellers (eds N. Duxbury, S. Albino, and C. Pato de Carvalho)
DOI: 10.1079/9781789243536.0020

strengthen relations, and communicate with their audiences. In particular, social media enables businesses to reach customers in different parts of the world and receive feedback in real time, employing minimum resources.

The strategic use of social media in creative tourism marketing

In the digital (social media) age, the four Ps of classic marketing (product, price, place, and promotion) continue to shift in favour of the four Cs: creating, curating, connecting, and culture. This is reflected in the paradigm change that continues to affect not only social media marketing for cultural and creative tourism, but also society as a whole. The basic notion is that we create something, then share it, link it, network it, and moderate discussions about it, instead of trying to convince people of its benefits with traditional advertising. As travellers have the possibility to write reviews, leave comments, and present different perceptions of destinations through a countless amount of review sites and social media platforms, they become prosumers in social media, meaning that travellers are benefiting from marketing activities initiated by companies but are also getting involved in promoting and improving the very tourism products and packages they consume. Moreover, tourists trust other tourists who have had similar cultural tourism experiences more than nice, eye-catching presentations from companies' websites, brochures, leaflets, television, and so forth. Attending to social media comments, ideas, and dynamics can contribute to improving tourism service quality and enriching visitor satisfaction.

Among the most popular types of social media are blogs, as they strongly influence a traveller's decision-making. Blogs easily integrate all other types of social media, and still allow the blog creator to maintain a personal platform with owned content. WordPress is the most used blog software currently and setting up a blog takes less than one hour.

The following are some tips for successful online publishing and communication with followers on social media (EuropeTour, 2018):

- Structuring the content can help in measuring its target results and relieve the pressure of 'having to post something now', over being more savvy in the way of presenting and sharing information.
- In order to 'organize, inspire, and communicate', focus should be given to important dates such as the World Tourism Day/International World Heritage Days, in order to build a content strategy around it. This helps to leverage off frequently used hashtags and search engine requests around such dates.
- As a general rule, it is important to think about what to write, but also when best to publish it.
- Google calendars are one of the best tools available in terms of a quick, easy-to-use and collaborative editorial calendar to use.
- Starting a series of regular features, such as blog posts, video posts, photo contests, and so forth can help maintain a timeline of publication, and a better overview, about it.
- All of the above-mentioned efforts will help in being perceived as far more strategic and professional concerning digital publications. They also help you to unify brand message and tone of voice across platforms, and to become a stronger negotiator and more solid about pitches with brands and other cooperation partners.

Influencer marketing

As cultural travellers plan their trips and gather information using social networks, collecting tips from friends and other virtual contacts, they rely on information provided by so-called 'travel influencers' in order to show the authentic side of the destination visited. 'Influencers' are relatively new on the market and are considered 'digital opinion leaders' with a wide reach among their (often very specific) target groups (influencers = potential guests and visitors!). As travel influencers have become more established, many have built up high levels of loyalty and engagement among their followers, who, in turn, take their opinions and guidance seriously. Communicating with (potential) target groups may therefore be done by hosting these influencers on public relations campaigns, digital storytelling contests, or other forms of professional cooperation.

Best practice example: Kreativ Reisen Österreich – Creative Tourism Austria

Kreativ Reisen Österreich – Creative Tourism Austria was founded in 2011 as an association to promote and highlight the creative tourism offers across all of Austria's nine provinces. It works in close cooperation with the Austrian National Tourist Office, particularly with regards to social media marketing, networking, and cooperation. The website www.kreativreisen.at (accessed 11 May 2021) matches these offers with an increasingly mobile user group, and continuously propagates newsletter and social media marketing. Given its market focus on the German-speaking areas of Europe, an English translation has yet to be added to the site.

From the beginning, the association Kreativ Reisen Österreich has been innovative in managing a notable number of successful (social media) cooperation initiatives. Every week, several updates are published on the association's Facebook page (www.facebook.com/kreativreisen; accessed 11 May 2021), using the growing Facebook community to highlight members' offers and interesting national or international events. In addition, Pinterest, Twitter, Flickr, and other social media channels are used, though Facebook remains the primary channel for community building among Austrian and German audiences. The latter is used for internal as well as external communication, moderating discussions with members of Creative Tourism Austria as well as creative travellers themselves. Facebook is also used for campaign building around prize draws, annual events, and member workshops.

Every year, at least one or two media conferences, press visits, or blog trips are organized for journalists and digital influencers who wish to experience creative tourism in Austria. Funding is covered by participating institutions, as well as local and regional tourism boards, with the management team providing the expertise and contacts. These trips have been very successful over the years and follow the hashtags #KreativReisen, #CreativeTravel, and #inAustria (the latter is also promoted by the Austrian National Tourism Board). Some of the results, including videos and storytelling done by the bloggers, can be seen here: www.kreativreisen.at/social-media (accessed 11 May 2021).

Conclusions

Creative tourism, due to its interactive nature and lively cultural exchange between locals and travellers, particularly lends itself to the opportunities presented by social media marketing. This is especially true when it comes to blogs, video posts, or (live) story updates on social media. Care should be given to always have a professional strategy in content marketing: choosing one social media platform over another has to be in line with your company's strategic communication goals and may be aided by other competent partners, such as professional digital content creators. Further, establishing a publication guideline in the form of an editorial calendar can help creative tourism managers better reap the benefits of social media marketing. This has been demonstrated in highlighting the example of Kreativ Reisen Österreich – Creative Tourism Austria, which over the years has successfully built up an online community of creative travellers in the German-speaking parts of Europe.

Regarding future trends in digital marketing, suffice to say that good content, and now context, will always matter and interactive content will become increasingly mainstream. An interesting context creative tourism could seek for the purpose of its continued proliferation worldwide is an association with destination video storytelling. As virtually all tourism destinations employ various video platforms, and live video formats, to highlight their travel offers, creative tourism is especially likely to head to the forefront of this type of social media marketing tool.

References

EuropeTour (2018) *Social Media Guidelines for Cultural Tourism in Rural Areas: A Practical Guideline for Decision-makers and Stakeholders in Cultural Tourism, Rural Municipalities, and Cultural Institutions*. EuropeTour project, supported by European ERASMUS+ funding scheme 2015 – 2018. Available at: www.europetour.tips/results/social-media-guidelines/ (accessed 30 April 2020).

Expedia/Egencia (2014) *2014 Expedia/Egencia Mobile Index.* Available at: http://viewfinder.expedia.com/news/2014-expedia-egencia-mobile-index/ (accessed 20 April 2020).

Paschinger, E. (2015) *The Creative Traveler's Handbook.* Full Flight Press, Toronto, ON, Canada.

Poria, Y., Butler, R., and Airey, D. (2004) Links between tourists, heritage, and reasons for visiting heritage sites. *Journal of Travel Research* 43(1), 19–28.

Richards, G. (2013) From culture to creativity? In: Smith, M. and Richards, G. (eds.) *The Routledge Handbook of Cultural Tourism*. Routledge, New York, New York, USA, pp. 297–303.

Surugiu, M.-R. and Surugiu, C. (2015) Heritage tourism entrepreneurship and social media: Opportunities and challenges. *Procedia—Social and Behavioral Sciences*, 188, 74–81.

21 Loulé Criativo: Our creative tourism development story

Marília Lúcio

Municipality of Loulé, Portugal

Loulé Criativo is a leading creative tourism initiative developed by the Municipality of Loulé in 2014, and the only structured and long-standing initiative in Portugal whose focus is creative tourism as a strategic approach to local policy development and planning towards sustainable local development.[1] Loulé Criativo focuses on enhancing the identity of the territory, with creativity and innovation as the driving force; supports the training and activity of artisans and professionals in the creative sector, contributing to the revitalization of traditional arts and the promotion of new approaches to intangible heritage; and offers an attractive annual programme with residencies, workshops, and creative experiences.

The various thematics of Loulé Criativo are tightly connected with the territory of Loulé, and have its culture, tradition, and identity as starting points. This is a way of proposing an authentic and high-quality tourist product, economically developing the interior areas, supporting small- and medium-sized companies linked to creativity, and combating tourism seasonality. The cultural, natural, and geographic characteristics of the territory of Loulé provide both material and immaterial resources that have inspired and been used to develop creative tourism initiatives.

The main intervention areas of Loulé Criativo are: *Creative Tourism*, the *Network of Workshops*, and the *Loulé Design Lab*. Through *Creative Tourism*, the Municipality of Loulé offers travellers the possibility of immersing themselves in the local culture, providing them with unique challenges through experiences and learning. The *Network of Workshops* currently includes five workshop places, which were acquired and rehabilitated by the municipality, in which artisans are working live, and exhibiting and selling their products. The spaces are themed and aim to revive old crafts that had a preponderant role in the economic life of families for many years, and nowadays can take on a contemporary use. The *Loulé Design Lab* is a laboratory for creation, research, and experimentation, where conditions are provided for education and the establishment of a creative community.

Setting the terrain

Loulé is a municipality located in the Algarve region, in the south of Portugal. Being one of the largest municipalities in the Algarve region, it combines the diversity and scenic beauty of the sea, the *barrocal* subregion, and the mountains with a cultural richness that testifies to a millennium of history of Mediterranean influence. In the city of Loulé and its surroundings, the diverse manifestations of its rich cultural heritage, as well as more contemporary artistic expressions, comprise unique experiential and creative resources.

Loulé is included among the municipalities of the Algarve that, until the end of the 20th century, presented a great diversity of arts and crafts, the result of its extensive territorial area and, above all, of its great landscape, demographic, and economic heterogeneity. It was once known as the 'land of handicrafts' (Martins, 2014). However, many of these craft activities have disappeared over in recent years. The ancestral transmission of knowledge that traditionally took place in the master–apprentice relationship within the workshop space, through participatory observation and practice, contributed to the gradual and cumulative execution of techniques.

Corresponding author: loulecriativo@cm-loule.pt

DOI: 10.1079/9781789243536.0021

Today, some of these traditional activities are practically extinct in the Algarve or are practiced by very few people who are already at a very old age, as is the case with the coppersmiths, boilers, brokers, barbers, and blacksmiths.

With the regional economy essentially based on the tourism sector, this is the sector that serves as a lever not only for the development of the offer of goods and services but also for activities related to civil construction, business, and fishing. Most tourists visit Loulé during the summer for its beaches in the coastline area. Creative tourism provides an opportunity to offer new products and new activities that promote an immersion in the territory all year long. It is in this context that the importance of a creative tourism project, such as Loulé Criativo, is confirmed.

The concept of creative tourism was defined at the beginning of the 21st century (Richards and Raymond, 2000) as a potential driver of local development in cities, regardless of their size, and a possible driving force for the revitalization and preservation of traditions, arts, and ancestral crafts that are at risk. The Municipality of Loulé initiated a study to assess the differentiating potential of its territory and its culture, as a way of valuing know-how, disseminating its culture and local heritage, and promoting sustainable development.

In order for this to effectively occur, there was a need to establish local partnership networks, or clusters, around the exchange and sharing of knowledge. This effort supports innovation and entrepreneurship, enhances the organization of resources, and helps to structure the offer and dissemination efficiently.

Thus, the Loulé Criativo project integrates different aspects that are articulated and that as a whole make it unique in the region and even in the country. The project, installed in a network of spaces, properly articulated, is an anchor that can attract talents, professionals, students, and researchers; functions as an incubator for new economic activities based on arts and crafts; and serves as a great attraction and centre of dynamism in the municipality, including in the field of creative tourism.

Loulé Criativo: first steps

The foundations of the project 'Loulé Criativo' in 2014 were the existence of ancestral know-how in the area of arts and crafts that was being lost, the need to have a new economic opportunity to preserve and transmit that knowledge, the existence of material and immaterial heritage, and a tourism sector eager to have new experiences of immersion in the local culture. In this initial phase, the Municipality of Loulé hired a consulting company in the field of tourism and cultural experience, ProActiveTur, to outline, with the municipality's team, strategic points of guidance for the creation and implementation of the project. Close cooperation was developed with regional, national, and international projects and associations, and benchmarking experiences were undertaken. Several surveys were carried out focusing on the material and immaterial heritage of Loulé, as well as several national and international visits, to outline and plan a structure for the Loulé Criativo project.

In the course of this work, the project's main objectives were defined:

- Create and promote a brand that mobilizes agents for good practices in creative tourism.
- Organize and structure a dynamic and attractive offer of creative experiences that put tourists in contact with the unique aspects of regional identity and heritage based on 'hands on' activities.
- Implement an artistic and creative residency programme that mobilizes international critical mass and introduces a contemporary vision to the sector.
- Develop a cultural programme that includes permanent and temporary exhibitions and events related to the theme of creativity, art, heritage, arts, and crafts.
- Develop an innovative system in Portugal for the transmission of knowledge between masters and apprentices.
- Implement adequate and updated training in traditional arts and crafts.
- Foster support for the installation of business of artisans and professionals in the creative sector adjusted to their needs.
- Create conditions for research in arts and crafts and related topics.
- Promote the continuous stimulation of innovation in the products and work processes of professionals.

The planning of the different phases of project implementation are outlined in Table 21.1. A set of areas was structured, which become the operational functions of Loulé Criativo and which would work in synergy, namely: creative tourism, creative

Table 21.1. The planning phases of Loulé Criativo

Phases	Components
Phase I	Partner identification Space allocation
Phase II	Structure of the offer of creative tourism Creation of an artistic residency *valencia* (programme)
Phase III	Project preparation

and artistic residencies, training in arts and crafts, and the incubation of designers and makers in the Loulé Design Lab.

Loulé Criativo: Implementation

Creative Tourism

In order to develop a consistent and integrated strategy for this project, its implementation was informed by the earlier study about the potential to develop a creative sector anchored in Loulé's cultural and territorial resources. Within the Creative Tourism stream, Loulé Criativo aimed to develop an organized tourism offer that allows the 'new generation of tourists' to participate actively in the culture, traditions, and local way of life, in interaction with residents and the unique character of the tourist destination. The Loulé Criativo creative tourism initiative lives and interacts with the other areas within Loulé Criativo, namely, the *Network of Workshops* and *Loulé Design Lab.*

Building a network

In March 2014, the first steps were taken towards the creation of a network in Loulé that aggregates cultural, economic agents, and individuals from the world of the arts and crafts and the gastronomic sector to create a Creative Tourism programme. It was considered that it would be advantageous to carry out this programming through a network dynamic, instead of individually requesting the intervention of each service provider, as there would be clear gains in the critical analysis of the process, its results, and even in the field of operations. At the beginning of the project, the municipality organized several meetings and training actions with partners to learn about their expectations and also to present other creative tourism practices. Several local agents were invited, including shops, companies, traders, artists, artisans and cultural institutions in order to develop the creative tourism component. The main areas for the integration of creative tourism activities were defined as: the arts, crafts, gastronomy, and heritage.

The primary goal is that the local creative tourism network grows and new creative tourism products are developed in the territory, including the integration of new partner agents. Thus, the programme is permanently open to the integration of new local agents, artisans, companies, and artists who want to open their doors to visitors and provide intimate and interactive opportunities to discover the culture of our territory. The Loulé Criativo project encourages and supports the burst of new activities and new entrepreneurs who incorporate the economic chain of creative tourism and contribute to maintaining and innovating traditions, celebrating identity, and creating an ecosystem favourable to creativity. The creative tourism network developed in Loulé has the active involvement of an array of local agents, including stores, companies, traders, artists, artisans, and cultural institutions, among others, who operationalize the network. At the moment, the project has 28 partners integrated into the local economy, some of which include their own network of internal employees and sub-networks of partnerships.

This local network of creative tourism partners is equipped with a simple, agile, and fast operating model, not only to respond to requests that arise from tourists and others interested in activities, but also to follow-up on other initiatives related to strengthening the network and promotion, among other aspects. The overall structure is coordinated by the Loulé Criativo project. Its management and operation include functions such as:

- call and mark moments of discussion;
- aggregate and work on proposals for creative tourism programmes;
- manage information requests and registrations;
- organize and monitor activities on the ground;
- articulate actions and exchange information with strategic partners;
- represent the network at publicity and promotion events;
- propose the admission or removal of elements in the localized network; and
- promote the exchange and sharing of information between network elements.

The partners meet whenever necessary and are in daily contact with the project's co-ordinating team in order to participate in its dissemination actions and thus contribute to the greater success of Loulé Criativo tourism. We have created strategies to regulate and promote the development of good practices, and a structure to monitor the impact of the network and its evaluation, involving local agents, the university, trade associations, and the municipality, among other entities.

Following some local dissemination events in 2014, the annual organization of a diversified programme began in 2015 with workshops, short courses, thematic weekends, and other activities, thus creating a rich environment for the visitor be part of the spirit of the place, meet its people, and participate in enriching personal learning experiences. An annual calendar is organized, with specific dates for each activity to be carried out, so that those interested in participating can register according to their availability. There is also a set of activities that are available by appointment, upon request from individual participants or groups, and according to the availability of the partner who provides the respective activity.

The strategy is dynamic and constantly adapting to the needs raised each season. There is a direct correlation between the seasonality of the agriculture and food, festivities and local traditions, and other endogenous materials that inspire our artisans and the creative experiences. Between 2016 and 2020, the planning of creative experiences, activities, and exhibitions throughout the year was strategically developed to link the seasonality of orchards and agriculture to local traditions, crafts, and gastronomy. As well, the festivities and municipal events are accompanied by unique creative experiences organized specifically for these celebrations. Table 21.2 presents examples that illustrate how some creative experiences are integrated into seasonal features and events.

Table 21.2. Examples of the year-round integration of creative experiences by Loulé Criativo

Month	Seasonal features and events	Experiences
February	Blossoming almond trees cover the fields of white and pink	Almond blossom route: walking through these fields is an unforgettable experience, available only once a year
	Carnival in Loulé, the oldest in Portugal	An excellent opportunity to remember or get to know the ancient traditions associated with this festival: Paper Flowers: learn how to create carnival decorations Orange 'Filhós de Carnaval' Pastries explore the gastronomy related to this festivity Photographic tour during Carnaval day
March/April	Easter	Easter Folar at Tacho: Easter folar (Easter bread) with figs and alfarrobeira (locust beans/carob)
May	Cob day	Espiga Branch: Spring invites you to stroll through the colourful fields full of wild flowers, this is the day of the traditional harvest of five samples from five plants that represent this branch
	Snail season	Snails à Algarvia: gastronomic experience
July/August	Popular saints Season of sardines	From the Market to the Kitchen – Grilled Sardines: gathering family and friends or celebrating the 'Popular Saints' celebrations
September	Grape harvesting Almond, fig, and carob harvest	Participative harvesting at Quinta da Tôr: Help with the harvest and discover the Pomar de Sequeiro on a morning that stimulates memories of other times, where the harvesting of almonds, figs, and alfarrobeira (locust beans, from which carob is derived) was part of the life of the people of this region
October	Olive picking	Discover the local tradition of olive picking, olive oil production in the mill, and preparation of olives for preserving, respecting traditional techniques
November	Catch of the Medronho	The Medronho tree is an important element of the landscape of the Algarve interior. This harvest is done by hand and the processing is done on a small scale (distilleries), in order to obtain the perfect taste of the Liqueur of Medronho

External partnerships and networking

Successful synergies have also been developed with national tourism agencies and accommodations businesses in order to attract tourists who want to experience their trip creatively and discover Loulé in a genuine and sustainable way. The success of these experiences and synergies cultivates a significant impact on local businesses, job creation, economic growth, and sustainability throughout the year.

Another important dimension of this development process has been pairing the on-the-ground development activity with actively participating in and learning from broader networks. From the beginning of the project, the international Creative Tourism Network (CTN) was involved in order to advise on the goals and lines of the project. Loulé was the first Portuguese municipality to join CTN in 2015. CTN aggregates and promotes this new generation of tourism worldwide. A very close relationship has been maintained, involving the exchange of suggestions and experiences. Several joint promotion and dissemination actions have been organized online and at international tourism fairs.

In 2016, Loulé Criativo became part of CREATOUR® (Developing Creative Tourism Destinations in Small Cities and Rural Areas), a multidisciplinary research project based on a combined theoretical-practical approach to collaborative processes. This project aims to contribute to the development of a sustainable creative tourism sector, in small cities and rural areas in four regions of Portugal, namely, North, Centro, Alentejo and Algarve. With this new platform, the first steps of the Network of Workshops were taken, incorporating several creative experiences interconnected with local crafts and crafts. The CREATOUR® project has helped the municipality to develop a more consistent project, helping with the quest to know better the public who does this type of tourism, to learn from other experiences, and to benefit from the exchange between the CREATOUR® partners.

Network of Workshops

In response to the need to create a network for the transmission of knowledge and learning in traditional arts and crafts, it was necessary to structure a network of workshops with the main objective of maintaining tradition and disseminating the knowledge of techniques of inestimable cultural value. The Network of Workshops, created by the Municipality of Loulé, aims to maintain traditions, disseminate techniques and knowledge of significant cultural and aesthetic value, and facilitate the commercialization of objects produced with materials and techniques typical of the region. In addition to promoting the production, execution, and marketing of pieces using traditional techniques, the workshops also aim to promote creative experiences and training between masters and apprentices or visitors. The profits generated from these activities go entirely to each of the artisans, maintaining a strong relationship of trust and reliability about the project and positive socio-economic results.

The main goals for the Network of Workshops are:

- train new artisans;
- prevent the extinction of some arts and crafts;
- contribute to the sustainability of cultural heritage;
- promote job creation among the young and unemployed;
- diversify business opportunities;
- encourage artisanal and crafts production in the region; and
- reaffirm the region as a generator of specialized knowledge and high-quality products that create notoriety in the sector.

A survey of the arts, crafts, and artisans who were active or who would have been apprentices of almost-extinct crafts was carried out. An assessment was also made of available municipal spaces that would be appropriate for the implementation of the workshops, considering the history and location of the respective crafts. In this initial phase, several spaces were selected in the historic area of the city of Loulé and the necessary requalification works were carried out.

From 2017 to 2020, five workshops have opened involving 20 artisans:

- *Casa da Empreita:* This workshop was re-oriented in a current way, located in what a century ago were the houses of the enterprise, which existed all over the Algarve, with Loulé standing out for its importance. Casa da Empreita consists of a collective of artisans who work and market what they produce, interweaving the palm (leaf from the dwarf palm), sewing and connecting each of its branches, and producing various pieces, such as bags, baskets, mats, and other pieces in both traditional and innovative formats (see Fig. 21.1).

Fig. 21.1. Palm-weaving workshop at the Casa da Empreita, Loulé, Portugal. Photo courtesy of the Câmara Municipal de Loulé, used with permission.

- *Oficina de Caldeireiros:* The sounds of hammering on copper and brass were heard again in the streets of Loulé. Three boilermakers work in this workshop (to date), in the space that housed the 'Caldeiraria Louletana' in the past. In this workshop, pieces such as pots, cataplanas, bracelets, lamps, bowls, and other items are produced manually.
- *Oficina do Barro:* Domingos Gonçalves 'Xavier', who remained active in this workshop until the building went into ruin, returned to mould and demonstrate how to work the clay, as well as to commercialize its most typical pieces, the Algarve chimneys, in the left wing of the workshop. In the other (right wing), within the scope of the Loulé Criativo project, other potters are welcomed during incubation periods, allowing experimentation and their launch and integration into the workshop activity. In 2019, there were three resident potters. The workshop also hosts training activities for professionals in the sector and, at the same time, works to bring the activity closer to school students and other audiences.
- *Oficina de Cordofones:* In this workshop, the luthier Manuel Amorim is dedicated to the construction and repair of stringed musical instruments, ranging from Portuguese guitars to classical guitars, bell violas, ukuleles, mandolins, and cellos, among other stringed instruments. The workshop aims to demonstrate to younger generations the entire manufacturing process, the tools involved, the different types of wood, as well as the history of the instruments to foster a taste for the activity and promote its continuity.
- *Oficina de Relojoeiro:* In this space, the master watchmaker José João Guerreiro applies his knowledge, carrying out maintenance and other arrangements, and keeping alive an almost extinct trade in the region.

The way the Network of Workshops does its work has a great impact on local arts and handicrafts that have dissipated or are about to disappear, creating jobs for artisans and expanding and sustaining traditional skills and local heritage, which otherwise could disappear completely. The boilermaker is an example of a profession that the Municipality of Loulé, through the Loulé Criativo project, managed to save from extinction. Through a creative residency, it was possible to find a craftsman who had worked in the area about 30 years ago, but who later opted for another profession, then, being retired, came to support this residence. Since the first training course, held in 2017, two of the trainees joined the copper shop where they have been developing this craft, together with their 'master'. The second Coppersmith Course is currently underway (2020). This is the vital energy of the project, a continuous effort to revitalize new strategies related to the mobilization of endogenous natural resources, promoting skills, knowledge, and capacity for innovation.

The knowledge of these traditional techniques is also disseminated through the exhibition of handicrafts work, creative experiences, and the sale of handicrafts with both traditional and contemporary lines. In the five workshops of the network, visitors have a unique opportunity to get to know the 'person behind the craft', to know their history, to experience their passion, and to learn traditional techniques directly through their hands and ancient wisdom, allowing a unique opportunity for involvement in this strong component of Loulé's identity.The Network of Workshops is in constant development, with plans to promote and support future projects, encompassing other crafts that produce handicrafts with indigenous materials. The training, experiences, and master–apprentice contact is continuous throughout the year, with several training courses aimed at beginners as well as master classes. In Loulé Criativo, there is space and an answer for everyone who wants to learn, know, and develop these arts and crafts.

Loulé Design Lab

Since the beginning of this project, one of the areas that was defined as essential was a laboratory for experimentation and design as well as the incubation of enterprises and entrepreneurs. A very important aspect associated with arts and crafts is permanent investigation and the constant introduction of innovative factors. Consequently, several guidelines were established to frame the operation of the Loulé Design Lab, in particular:

- The establishment of agreements with universities and reference institutions in the areas of creativity, research and innovation.
- Support for research and development in the arts and crafts sector and incorporation of their contributions in the projects of users or beneficiaries: inputs in terms of product design, packaging, visual communication, research in the area of materials and components, innovation in production processes, and strengthening of knowledge and connection with cultural heritage, anthropology, and other disciplines; and
- Provision of properly equipped space, working conditions, access to information and sources, and support for the dissemination of results to researchers, interns, and master's students who wish to develop their work.

The Loulé Design Lab started in 2017, defined as a laboratory for creation, research, and experimentation, focused on supporting and promoting design specifically linked to local culture. This creative community participates actively throughout the year in exhibitions, fairs, events, residences, and showcases, diversifying its projection, interaction, and partnerships with local, national, and international activities.

One of the greatly anticipated moments is the participation of the projects integrated into the Design Lab in the 'Algarve Design Meeting', an annual event that promotes the dissemination of Portuguese and international design through an extensive programme of activities. The partnership between Loulé Criativo and this great regional event allows the creative community integrated in the Loulé Design Lab to have the opportunity to publicize their creations and develop synergies with the different collectives of our society.

In 2019, four members of this creative community partnered and opened their store 'Coletivo 28' in the historic area of Loulé, an impressive and very successful step.

In 2020, the Loulé Design Lab hosted 29 resident projects that received business incubation support from a team the municipality contracts. The designers and makers have co-working spaces. The Loulé Design Lab business incubation support includes:

- an Incubator Zone, with space for temporary installation of artisans and artists companies;
- a continuous training programme in the areas of design and innovation, product, project, management, marketing, communication, and sales;
- access to services provided within the scope of the research laboratory;
- experimentation and design that add value to the product, packaging, and business communication;
- a shop for the promotion and sale of artisan and artistic products created by the beneficiaries of the space; and
- support in the elaboration of business plans, projects, and actions for internationalization.

Promotion and communication

The initial promotional plan, developed to leverage Loulé Criativo, aimed to position Loulé internationally as a privileged destination for creative tourism, to broaden the target audience, and to more significantly reach the foreign market, especially that of northern Europe. This required the development of a two-phase communication plan, which is outlined in Table 21.3. A specific promotion and dissemination programme was designed, with a website, a Facebook page, and social media campaigns.

The main vehicle of promotion of Loulé Criativo is the website www.loulecriativo.pt. It includes information about the project and the municipality, and highlights the annual schedule of activities presented by the partners. All workshops are available on the website in the 'Experiences' section and each activity has its own form, which presents useful information about the programme, partners, price, location, and contact details for registration purposes. A Facebook page and an Instagram page for Loulé Criativo were also created, as tools that would permit the dissemination of the entire programme.

Since then, a communication plan has been drawn up, analysed, and presented every year, in which the promotional strategies to be developed over the months are outlined. Today, the creative tourism programme's dissemination strategy and the promotion actions include several channels:

Table 21.3. Loulé Criativo's two-phase communication plan

Phase	Components
Phase 1	Website production
	Production of informative and promotional materials about Loulé Criativo, such as flyers, roll-ups, leaflets, and posters, among other items
	Production of digital materials, such as newsletters and promotional banners
	Creation of promotional video
Phase 2	Conducting online promotional campaigns, with a strong focus on digital marketing and social networks
	Participation in international tourism fairs, articulating with the action of the Algarve Tourism Association (ATA), which annually participates in several of these events
	Participation in BTL, the largest national tourism fair, held in Lisbon
	Organization of educational visits with specialists in these subjects, such as journalists, opinion leaders, and tourism agents

- website and social networks in synergy with all partners, a monthly newsletter, and prominent regional publications available to ex-pats and local visitors;
- strategic publications, interviews and articles in national and international magazines, newspapers, blogs, and websites;
- dissemination through the Algarve Tourism Association, the Creative Tourism Network, and CREATOUR®;
- participation in strategic local, regional, and national events;
- participation in promotional fairs and in national and international events in which the Municipality of Loulé is present;
- reorganization of the annual planning of creative experiences, adjusted to the new areas of activity;
- promotion, dissemination, and participation in fairs and events, attracting new partners;
- monitoring, follow-up, and advice to network partners; and
- new activities and new partners.

Through these efforts, a high level of participants has been maintained in the low- and medium-tourism seasons and we have also achieved the objective of mobilizing tourists from the coast to the interior during the high season. In 2019, we reached the highest value of activities carried out and the highest number of participants to date, reflecting the exponential growth that the project has achieved (see Table 21.4).

Table 21.4. Number of creative experiences/events and participants in Loulé Criativo activities, 2015 to 2019

Year	Activities	Participants
2015	Promotion of 89 activities	Total of 419 participants
2016	Streamlining 89 activities	Total of 219 participants
2017	Promotion of 53 activities	Total of 353 participants
2018	Promotion of 67 activities	Total of 530 participants
2019	Promotion of 135 activities	Total of 853 participants

Loulé Criativo headquarters: Palácio Gama Lobo

In 2019, the new headquarters of Loulé Criativo was launched in the renovated Palácio Gama Lobo as 'Espaço de Conhecimento, Oficios e Artes' (Space of Knowledge, Workshops, and Arts). Located in the city of Loulé, this Palace is an emblazoned building, an old noble house, a landmark of Loulé's civil architecture, and an important element in the city's expansion and urban development in the second half of the eighteenth century.[2] The Gama Lobo Palace allows all areas of Loulé Criativo to be brought together in a single location, giving a new impetus to the project and serving as a motor for various activities, including exhibitions, creative experiences and residencies,[3] and workshops and training. In turn, the combination of these activities in one place fosters and supports dynamics of improvement and experimentation, innovation, creativity, articulation with other arts and technologies, new economic and creative activities, demonstration and cultural animation, support for commercialization, tourist visits, and so forth.

Closing remarks

Over the years, Loulé Criativo has grown and evolved into a major territorial development project, a unique project at the national level, and a reference at the international level. All the different areas incorporated in the project (Creative Tourism, the Network of Workshops, and the Loulé Design Lab) contribute to develop a strong programme around cultural heritage, traditional arts and crafts, and design.

The Loulé Criativo project aims to provide a continuous induction of innovation in the products and work processes of professionals; the necessary conditions for research in the arts and crafts and related topics; support for the installation and business of artisans and professionals in the creative sector, adjusted to their needs; and a cultural programme that includes permanent and temporary exhibitions and events related to the theme of creativity, heritage, and arts and crafts. It strives to be closer and closer to its communities, to create business opportunities for local agents, and to offer quality, differentiated, and authentic tourism so that anyone who decides to spend their vacation in Loulé can have an experience of immersion in the local culture.

Loulé Criativo was the regional winner of the 2016 municipality of the Year Award, organized by the UM-Cities platform of the University of Minho, Portugal. This award aims to recognize and reward good municipal practices that have a significant impact on the territory, the economy, and society, and promote growth, inclusion, and sustainability. In 2019, the project was a finalist in the 'Promotion for Public Entities' category of the National Crafts Award, promoted by the Institute of Employment and Professional Training (IEFP). This is an award that aims to encourage artisanal production, in its traditional and contemporary aspects, distinguishing artisans who stand out for their technical and professional skills and the excellence of the results achieved, as well as agents and public and private entities with relevant interventions promoting arts and crafts.

Numerous investigations and case studies have been carried out on Loulé Criativo (e.g. Carvalho *et al.*, 2016; Almeida de Sá, 2017; Remoaldo *et al.*, 2019; Carvalho, 2020), helping it serve as a reference for the creation and management of innovative creative tourism projects in other countries. Looking forward, Loulé Criativo will continue to strengthen the identity of Loulé, renewing and deepening the particularities of its local identities.

CREATOUR®

In addition to the activities that Loulé Criativo has promoted since 2016 through the Partners Network, we applied (and were successfully selected) to be part of the CREATOUR® project. We proposed to open several spaces dedicated to traditional arts and crafts that would function daily as venues for exhibitions, sales, and training in the areas to which they are dedicated. Throughout the CREATOUR® project, the Network of Workshops evolved and is today one of the main tourist attraction centres of the city.

At the same time, we also aimed to diversify our existing creative experience offer and thus organize activities linked to tradition and local culture in which several partners from different areas would intervene (e.g. art, crafts, gastronomy, rhythm, and heritage) and which could function as a route of local experiences, subordinated to a theme linked to local culture. A set of experiences connected with the interior of the municipality was developed mobilizing the local economy in this area and promoting the connection of the partners with local traditions and festivities, highlighting the experiences inspired by the Sequeiro Orchard and its connections with gastronomy, the Mediterranean Diet, and associated traditions. The diversity of experiences developed was exceptional, in 2019 reaching a record number of experiences and number of participants (see Table 21.4). The demand and interest of the media has also increased substantially.

For Loulé Criativo, CREATOUR® was an opportunity to exchange experiences, ideas, and opinions, a space for reflection on everything that has already been done and on the implementation of the new projects presented. CREATOUR® allowed us to develop new activities within the project that leveraged the economy of some points of the municipality by creating jobs, business opportunities, and training in specific areas. It also facilitated synergies, knowledge exchange, and strengthened our contents and objectives. CREATOUR® provided us with unique and exceptional opportunities, which helped us to assert ourselves as a quality creative tourism destination, diversifying the region's tourism products, boosting rural development, mitigating seasonality, and contributing to the affirmation of the local economy.

Advice and recommendations for other municipalities

Based on our experiences of designing, launching, and implementing Loulé Criativo since 2014, we highlight the following practices and strategies that proved to be effective and useful in our context.

Develop a policy framework:

- Municipalities should incorporate creative tourism policies in their action plans and strategies for local sustainable development.
- Furthermore, designing an integrated policy framework provides clarity and strength to guide the strategic development of initiatives interlinking culture/heritage, tourism, and local sustainable development.

Build a local knowledge base:

- Study and assess the differentiating potentials of the territory and its culture as a way of valuing local know-how, disseminating its culture and local heritage, and promoting sustainable development.
- Carry out several surveys, focusing on local material and immaterial heritage, as well as national and international visits to become informed by others' decisions and practices.
- Conduct a survey of the arts, crafts, and artisans who practice, have been active, or who were apprentices of extinct and almost extinct crafts.
- Assess available municipal spaces that would be appropriate for the implementation of the creative workshops.

Provide infrastructure:

- In order to develop a sustainable creative tourism product, a set of creative workshop venues should be created. Assess whether there are unused municipally owned buildings that could be used for this purpose.
- The spaces can be themed, as they intend to revive old crafts that had a significant role in the local economy, and ideally should be linked to historic areas for these activities if this is possible.

Develop a local network:

- Develop a local creative tourism network that actively involves an array of local agents, including stores, companies, traders, artists, artisans, and cultural institutions, among others.
- The creative tourism network should be permanently open to the integration of new local agents, artisans, companies, and artists, promoting opportunities and new discoveries.
- The local network of creative tourism partners should be equipped with a simple, agile, and fast operating model to quickly respond to requests that arise from tourists and others and to follow-up on other initiatives related to strengthening the network.
- Create a structure to monitor and evaluate the impact of the network, involving local agents, a university or research centre, trade associations, and the municipality, among other entities.

Communicate and promote synergies:

- An annual communication and dissemination plan for creative tourism should be drawn up, analysed and publicly presented in which the promotional strategies to be developed over the months are outlined.
- The planning of creative experiences, activities, and exhibitions throughout the year can be strategically developed in connection with seasonal festivities, municipal events, agriculture, and other local activities.

Sustain the continuance of local traditions and know-how:

- To create a network for the transmission of knowledge and learning in traditional arts and crafts, we recommend structuring a network of workshops with the main objective of maintaining tradition and disseminating knowledge of techniques.
- The training, experiences, and master–apprentice contact should be continuous throughout the year, with several training courses aimed at beginners as well as master classes.

Encourage innovation:

- Create enabling conditions for research in arts and crafts and related topics, to foster new knowledge and flows of information that can innovate working practices and perspectives.
- A creative laboratory, including research, experimentation, and design, can incubate companies and nurture a spirit of exploration, spearhead innovations, and foster connections in the creative community and more widely.

Acknowledgements

This chapter was developed within the CREATOUR® project (no. 16437), which is funded by the Portuguese Foundation for Science and Technology (FCT/MEC) through national funds and co-funded by FEDER through the Joint Activities Programme of COMPETE 2020 and the Regional Operational Programmes of Lisbon and Algarve.

Notes

1. In recent years, other Portuguese municipalities have also been following this line of action, such as São João da Madeira, Espinho, Bragança, Mértola, and Beja.

2. The architecture project for the restoration of the Gama Lobo Palace was prepared by the architect Luís Guerreiro of the Municipality of Loulé.

3. The Gama Lobo Palace encompasses the programming and operation of artistic, technical, and creative residences, offering a space for the accommodation of residents and for their research, production, and creation.

References

Almeida de Sá, J.M. (2017) Turismo criativo em Portugal: Reflexão sobre territórios turísticos alternativos no caso do Algarve. Master's dissertation in Leisure, Heritage and Development, Faculty of Arts, University of Coimbra, Coimbra, Portugal. Available at: https://estudogeral.sib.uc.pt/bitstream/10316/85574/1/Jos%c3%a9%20Miguel%20S%c3%a1.pdf (accessed 13 November 2020).

Carvalho, C.P. de L. (2020) Investigação de parâmetros de análise do design de equipamento enquanto ferramenta de impacto social. Doctoral thesis, Faculty of Fine Arts, University of Lisbon, Lisbon, Portugal. Available at: www.academia.edu/41961013/INVESTIGA%C3%87%C3%83O_DE_PAR%C3%82METROS_DE_AN%C3%81LISE_DO_DESIGN_DE_EQUIPAMENTO_ENQUANTO_FERRAMENTA_DE_IMPACTO_SOCIAL_Tese_especialmente_elaborada_para_a_obten%C3%A7%C3%A3o_do_grau_de_Doutor_em_Belas_Artes_na_especialidade_de_Design_de_Equipamento (accessed 13 November 2020).

Carvalho, R., Ferreira, A.M., and Figueira, L.M. (2016) Cultural and creative tourism in Portugal. *Pasos: Revista de Turismo y Patrimonio Cultural* 14(5). 1075–1082. Available at: www.pasosonline.org/Publicados/14516/PS516_01.pdf (accessed 13 November 2020).

Martins, L. (2014) *Para uma contextualização de Loulé medieval*. Centro Interdisciplinar de História, Culturas e Sociedades da Universidade de Évora, Évora, Portugal.

Remoaldo, P., Ribeira, J.C., and Carvalho, R. (2019) Turismo e descentralização: da perceção de alguns agentes sobre o turismo criativo à potenciação do desenvolvimento local. Paper presented at IV Conferência P3DT: Descentralização & Desenvolvimento, Ermesinde, Portugal, 11–12 April 2019. Available at: https://ler.letras.up.pt/uploads/ficheiros/17135.pdf (accessed 13 November 2020).

Richards, G. and Raymond, C. (2000) Creative tourism. *ATLAS News*, no. 23, pp. 16–20.

22 Santa Fe, New Mexico's Creative Tourism Initiative

SABRINA V. PRATT

SVPratt Creative Strategies, United States

The City of Santa Fe, New Mexico's Creative Tourism Initiative ran from 2009 to 2015. It began as a result of Santa Fe's membership in the United Nations Educational and Scientific and Cultural Organization (UNESCO) Creative Cities Network. Santa Fe joined the network in 2005, and that same year a UNESCO representative involved in forming the Creative Cities Network brought up the concept of creative tourism as an economic development tool. Crispin Raymond and Greg Richards had only recently defined it in late 2000, and the UNESCO representative suggested it was something the Creative Cities Network could look into (Wurzburger *et al.*, 2009, p. 44).

Santa Fe, population 84,683 (US Census Bureau, 2019), is known for its history, arts, and culture in a southwestern US state that shares a border with Mexico. As a crossroads for Native American, Spanish, Mexican, and European culture, plus a beautiful desert and mountainous landscape, tourism is one of New Mexico's primary industries. The arts are also a significant economic driver for Santa Fe and the state, from Native American arts to important painters such as Georgia O'Keeffe. Santa Fe is the site of several important art markets, and young though it is at only 400 years, the city has long attracted artists and art buyers.

As a tourism destination, Santa Fe frequently receives awards and accolades. Among many received in 2020 was a third-place ranking on *Travel + Leisure* readers' top 15 cities to visit in the United States (Clemence, 2020). The Western States Arts Foundation's Creative Vitality List (WESTAF) ranked Santa Fe as the #1 city among the 30 most creative US cities with a population under 500,000. WESTAF also noted that Santa Fe has the most creative economy gig workers of any city on the list (WESTAF, 2020).

Back in 2006, the City of Santa Fe hosted a small meeting of the nine members of the UNESCO Creative Cities Network to prepare for hosting a conference. The Santa Fe International Conference on Creative Tourism was held in 2008, in collaboration with the UNESCO Creative Cities Network. Creative tourism was discussed as a means of supporting artists economically and capitalizing on the interest generated by being a Creative City. Ultimately, the essence of Santa Fe's definition was that experiences should have a hands-on aspect and give insights into the local culture. The participants believed it was important that creative tourism offerings were designed and offered by, or in collaboration with, the local creative community.

The Creative Tourism Initiative, led by the City of Santa Fe, developed a robust selection of creative tourism experiences and promoted them. The City assigned staff members of its Arts Commission, the city's arts agency, to design and implement the programme. The primary resources used were:

- artists and other creative businesses with teaching capacity (the creative economy gig workers noted in the WESTAF Creative Vitality List);
- a co-ordinating and promotion partnership between the City of Santa Fe Arts Commission and the Santa Fe Convention and Visitors Bureau (CVB); and
- City of Santa Fe funding.

The investment in the 2008 conference helped launch the initiative. Artists and creative businesses involved in the tourism industry had been researched for the conference, providing a database

Corresponding author: sabrina@svprattcs.com

 Creative Tourism: Activating Cultural Resources and Engaging Creative Travellers (eds. N. Duxbury, S. Albino, and C. Pato de Carvalho)
DOI: 10.1079/9781789243536.0022

of 50 entities (Wurzburger *et al.*, 2009, p. 211). Many creative tourism workshop presenters were identified in the course of planning the conference sessions. Branding, graphic design work, and a website developed for the conference were adapted for the initiative.

In keeping with its mission to provide leadership in support of arts and cultural affairs, the Santa Fe Arts Commission drove the Creative Tourism Initiative (CTI). CTI offered an opportunity to promote an additional income stream for artists. In the City's budget and service structure, the CTI fitted into a group of programmes that supported the health of the creative economy through grants to arts organizations; art commissioned for public spaces such as buildings, parks, and streets; and a community gallery showing the work of local artists. Many of these arts and culture programmes were supported by an innovative hotel tax model, which made the arts–tourism connection a natural fit.

CVB staff provided advice, research, and assistance with marketing the santafecreativetourism.org website through their santafe.org site and by using their public relations contractor to wrap the initiative in their existing tourism public relations (PR) strategy. Existing tourism budgets enabled the programme to take advantage of press releases and writing services.

CTI launched in 2009 with the support of an intern added to the small staff of the Arts Commission (later hired as a part-time contractor). His role was to implement the programme and act as the central communication point for multiple communication streams and outreach efforts into the arts community. Like the artists his role was to support, his duties were entrepreneurial – from product and small business development, to building and operating a robust website, to promotion and marketing.

Product and small business development

The website was started with 40 artists and institutions. It was the job of the intern/contractor to locate more existing workshops and recruit more artists who were interested in expanding their businesses by offering creative tourism experiences. The City's support and resources helped develop and broaden the catalogue of creative tourism experiences to 235 in 2015 (Hanifl, 2019, p. 116).

Challenges included coalescing geographically dispersed workshops under the umbrella of a marketing and promotion programme. Other practical considerations were whether signage was needed at the experience locations, if a sign-up system was needed, and whether a regular schedule was important.

The City designed training programmes to help interested people develop creative tourism businesses. All programme participants were offered education on social media, search engine optimization, and internet marketing. In 2012, there were nine sessions on nine different topics. The training sessions were offered free or at a minimal charge.

Collaborations with other organizations and businesses brought in diverse expertise and kept the training programme fresh. WESST, a small business development and training organization that works with small businesses across New Mexico, provided some workshops. The Global Center for Creative Entrepreneurship, now known as Creative Startups, also lent their expertise.

To further develop the skills of creative tourism business owners, one-on-one consultations on internet marketing were offered to anyone interested in participating in the programme. Giving participants these skills helped strengthen the small businesses. It also helped grow the initiative's online presence and drove traffic to santafecreativetourism.org.

Website

The website provided a point of connection between tourists and creative tourism experiences. To reach the widest market of potential visitors to Santa Fe, it was co-ordinated with the primary city tourism website, santafe.org. The fact that the co-ordinating body for this effort was the City of Santa Fe meant that various resources were available. One was an in-house videographer, who made short videos of the artists demonstrating their work. The CTI contractor was very active with a blog and social media.

Promotion and marketing

The contractor was also responsible for developing and implementing promotion and marketing strategies. Some of the components of this work were to:

- identify and promote to new markets;
- determine and implement methods of selling workshops to tourists beyond website activity;

- develop special events with the artists;
- develop print advertising such as rack cards and distribute them;
- network with the tourism industry to inform them of creative tourism efforts; and
- create an evaluation plan and maintain monthly records of creative tourism activity.

An example of a seasonal promotion activity was a spring break vacation promotion with local hotels. Hotels set up tables in their lobbies, and members of the initiative signed up to hold demonstrations and sell their work. This was designed to bring attention to creative tourism at a time when hotels were fully booked, and families might require new engaging activities.

A diverse array of creative tourism experiences was promoted throughout the 2009–2015 programme time period. As can be seen in the images on the 'Santa Fe Workshops, No Experience Necessary!' rack card (Fig. 22.1), they included painting, photography, dancing, drumming, pottery, traditional tin smithing, printmaking, and cooking. Among this array, two creative tourism businesses stand out because of their longevity and purpose as creative, experiential services for visitors: the Santa Fe School of Cooking and the Santa Fe (Photographic) Workshops.

'Our success is because of our celebration of the local foods of the area', stated Nicole Ammerman, director of the 31-year-old Santa Fe School of Cooking, 'We focus on the cultural influences of the food, including the Spanish, Native American, and Mexican influences, and when you've mixed all the ingredients together, we call it contemporary Southwestern cuisine' (Ammerman, personal communication, 9 September 2020). Hands-on cooking classes and demonstration classes with food tasting draw visitors looking for a unique, communal experience Fig. 22.2). The retail shop features regional cooking supplies, condiments, dishes, and cookware. Walking tours visit downtown restaurants, and the cooking classroom is used for corporate gatherings, entertaining, and charitable fundraisers.

The Santa Fe Photographic Workshops, also established 30 years ago, has evolved into the Santa Fe Workshops and diversified from photography to include a Writers Lab and other art forms (Fig. 22.3). Santa Fe Workshops and Santa Fe School of Cooking are both business models that employ members of the creative gig economy to teach their classes and workshops.

Today, many individual artists continue to include workshops and demonstrations as part of their small business model. Artists teach encaustic painting, glassblowing, pottery making, skateboarding, and many other interesting topics. Some of the linkages to tourism are overt, such as the glass shop located at a tourist-focused handmade craft complex. Others may appeal to visitors but are found by chance. Opportunities include brief workshops for novices available with little advance planning as well as multi-day or week-long workshops for those seeking more in-depth instruction. The many individuals operating these businesses need support to generate leads for bookings, which is no longer provided by city resources.

A loss of financial and other resources ended the CTI in 2015. Randy Randall, Executive Director of Tourism Santa Fe as of 2020, does not use the term 'creative tourism'. He does, however, acknowledge that millennial generation travellers want an experience. He also believes that these travellers are interested in buying items that have meaning to them, so they might take a pottery class to get an understanding of the process and cultural context, appreciate the art form, and purchase a piece from the teacher that they then put on display in their home (Randall, personal communication, 1 September 2020). To serve these visitors, the city's marketing includes itineraries with hands-on experiences, and santafe.org features 'Classes and Workshops' listed in the 'Things to Do' section. A 2020 innovation is an extensive list of virtual experiences for visitors and locals.

However, a website per se does not connect creative tourism businesses to tourists, nor does it drive development and expansion of programmes. An artist who was interviewed was not aware that her workshops were listed on santafe.org and was certain that she does not get business from the listing. She sells her workshops through word of mouth from former clients, walk-in traffic, referrals from art galleries, partnerships with other artists, and referrals from high-end hotels that have guests looking for things to do. Social media platforms such as Instagram offer a more active way for artists to promote themselves in coordination with their websites, and may connect temporary visitors to classes, workshops, and experiences quite effectively.

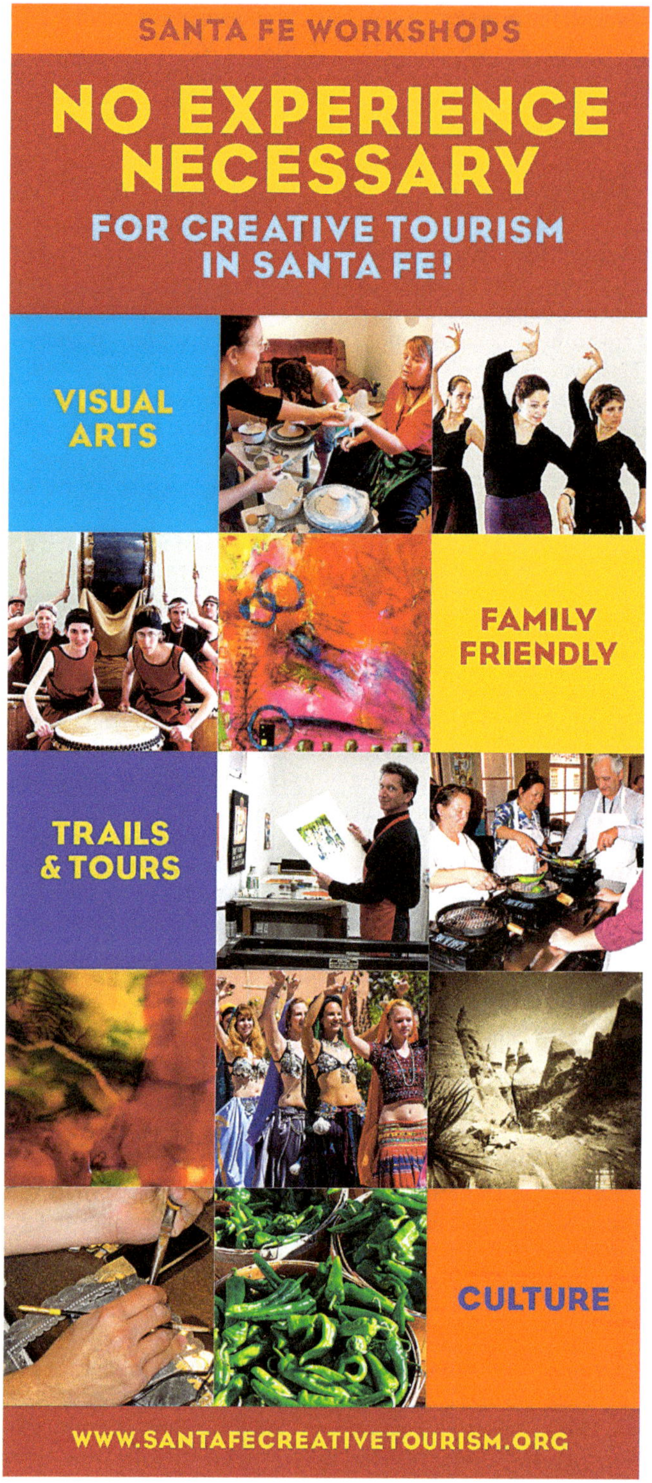

Fig. 22.1. 'Santa Fe Workshops, No Experience Necessary!' rack card. Photo by Michael Motley, used with permission.

Fig. 22.1. Continued.

Fig. 22.2. Santa Fe School of Cooking Chef de Cuisine, Noe Cano, sharing tips on roasting green chile. Photo by Eric Swanson, used with permission.

Fig. 22.3. A Santa Fe Workshops photography workshop in progress near Santa Fe, New Mexico. Photo by Stafford Squier, Santa Fe Workshops, used with permission.

Whether the CTI businesses could have collaborated to sustain a joint marketing and promotion programme is an unanswered question. Twelve years later, companies such as Airbnb and Travel Zoo are better positioned and resourced than a local government-driven initiative to advertise 'experiences' to their customers. Though perhaps an efficient way to sell creative tourism, large booking fees may be a deterrent for small creative businesses with slim profit margins.

My recommendations for a successful creative tourism programme are:

- Establish public–private partnerships:
  - Consider seed funding from government or grant sources.
  - Once the programme is established, consider charging the creative tourism businesses to be part of the website and for joint promotions. This revenue stream would sustain the programme when other funding sources dry up.
  - Collaborate with tourism industry partners such as hotels, restaurants, and tour companies to increase visibility and sales for all partners.
- Budget for strong social media expertise and networking.
- Encourage the city's tourism promotion entity to lead the initiative or co-ordinate the experiences to be integrated into tourism promotion.
- Build a community of creative tourism businesses and grow active leadership from within the community.
- Include training on small business issues and how to promote to visitors.
- Provide a central location if possible, so businesses without their own space have a place to offer workshops.

References

Clemence, S. (2020) The top 15 cities in the United States. *Travel and Leisure*, 8 July. Available at: www.travelandleisure.com/worlds-best/cities-in-us (accessed 1 September 2020).

Hanifl, B. (2019) Creative tourism in Santa Fe, New Mexico. In: Duxbury, N. and Richards, G. (eds.) *A Research Agenda for Creative Tourism*. Edward Elgar Publishing, Cheltenham, UK, pp. 113–121.

United States Census Bureau (2019). QuickFacts: Santa Fe, New Mexico. Available at: www.census.gov/quickfacts/santafecitynewmexico (accessed 22 September 2020).

Western States Arts Foundation (WESTAF) (2020) The 30 most creative small cities. *Creative Vitality Suite*, 29 May. Available at: https://cvsuite.org/cvlist-homepage/the-30-most-creative-cities/#06 (accessed 1 September 2020).

Wurzburger, R., Aageson, T., Pattakos, A., and Pratt, S. (eds.) (2009) *Creative Tourism, A Global Conversation: How to Provide Unique Creative Experiences for Travelers Worldwide*. Sunstone Press, Santa Fe, New Mexico, USA.

23 Creative tourism development and innovation in Kenya: An entrepreneurship perspective

Lénia Marques

Erasmus University Rotterdam, The Netherlands

As the field of creative tourism is becoming more established, studies on the subject have been growing exponentially in the last decade. This growth has also expanded the range of conceptualizations of creative tourism, leading to a landscape of interpretations (Richards, 2011, 2018; Duxbury and Richards, 2019). On the one hand, this diversity has enriched the field; on the other hand, the concept seems to be subject to the pressures of its popularity, which leads to disparate forms of understanding. One positive aspect of this array of interpretations around the concept of creative tourism relates to its flexibility and adaptability, which allows it to fit into different contexts. This fluidity can be a great advantage since it embraces a global activity with local expressions. It is in this sense that the boundaries and interpretations of what creative tourism is and can be are different, as they can be adapted to local contexts.

This is particularly important when considering creative tourism in African developing countries. Despite the recent boom in creative tourism literature, Africa as a geographic and relational territory plays only a very small role. In the African continent, South Africa has been the most studied destination from a creative tourism and creative industries perspective (Rogerson, 2006, 2007; Booyens, 2012, 2015; Booyens and Rogerson, 2015, 2016), with the occasional other study on Mali (Marques, 2012), Kenya (Langejan, 2018), or to a certain extent Nigeria (Andrew and Andrew-Essien, 2013) and Zambia (Seshamani, 2014). In such a diverse continent, with wide cultural diversity, storytelling, sense of community, and craftmanship, further studies are needed to understand creative tourism in non-Western contexts. In order to do so, understanding the structural role of tribes in the African context, both in their communities and in relation to visitors, is fundamental to gain insight on the creative tourism initiatives and developments in the continent. In East Africa, Kenya is one of the most visited destinations, and the Maasai tribe in particular has been used to successfully attracting visitors. The question remains, however: How does creative tourism come about in such a particular context?

This chapter is a contribution to expand the body of knowledge of creative tourism in Africa by providing some insights on creative tourism in Kenya. It reports briefly on research performed in 2018, which had the aim of gaining a deeper understanding of creative tourism in Kenya, starting off from the supply perspective. Fourteen interviews with entrepreneurs of different type of creative tourism related-products and tourists were held in Kenya, mainly around the region of Nairobi. The aim of this chapter is to provide an overview of entrepreneurial innovations in creative tourism using Kenya as a case study, in particular creative tourism activities developed with(in) the Maasai tribe.

Kenya and creative tourism

Kenya is becoming a popular international tourism destination in Africa, with tourist numbers increasing, most significantly within the African market.

Corresponding author: marques@eshcc.eur.nl

DOI: 10.1079/9781789243536.0023

The Maasai people are one of the country's most known assets, with their culture, beliefs, traditions, dance, music, crafts, and food. This popularity has also been presenting some challenges, such as the exoticization of the Maasai or the relations of power, in particular within Western imaginaries and discourses (Toogood, 2012).

As with other African ethnic groups, as well as with other local traditions in Western countries, the Maasai have also been involved in performances, making them a clear example of staged authenticity (MacCannell, 1973) – performed, as well as perceived. Although these activities are integrant part of their culture, the enactments for tourists started as a response to a waning agricultural lifestyle. The regular touristic enactment of their traditions, such as costumes, music, and dances, has become an important alternative source of income for Maasai families and communities.

In particular, Western tourists travelling to Kenya wish to get acquainted with different (*exotic*?) cultural forms. Tourists are increasingly motivated to have a meaningful, 'authentic', 'unique' experience while travelling, leading to more interaction and co-creation. Although built cultural heritage is still very important and provides the geographical site to visit, many of the cultural expressions that tourists are increasingly interested in relating to intangible heritage. The way to relate to the Maasai's intangible heritage can be done in different ways and with different levels of engagement. Viewing an enactment of rituals is a more passive form of getting acquainted with unfamiliar practices, traditions, beliefs, and rituals. This could be considered a form of creative tourism in the sense that there is an active search for culture under a creative form, both for the producer and consumer, with a sense of embeddedness caused by co-presence in a meaningful place (where the local culture is situated). It can be argued, however, that the learnings related to such performance do not make a long-lasting impression, or profoundly transform the creative tourist. For deeper levels of engagement and active local cultural learnings and bonds, interaction and co-creation should be enhanced, and therefore new forms of (creative) tourism should emerge to allow this to happen.

In the context of the Maasai, innovations in creative tourism have been happening, not only in an informal way, as Marques (2012) pointed out in the case of Mali, but also in a more structured and formal manner. Entrepreneurs have been emerging who work with innovative approaches to creative tourism related to the Maasai culture. Barriers and sometimes tensions exist, as it is a big challenge for entrepreneurs to convince local actors that there is more touristic potential beyond exoticized enactment.

Creative tourism entrepreneurs and innovation

In this context, the question then becomes 'How can innovation in creative tourism be introduced in a successful and sustainable manner, that contributes not only to the economic welfare of people, but also to their social and cultural empowerment?'. Approaches to innovation are manifold, and very often they focus on organizations. However, they also apply to entrepreneurs. If creativity might be of abundance within creative tourism, innovation provides the necessary competitive advantage to survive in competitive markets.

Innovation can be perceived in a range of ways and relate to different practices. As Van de Ven and Engleman (2004) state, 'an innovation is a new idea that may be a recombination of old ideas, a scheme that challenges the present order.... As long as the idea is perceived as new to the people involved, it is an "innovation", even though it may appear to others to be an 'imitation' of something that exists elsewhere' (p. 48). Baregheh *et al.* (2009) go further, emphasizing not only an idea, but also its implementation: 'Innovation is the multi-stage process whereby organizations transform ideas into new/improved products, service or processes, in order to advance, compete and differentiate themselves successfully in their marketplace' (p. 1334). In the case of entrepreneurs, there is not necessarily a formal organization, although there are relationships (a network) that often provide a starting point.

In tourism and leisure, innovation has been seen through different lenses (e.g. Hjalager 2010; Marques and Biscaia, 2019; Booyens, 2015; Booyens and Rogerson, 2016, 2019; Rodriguez *et al.*, 2014; Hall and Williams, 2008). As pointed out by Marques and Biscaia (2019), following on Hjalager (2010), innovation can be of different natures: product or service innovation, process innovation, managerial innovation, management innovation, and governance innovation (see Table 23.1). These types of innovation are not mutually exclusive. On the contrary, they often go hand-in-hand with each

Table 23.1. Innovation in creative tourism. Adapted from Marques and Biscaia (2019) and Hjalager (2010)

Innovation type	Description
Product/service innovation	A new product is invented or put together. A new service is put into the market. It can also be a new addition or new way of presenting a certain product or service.
Process innovation	Innovations brought into activities that are on the backstage of a product/service offered, for example, related to access and logistics.
Managerial innovation	Related to human resources, mainly internal collaboration and knowledge sharing.
Management innovation	Changes and experiments in different areas of management, such as organizational design, strategy, marketing, finances, leadership, and human resources management.
Governance innovation	Innovative ways of governing leisure at central, regional, and local level. This can include decision-making in terms of infrastructures or policy.

other. Innovation occurs within a certain context, and it is dependent on contextual factors such as culture and place. In the Kenyan context, specific tribal culture, such as the Maasai, together with geographical and territorial characteristics play an important role in the way creative tourism entrepreneurs innovate.

Entrepreneurs and innovation in creative tourism in Kenya

From the findings, several types of innovation that support the creation and development of creative tourism in Kenya emerge. These innovations are important even when they are micro-innovations and are extremely simple, as they can have a real impact on individuals and their communities. Creative tourism is a more sustainable and responsible form of tourism, contrary in many ways to mass tourism, because it functions on a small scale, it focusses on local actors, and, in this ethnic context, it requires community commitment and involvement. Creative tourism presents opportunities for innovations with real impact for a number of people. However, it is not always that simple, since countries like Kenya and many other developing countries face challenges in terms of wealth distribution, including income from tourism products (Kareithi, 2004; UNCTAD, 2013), with high levels of corruption and also foreign companies leaking benefits out of the country. The commitment to the community depends, in the last instance, on the individual and their professional ethics. As one interviewee mentioned, 'if everything goes well it would be an awesome project, it would be a completely new way of travel you know. Where you can make the tourism beneficial to the country upliftment because we can involve local communities. So, this is a way of community development, way of training and sharing knowledge' (interviewee 7).

In order for that to happen, some conditions are necessary. In the interviews, three major challenges for the development of creative tourism in Kenya emerge: (i) innovative and creative thinking; (ii) appropriate skills; and (iii) distribution, access, and communication. These challenges are most likely be similar in other regions of Africa and of the world. Responding to these challenges becomes, therefore, the key success factor to develop creative tourism. Innovation can contribute to address these challenges in a range of different ways, such as exchange of skills, synergies, bridging benefits, or access to markets. The different types of innovation responding to these challenges create the necessary competitive advantage to survive and thrive in a competitive market.

A range of innovation types becomes apparent in the Kenyan-Maasai context which relate, firstly, to product/service development. Product or service innovation is the most immediate and usually the first step for a creative tourism entrepreneur. Several interviewees mentioned different innovative situations, such as, for example, the start of a business offering personalized packages or platforms (by Westerners living in Nairobi); tours in a Nairobi slum, with local stakeholders and stories; tours to remote places, quite inaccessible without local knowledge; and applying new means of transportation to an existing product.

Innovations can also be made in terms of the processes involved in making, distributing, and co-creating the product. In creative tourism entrepreneurship, co-operation is an idea that often surfaces. This can be made between entrepreneurs (Kenyans and foreigners; from different communities or within the same community, etc.), or between the tourist and local creative entrepreneurs. Another important aspect of process innovation relates to the introduction and use of technology, in particular, digital environments and social media.

Managerial innovation also happens through skills development, although, according to our findings, this occurs mainly in an informal way. It becomes clear that for many interviewees, involving and training other people in the community as the business evolves is part of their work and living ethos. Management innovation does happen, however, in a more subtle way. Business models and vision are words that gain relevance mainly in the discourse of Westerner entrepreneurs.

Contrary to the other types of innovation, entrepreneurs perceive a lack of governance innovation. In fact, the very few comments on governance structures and organization were negative, pointing to the leak of resources outside of the places which attract tourists as well as problems with corruption. The gap between innovation that depends directly on the entrepreneurs and that of governance structures seems to be quite substantial.

In the Kenyan-Maasai context, innovation is, for the most part, inherent to the creative tourism entrepreneurs' activities and has allowed them to open different doors for business. In this context, innovations do not come alone. Along with innovativeness, some necessary elements seem to be common and recurrent: collaborations, knowledge, skills, synergies, and empowerment.

What's next?

As with other initiatives around the world, discussions and studies about creative tourism should take into account that creative tourism started as a Western concept, for Westerners, about a certain more 'exotic' (different, unfamiliar) reality. Therefore, a critical and cautious approach is needed when trying to frame different cultures for the Western gaze, as the African tribal context might be.

With this in mind, it is clear that, as in many other countries in the African continent, creative tourism in Kenya has immense potential. Creative tourism can provide a partial answer to the challenges faced by communities in which tradition and local embeddedness play an important role in their sustainability. Similar to previous studies, this research suggests that the connection to the community is central, even more so in African countries than in Western contexts. This is mainly seen as an advantage, although it can also present additional obstacles to creative entrepreneurialism in tourism, as often it depends on individuals and it involves some form of innovation. The processes tend therefore to take time. In Kenya, creative tourism initiatives have been developing in recent years, bearing witness to the innovative capacities of entrepreneurs, either originally from the local communities or foreigners who have chosen to establish themselves in Kenya. Creative tourism is taking its first infant steps in the country and in the African continent, and the dynamic relation among creative tourism entrepreneurs has been slowly evolving.

These creative tourism entrepreneurs have identified opportunities and implemented innovation at different levels. In this framework, even what could seemingly be considered a small innovation might have a significant local impact and open new doors for development. With the potential resources that Kenya and the Maasai communities have, innovations in terms of product are seemingly the most immediate. However, these can hardly occur without process and managerial innovation, in particular, related to capacity building. To become more structural, managerial and governance innovations are important for the future of creative tourism because they facilitate these processes and provide the space for continuous innovation to occur. In the case of Kenya, and more specifically of some Maasai communities, creative tourism activities have a significant local economic impact. They enable certain community values and practices to endure in time, while allowing the community to develop tourism activities beyond framed ritualistic enactment.

Creative tourism is a more sustainable way of travelling based on local values, cultures, and lifestyle (Richards and Marques, 2012). Sustainable, and yet very fragile. While creative tourism presents an opportunity for contributing to the UNESCO's sustainable goals as well to alleviating poverty (UNCTAD, 2013), it is not a panacea to all problems – not on paper, and not in the field. Facilitating entrepreneurs to innovate together and with the tourists in processes of value (co-)creation can form a solid platform for local, regional, and even national development.

Acknowledgements

The author would like to express her gratitude to Simone Langejan for sharing her insights on Kenya, as well as being a critical reader of this chapter. The chapter would not have been possible without the collaboration of Charlie's Travels and the kind openness of all the other creative entrepreneurs interviewed – to all, a big thank you.

References

Andrew, E.H. and Andrew-Essien, E. (2013). Arts, tourism and the environment: An overview of critical intersections in Nigeria. *Mediterranean Journal of Social Sciences* 4(6), 275–275.

Baregheh, A., Rowley, J., and Sambrook, S. (2009) Towards a multidisciplinary definition of innovation. *Management Decision* 47(8), 1323–1339.

Booyens, I. (2012) Creative industries, inequality and social development: Developments, impacts and challenges in Cape Town. *Urban Forum* 23, 43–60.

Booyens, I. (2015) Innovation and networking in tourism for the competitiveness of the Western Cape regional tourism economy. PhD thesis. University of Johannesburg, Johannesburg, South Africa.

Booyens, I. and Rogerson, C.M. (2015) Creative tourism in Cape Town: An innovation perspective. *Urban Forum* 26(4), 405–424.

Booyens, I. and Rogerson, C.M. (2016) Unpacking the geography of tourism innovation in Western Cape Province, South Africa. In: Szymańska, D. and Rogatka K. (eds.) *Bulletin of Geography. Socio-economic Series*, no. 31, Toruń: Nicolaus Copernicus University, Poland, pp. 19–36, doi: 10.1515/bog-2016-0002.

Booyens, I. and Rogerson, C. (2019) Creative tourism: South African township explorations. *Tourism Review* 74(2), 256–267, doi: 10.1108/TR-12-2017-0200.

Duxbury, N. and Richards, G. (2019) Towards a research agenda for creative tourism: Developments, diversity, and dynamics. In: Duxbury, N. and Richards, G. (eds.) *A Research Agenda for Creative Tourism.* Edward Elgar Publishing, Cheltenham, UK, pp. 1–14.

Hall, C.M. and Williams, A.M. (2008) *Tourism and Innovation*. Routledge, London, UK.

Hjalager, A.M. (2010) A review of innovation research in tourism. *Tourism Management* 31(1), 1–12.

Kareithi, S. (2004) Tourism and development: Using tourism as a strategy for poverty reduction in Narok District, Kenya. PhD Thesis. University of Luton, UK. Available at: http://uobrep.openrepository.com/uobrep/handle/10547/293249 (accessed 22 September 2020).

Langejan, S. (2018). An inspiring symbiosis: value co-creation as a powerful tool in the development of creative tourism. MA Thesis, Cultural Economics and Entrepreneurship. Erasmus University Rotterdam, Rotterdam, The Netherlands.

MacCannell, D. (1973) Staged authenticity: Arrangements of social space in tourist settings. *American Journal of Sociology* 79(3), 589–603.

Marques, L. (2012) Boosting potential creative tourism resources: The case of Siby (Mali). *Journal of Tourism Consumption and Practice* 4(2), 111–128.

Marques, L. and Biscaia, S. (2019) Leisure and innovation: Exploring boundaries. *World Leisure Journal* 61(3), 162–169. doi: 10.1080/16078055.2019.1639257

Richards, G. (2011) Creativity and tourism: The state of the art. *Annals of Tourism Research* 38(4), 1225–1253.

Richards, G. (2018) Cultural tourism: A review of recent research and trends. *Journal of Hospitality and Tourism Management* 36, 12–21.

Richards, G. and Marques, L. (2012) Exploring creative tourism. *Journal of Tourism Consumption and Practice* 4(2), 1–11.

Rodriguez, I., Williams, A., and Hall, C.M. (2014) Tourism innovation policy: Implementation and outcomes. *Annals of Tourism Research* 49, 76–93.

Rogerson, C.M. (2006) Creative industries and urban tourism: South African perspectives. *Urban Forum* 17, 149–166.

Rogerson, C.M. (2007) Creative industries and tourism in the developing world: The example of South Africa. In: Richards, G. and Wilson, J. (eds.) *Tourism, Creativity and Development*. Routledge, London, UK, pp. 229–239.

Seshamani, V. (2014) The business of creativity and creative arts for business: Perspectives on the potential for Zambia's tourism promotion. *Journal of Tourism and Hospitality Management* 2(4), 155–160.

Toogood, M. (2012) Placing the Maasai. In: Convery, I., Corsane, G., and Davis, P. (eds.) *Making Sense of Place*. Boydell, Woodbridge, UK, pp. 261–270.

UNCTAD. (2013) *Creative Economy Report 2013, Special Edition: Widening Local Economic Pathways*. United Nations, Geneva, Switzerland.

Van de Ven, A.H. and Engleman, R.M. (2004) Central problems in managing corporate innovation and entrepreneurship. In: Katz, J. and Corbett, A.C. (eds.) *Advances in Entrepreneurship, firm Emergence and Growth, Vol.* 7. Emerald Group Publishing Limited, Bingley, UK, pp. 47–72.

Part IV: Embedding creative tourism activities within a community

Part IV presents contributions from a variety of contexts on how creative tourism organizers can promote resident engagement and work with their broader community. The chapters provide insights from different geographical perspectives and both academics and practitioners. The two 'core' chapters (the longest contributions) are based on the authors' experiences in Brazil and Namibia, with three shorter chapters featuring contributions based on situations in Spain, Denmark, and Portugal. A community-embedded approach to creative tourism can provide the 'link to locals' that the new generation of travellers is craving. It helps foster meaningful community involvement in conceiving, designing, and implementing creative tourism initiatives, and ensures a wider distribution of the benefits of creative tourism in the community. More broadly, community-engaged creative tourism initiatives can generate new perspectives on a locale's links with the broader world and, as Melanie Sarantou and Satu Miettinen note, can be applied as a strategy for inclusion, participation, and decolonization.

In 'Creative tourism as a local development strategy' (Chapter 24), Larissa Almedia outlines the community-embedded approach to creative tourism she has developed over 15 years' experience of designing and implementing community-engaged development projects in peripheral communities in north-eastern Brazil. Placing the community's involvement and well-being centrally in these initiatives, Almeida details the planning and operational components that are essential for the development of successful creative tourism initiatives and discusses some of the challenges she has encountered. The chapter presents a thorough description of the different steps that are part of her process of creating community-embedded creative tourism products in peripheral communities: mapping potential attractions and intangible assets, forming a working group and aligning expectations, designing a training and qualification route, product development, and training local agents for the market. The chapter calls attention to the importance of offering open and accessible training to all interested community members on how to develop creative tourism products, the inclusion of local residents in planning and decision-making processes, and making efforts to link creative tourism to local development strategies and governance. Highlighting the human potential embodied in peripheral and vulnerable communities, the author notes that they frequently suffer from negative external perceptions. The chapter points to ways in which community-based creative tourism can transform the image and potential of these communities.

In Chapter 25, 'Why creative tourism won't work if residents are not involved: A tale of two cities', Kathleen Scherf highlights the need to develop relationships between visitors and residents in order to develop a sustainable tourism situation. The chapter presents us with two very different experiences in Europe: the El Raval neighbourhood of Barcelona, Spain, and Copenhagen, Denmark. The Barcelona example is embedded in a context where over-tourism is a reality and, according to the author, the idea of visitors engaging with local residents is almost a mirage. The results of Scherf's surveys of residents and visitors in Barcelona's El Raval neighbourhood regarding relationships between residents and visitors are sceptical of integration. With this situation in mind, Scherf provides a series of suggestions on possible measures to embed tourist activities within the community. The Copenhagen experience, a programme that aims to engender and promote a new approach to 'localhood' that can help build bridges between resident and tourists, is introduced in this context. This programme promotes shareable moments between locals and visitors and acknowledges the importance of collective local knowledge to the creation of a more inclusive situation that aims to avoid unsustainable over-tourism in the city.

Next, further reflecting on this relationship between local residents – especially local youth – and creative travellers, Fernando Almeida points to the generative potential in fostering and enabling these connections in a context of contemporary communication flows and influences. In Chapter 26, 'Youth, ways of production, and community', Almeida brings attention to the emergence of the virtual global community, a situation that offers new possibilities for dialogue and the expansion of social relationships in new ways. This global-scale community is generating participatory processes in which local communities are being reframed. In this context, the author introduces a creative, community-engaged, and participatory artistic festival, Encontrarte Amares, which stimulates dialogue through cultural participation. In this biennial event, viewed as a type of creative tourism initiative, artistic activities promote spaces of co-creation between arts collectives and artisans in Amares, enabling intersections between contemporary artistic thought and traditional local heritage, intergenerational efforts, and linking the local to the international scale.

In 'Creative tourism as a strategic approach for decolonial thinking and doing in Namibian tourism' (Chapter 27), Melanie Sarantou and Satu Miettinen point to important implications of sensitively conceiving, designing, and implementing community-engaged creative tourism initiatives. The authors examine and critically reflect on how creative tourism can be applied as a strategy for inclusion, participation, and decolonization, and thus contribute to sustainable community development. In line with the critical turn in tourism, which promotes the decolonization of tourism research, this chapter suggests that creative tourism, narrative, and creative practices can enable such a turn to be practically implemented through the inclusion and participation of local communities. According to the authors, creative tourism (by rethinking and redoing tourism products and services) can promote participatory, tolerant, and inclusive collaboration within tourism communities, thus counteracting dominant and implicit narratives. The chapter draws on two Namibia-focused doctoral thesis projects (linking to ongoing research): one investigated creative tourism in Namibia, the other mapped postcolonial Namibian craft and design narratives and identities. Both research projects analysed, from different perspectives, how crafts production can support decolonial thinking. The case studies discussed in the chapter take place in Grashoek and Penduka in Namibia. The cases demonstrate the capacity for resistance through designing creative tourism initiatives to articulate alternatives to dominant narratives, the possibility of innovation in products based on local knowledge, and the importance of focusing on the impact of creative tourism in the sustainability of local communities.

Closing Part IV, Fiona Bakas' chapter 'Connecting local artisans to tourism in extra-metropolitan areas: The importance of social embeddedness' (Chapter 28) presents a closer look at the entrepreneurial agents who organize and propel creative tourism initiatives in many smaller communities. The chapter is based on recent research into the activities of artisan entrepreneur-mediators in small cities and rural areas of Portugal, conducted within the CREATOUR® project. The artisan entrepreneur-mediators take on leadership roles in identifying local resources and people, producing and locally managing activities, and investing in social relations. The research revealed that the success of these artisan entrepreneur-mediators is linked to their social embeddedness in the locale in which the creative tourism initiatives are organized, drawing attention to the importance of community engagement and trust as fundamental for the success of the artisan entrepreneur-mediator. Allied with the insights in Larissa Almeida's opening chapter about strategies she has taken to build trust with community members, Bakas suggests an array of actions that entrepreneurs can take to foster their social embeddedness in the community.

— N.D. and C.P.C.

24 Creative tourism as a local development strategy

LARISSA ALMEIDA

Recria – Brazilian Experiences and Creative Tourism Network, Brazil

Creative tourism can be a platform for many purposes: as you change your objective and the elements in it, you obtain different results. Using creative tourism as a competitive advantage, one can provide clients with a memorable experience whose result is an enthusiastic, engaged promoter. It can be used as a strategy to teach many abilities to people at any age, developing creative capacities as well as notions of history, language, biology, and even teamwork abilities, empathy, and other social skills. A creative tourism strategy can also be used to improve innovation in some places, encouraging people to take part in creative activities throughout the city and providing another perspective about the locale. I believe that, in any application, creative tourism can promote transformation, but when it is employed for local development, the potential of this platform blossoms.

According to the Brazilian Institute of Spatial Geography (IBGE), the Brazilian population in 2019 was about 210 million people, of which approximately 13.5 million live in poverty conditions with monthly per capita incomes below US$60.00 or US$1.90 per day – an indicator of extremely poor living conditions according to the World Bank. The country's social reality is marked by a stark social inequality, low education of the population, and precarious labour relationships. However, there is also plenty of wealth: Brazilians' best-known festivals, such as Carnaval and the Parintins folklore festival, and Brazilian music and dances are the expression of people's identities, talents, stories, and skills. These reflect the country's cultural, environmental, and social richness – the fabulous outcomes of Brazilian creativity are truly remarkable.

As for tourism, the territories provide the raw material – their intangible elements – for the elaboration of tourism products aligned with contemporary consumer demand. Tourists nowadays are searching for deep and significant experiences and, therefore, there is a growing interest in narratives and atmosphere that helps to stimulate creativity. Tourism offers only make sense if they cause feelings and sensations. Creative tourism matches those demands and, at the same time, it does not require large infrastructure, thus presenting itself as a viable local development strategy. In this context, I must highlight the human potential embodied in cultural entrepreneurs, who seem to be the driving force behind community development. They act locally and are often the hosts of local culture, coordinating cultural groups and striving to keep traditions alive. However, due to low schooling and precarious living conditions, it is often difficult for them to understand how they can transform local cultural assets into wealth through creative tourism products, connect to networks, and generate local development. Such entrepreneurs lack training and easily understandable, applicable information.

Facing this scenario and acting professionally in the non-governmental sector since 2007, it disturbed me to observe so much wealth in the communities where I worked while, outside the peripheries, people only saw those places through the lens of poverty. The news media always emphasized their 'lacks'. This concern inspired me to launch a tourism company whose mission is to take people to these places and introduce them to the affluence of the peripheral territories. My idea was that people could then know the places through their wealth and, from that encounter, they could reframe their perceptions

Corresponding author: larissal.almeida@gmail.com

 Creative Tourism: Activating Cultural Resources and Engaging Creative Travellers (eds. N. Duxbury, S. Albino, and C. Pato de Carvalho)
DOI: 10.1079/9781789243536.0024

of those places and, especially, of the people who lived there. My dream has been to connect cultures and knowledge, and to break down the walls that separate the centre and the periphery. My image of the future is a scenario in which everyone is connected through their knowledge.

In 2020, my work with social development spans 14 years, including 6 years spent on social development using tourism as a tool. Over that time, I have worked in several communities that shared the characteristics of being peripheral, suffering prejudice because of poverty, and with a low level of urban infrastructure, but which also have great cultural diversity, people who work together, who help one another, and who believe in a positive future. I have found that tourism is a powerful tool to generate local development and to engage people collectively. Within this experience, I developed a framework that enabled the mobilization of people to transform cultural and social events into a tourist product without losing their originality and with a 'package' desired by the market. Although that has been the groundwork that I have used for 6 years and applied wherever I worked, I made no effort to systematize and measure the results, so I cannot label it as a methodology. Moreover, it is still under construction and subject to ongoing experimentation.

This chapter is based on my professional practice in interventions in peripheral communities in north-eastern Brazil. It aims to present a way to stimulate social development using creative tourism to transform cultural assets into products in low-income communities. It also discusses how this intervention can generate a business network that supports local development and reflects the socio-economic progress of the territories through visibility and as a basic dimension of dignity. The chapter presents a way of achieving this in which I strongly believe.

Creative tourism and territorial development

Although the concept of creative tourism itself does not contain sustainable local development as a premise, my experience shows that territories with low development rates harbour the most authentic manifestations, perhaps because creativity is a need in such spaces. I find it impossible to think about the development of creative tourism detached from local development strategies.

Local sustainable development is a strategy based on the principle that it is necessary to reinforce local potentialities in an integrated manner, creating the economic, social, and educational opportunities necessary for human development and ensuring that these development conditions are sustainable (Parente, 2014). Local development is impossible without human development, using capacities that people develop thanks to the material and human resources and the culture(s) of their territory, which stimulates economic development. Development is, therefore, a process of continuous transformation of the economy and society based on the expansion of the potential and creativity of individuals, which affects all types of territory, regardless of the level of development (Sen, 2000).

Distinct from local actions, local development acts upon the latent potential of the territory from the inside out in order to generate transformation. It is necessary to activate three dimensions: the development of social capital, the stimulation of productive activities, and the establishment of governance (Parente, 2014). Social capital is comprised of the elements that provide meaning to a community. It answers the question 'Why are we together?' It is about social ties and cultural identity; once you share identities and recognize yourself in others, you tend to trust. Therefore, social capital is like a 'glue' that maintains community connectedness. It is directly linked to developing human capital, which translates into educating people not only in terms of technical things such as business or a specific profession, but also improving their general knowledge in order to empower them as societal agents. The stimulation of productive activities is related to improving local products and processing and diversifying economic activities. The mobilization of stakeholders and public policies that can support small entrepreneurs and strengthen their business is part of this strategy to stimulate productive activities. Finally, establishing governance stimulates the establishment of an institution to discuss governance inside the community and articulate various stakeholders, such as state governments, economic enterprises, and civil organizations, in arrangements that favour the territory. The activation of governance is important as a way to guarantee the sustainability of the actions developed during the process.

Creative tourism stands as a real possibility for community development since its raw material is local culture, and its ideal operationalization involves people who live what they are willing to share. Touristic interest is facing a transition from the desire to visit great monuments to the desire to undergo great experiences; for that reason, any place has the potential to thrive in creative tourism (Richards, 2009). Narratives create 'stories' about people and places that make specific destinations attractive, and communication largely concerns cultivating and distributing this image.

For Zapata (2014), the territory's characteristics (cultural identity, memory, talents) are the raw material for creative tourism since they gather all the pre-existing elements in the territory and transform them into suitable products for delivery to visitors. It is a low-cost solution to foster local development and it can revitalize the community's self-esteem, stimulate economic development, increase knowledge by encouraging schooling, lessen prejudices, and generate integral development. There is a common perception that development is about economic metrics and market growth; however, seeking integral development is to consider strategies that improve elements besides objective economics, such as cultural, social, environmental, and political dimensions, and acknowledging the interdependence of all of these dimensions. Molina (2011) corroborates this vision, highlighting that creative tourism is oriented to creating, sharing, and expanding well-being; to the development of unique skills and values that generate income; and to re-shaping the space and relationships expressed in the community and turning its members into protagonists.

Creative tourism works as a platform that meets all the requirements to stimulate development. When creative tourism is used as a strategy, it is possible to invite different types of organizations to act on behalf of the locale for the benefit of the area. It unites the city, creatives, and travellers through exchanges forged on respect for people, for knowledge sharing, and for the place itself. It encourages the integration of several microentrepreneurs and increases financial in-flows, enhancing local sustainability.

A way to implement creative tourism

On a personal level, I have always been passionate about people and culture. As an entrepreneur in tourism, I wanted to present unique, surprising products. However, I am not originally from the places I want to show, and I worried about how to avoid turning it into a safari-like experience. I was looking for charming elements that could tell stories about the way of life in that place, and intended to introduce visitors to my understanding of culture and the sensations the city could produce. However, the things I wanted to show were hard to access and surrounded by prejudice.

As I mapped elements with the potential to be tourism products, I realized that turning them into marketable products would require a qualification process that prepared them for the market. So, intuitively, the work process was built, and the peculiarities of the work was developed as it progressed. Furthermore, considering that its purpose is to train autonomous operators, my constant concern in this work has been to include people into decision making within the process in a systemic and integrated way. Thus, the steps of the process involve mapping the potential elements that have attractiveness for tourism, both potential attractions and intangible assets; ensuring people are interested in engaging in the process, forming a working group, and aligning expectations; training entrepreneurs to deliver the product with quality through the development of a qualification route and product development process; and training local agents to commercialize and market it. Presently, some colleagues and I are experimenting with stimulating this work within a network to boost learning and improve opportunities (described later on).

Step 1: Identification of potential attractions

This step consists of mapping everything interesting about a territory. The kick-off step of the mapping is document research to discover those elements that are already known. We search within companies, government institutions, and communication agencies, and we also conduct an historical investigation to find any potential attraction, a step that demands a broad, attentive analysis. Subsequently, we visit points of interest such as squares, bars, restaurants, marinas, and any other element that is relevant for the conceptualization of an attractive product. In the mapping, we point out all interesting places, whether they are in operation or not, whether they are public or private. We also walk through the territory and talk to people to listen to them and find out if there are any points of local interest, regardless of whether those spaces are not recognized outside the community, if they are gathering places, or if there is a story behind them.

The second step of the mapping is to identify in the previous research which points are in operation or not. If they are in operation, we approach owners to learn about their interest in integrating a tourism action; if they are not in operation, we check if there is an intention or possibility of operation. When it is a public place without operational conditions, it is necessary to consult the responsible public institution to verify the possibility of revitalization and operation.

In the initial moments of this phase, it is particularly common that people show distrust and unwillingness to open their spaces for activities. Some people can be resistant, especially well-established entrepreneurs. Oftentimes, they seem to perceive themselves as sufficiently successful for the background they came from, and they do not perceive the proposed project as an opportunity for growth. In short, they do not want to be bothered. This group seems to be individualist, tend to lack a broad vision of growth, and are unwilling to dedicate time and engage in a collective project unless there is explicit personal benefit.

As results appear, however, people start to believe in the process and show openness to dialogue and possible partnerships – but which are not necessarily confirmed. The process maps and catalogues all initiatives, but only those in which decision makers are receptive to the process will have products developed. Those who do not display interest initially are not involved, but are continuously invited to participate in integrative moments, events produced by the group, lectures, and other project activities as attempts to engage them.

Step 2: Identification of intangible assets

The purpose of this stage is to discover the stories, and the *borogodo*[1] of the place. Here we seek the icons, the leaders, the articulators, the people relevant to the daily work of the place. We look for passionate residents who know the local history and can speak about the local identity while building a bridge between yesterday and today. To access the community core, a mediator is essential, whose role is to facilitate the relationship with local residents and encourage trust; we are outsiders and local people will protect their group as their first reaction. It is similar to the experience of an anthropologist: as much as they might live among Indigenous populations, they will never be regarded as one of them, and that is why a mediator is so important. Additionally, there is usually little time to establish a relationship of trust and the mediator lends us his reputation so that we can move through the community. The mediator is someone from the community who enjoys a good reputation and is reliable, who has access to people. However, they should not be a strong leadership, because generally those people usually face restrictions from some groups within the community and this can hinder the process.

Next, a sample of residents is selected, based on suggestions from the mediator and also from people on the street and in commercial establishments. To compile this group, we ask who they would recommend for a conversation, about interesting people, local leaders with whom we can talk, and so forth. We then conduct in-depth interviews with each selected person to understand what connects the person to that place and what stories they tell. At this point, we begin to understand the disputes, the oppositions, and the dominant groups in the community.

This step is especially important because it is when we establish relationships and build trust. We listen to people and we tell them briefly about our intentions. The local mediator plays an important role in building trust, as these connections are established based on their reference. The local agent is the key element for interviewees to feel comfortable talking. Mediators attend the interviews with the consultant, and continue to play this role throughout the process, as they are usually hired as assistants. Their activities include participation in the planning of actions (i.e. we consult them to know whether specific actions are viable due to community openness), they mobilize community participants, and they organize meetings. Ultimately, they are the consultant's ears and eyes during the process.

This mapping process ends when the nominees or stories begin to show repetition. At this point, it is possible to compare the information obtained in the initial mapping and the intangible mapping stages to speculate about the most interesting elements of the territory, and to identify those with greater potential to become a tourist attraction, the ones with most attractiveness, the easiest options in terms of operationalization, and so forth – the potential of the territory gradually takes shape.

Step 3: Formation of a working group and alignment of expectations

Once we know the territory, its nuances, and its peculiarities, and understand the local potential more closely, it is time to gather people and form a working group. Initially, we make an open call for anyone and everyone who wants to participate, whether they have already been contacted or not. In my experience, I have always entered the community with a particular interest in mind, which has been either the desire to develop an innovative tourism product that could be a competitive advantage or an initiative of a social or governmental organization that wanted to invest in local development. This stage is a moment of alignment between the interests of the proponent and the community (ensuring transparency in all steps of the process and keeping records during all of the intervention are important for the success of the project). This is the proper moment to assess the community's interest in receiving a project like this and, if so, how it will be developed.

As a consultant or entrepreneur entering a community, we must always keep in mind that we are neither the saviours nor the holders of all knowledge and answers. Before our arrival, those places have been through a lot of history and movements, and there is a knowledge and wisdom that must be considered. Keeping this in mind, we present the results of the two previous stages and the intervention plan that we have designed. After that, we listen to the wishes and expectations of the group assembled and confirm their interest to join the programme. Together, we make agreements on how the work will proceed and we conduct checkpoint meetings during the intervention to review the agreements and plans. It is important to highlight that financial exchanges with the community are not an adopted practice – there are no subsidies for residents to join the programme, so all participation is voluntary and unpaid. Once these agreements are defined, the technical team meets to draft the intervention plan.

Step 4: Creation of a qualification route

Any intervention focused on emancipation and sustainable development requires the training of those involved so they do not depend on whoever is proposing the intervention. Training is a crucial point for people to develop and become able to manage activities by themselves in the future with no tutoring from outsiders – this step aims to implement the social and human development mentioned earlier. The qualification route we design covers technical, social, and behavioural skills in an integrative way.[2] Each territory presents a different challenge, but technical aspects usually include topics such as customer service, group management, languages, administrative and financial management, event organization, public speaking, and sales. As for social skills, aspects such as empathy, communication, civility, networking, trust, collaboration, and issues related to behavioural skills are addressed, such as proactivity, creativity, planning, and organization, among others. The training content combines lectures, workshops, technical visits, a technical mission, and collective projects (see Table 24.1).

Collective projects have proven to be one of the most efficient ways to conduct training. Besides being a good teaching–learning strategy, it is an opportunity to attract and engage new people in the project and to involve everyone in the process. It also keeps people interested and builds trust. As they perceive positive results, the confidence of the group in itself and in the process increases, which encourages further engagement in the proposed activities. The group is responsible for the activity and results; however, the consultant needs to be aware that conflicts, leaderships, interests, and personal values usually emerge and, if not addressed, they can lead to breakdowns and dissent in the process. At the same time, it is a great time to learn about the roles that each one plays in the territory and how it is possible to use each skill to obtain the best results.

I highlight here the power of working on the ambience of the place. I have observed in collective projects that improving some aspects of the ambience of the place, such as adding colours, plants, and beauty to a community space has a great impact on group morale. If people live in a rundown place, their tendency is to intensify those characteristics and worsen local conditions. Practice shows that the opposite is also true: when we care, we make the place pleasant and beautiful, and this encourages further care and increases the group's self-esteem in living in that place. People become more zealous and prevent those places from being plundered again, and the community tends to begin to take care of other spaces as well. A more beautiful space is not only better for the community, but

Table 24.1. Training components

Type of session	Description
Lectures	3- to 4-hour sessions of passive learning, with no practical activities
Workshops	6 to 16 hours of theoretical and practical activities. These events allow for deeper work on the themes
Technical visits	Participants visit places and/or experiences that stimulate insights and learn about an entrepreneur's practice. They are widely used for participants to be exposed to other possible practices. There is not a defined workload
Technical mission	An immersive activity that combines benchmarking with training. It generally lasts at least three days, and mixes lectures, workshops, and technical visits
Collective projects	In this dynamic, the group chooses an activity to revitalize or creates a relevant place or cultural activity (selecting a potential attraction among those mapped) such as an event, a vegetable garden, a façade painting – something that encourages the community to develop together, something that will later stimulate or represent an improvement for the community. Once the action is chosen, the group creates an action plan in which they will need to obtain the necessary resources, preferably within the neighbourhood

it also helps to minimize the resistance of visitors. The choice of projects in such approaches is defined based on reconciling the group's need, their financial availability to subsidize the content, and the opportunity to experience something unique and enriching.

Step 5: Product development

To develop the product, we compare the information from the initial cultural mapping with market information on the level of attractiveness, competition, and originality, and with the interests of the local entrepreneurs (feasibility depends on people willing to develop and operate the experiences professionally). There is no specific timeline for product development – the products are shaped as the work goes on. Of course, this stage includes considerations from experts to identify what is more marketable and would have greater appeal, but the product will only be developed if a local entrepreneur is willing to carry it out. As previously mentioned, the objective is that the operation happens autonomously, independent from the agency or consultant.

Once market information and issues have been identified and local operators have embraced the project, we structure the technical aspects of the experience. Table 24.2 outlines the dimensions to be determined and planned. Complementing this, Table 24.3 presents a series of additional aspects that also need to be observed in the development of creative tourism activities.

Bearing all this in mind, we design the product and test it, initially with a group from the territory, then with a group of random guests, and then with a specialized group of travel agents, tourism, and promotion agents – we simulate a FAM (Familiarization) tour. At each stage, we work on the product until it is suitable for commercialization. This stage also includes alignments on operation management, logistics, pricing, and communication.

Step 6: Training of a commercial agent

Selling community-based products requires sensitivity to communicate to the buyer all the product qualities and to settle all questions without reinforcing prejudices and devaluing the product. Also, tact is required to deal with community-based entrepreneurs who often demand more attention due to relationship dynamics in the community.

The ideal scenario is having someone from the community who acts as a local operator. This person must have good relationships and the trust of the community. There have been cases in which people who engaged in the overall process but did not have a tourism product have become a local operator. In other cases, people with previously developed tourism products have become local operators, but the results were not optimal. Usually they are strongly dedicated to their own operations and unable to meet the needs of the group. Based on the availability and desire of the local operator, we work on the development of management and marketing skills. This local operator training occurs as a mentoring process. The objective is to set up a business that operates experiences in the community.

Table 24.2. Technical aspects of the creative tourism experience

Technical aspect	Considerations
Location of the operation	Checking conditions of accessibility, security, infrastructure, and comfort
Duration	Checking the duration of the experience. Visitors should stay long enough to immerse in the atmosphere of the place without hurry, but also short enough so they do not feel bored by the time spent in that place
Number of visitors	Paying attention to the service capacity. These are experiential activities in which the exchange between host and visitor are precious, so the number of visitors is generally reduced, not exceeding 30 people
Proposed activities	The activities should build a narrative with an upward curve of interest and that allows the visitor to have a broad understanding of the place and activity experienced
Co-creation	Identifying which activities can be conducted within the time available, whether they are emblematic of the place visited, and whether they can be carried out by someone who is learning about it for the first time
Experience providers	The person who conducts the experience must know not only about the product history but also about the relationship between this product and the territory. This person needs to have good speaking skills and quick thinking to understand the group and conduct the experience properly
Service	Ensuring that the activities are conducted in a punctual, polite, and friendly manner, as previously organized
Space and materials organization	Verifying if there is adequate space and enough equipment for the number of visitors, as well as their maintenance, safety, and hygiene

Table 24.3. Aspects to be fostered in the development of creative tourism activities

Qualities/aspects	Description
Originality	Presenting the visitor an experience that cannot be replicated elsewhere, as it relates to the uniqueness of that territory
Enchantment	This is the quality of involving visitors in an atmosphere that stimulates all senses and leads them to a complete immersion in the experience. Enchantment is a subjective concept; however, it is possible to count on the creativity of the agents who constantly try to find new ways to captivate visitors. Each experience is unique
Valuing local identity	Activities should refer to the identity of the territory in which the experience takes place, telling local stories and their influences as well as showing how the activities the group develops contribute to strengthening the community and surroundings, if applicable. The narrative should focus on origins or interesting and curious aspects for the visitor
Alignment with other proposals in the roadmap	Observe how this activity matches other activities offered in the territory and the market in general. If activities are complementary, they become easier to commercialize since the product can be combined with others when designing routes
Posture	Treat the activity in a serious, professional manner. This also refers to being pleasant and surprising. Posture is the quality of being attentive to understand the visitor and the points for improvement in one's performance

Aspects such as business planning, strategic planning, financial management, communication, and relationship management are addressed.

After the development of the business strategy, it is also necessary to support the construction of the operator's network. This process involves scheduling meetings with potential buyers and stakeholders, supporting their participation in fairs, and encouraging participation in debates, seminars, and any place where community-based creative tourism products can be promoted. This is the most challenging stage, and the commercialization usually takes place through a partnership between the local operator and a more traditional agency, preferably those more sensitive to sustainability issues. Unfortunately, the traditional market in Brazil is not yet prepared to deal with the peculiarities of the services offered by communities so their access to traditional marketing spaces is still not easy. Furthermore, reaching a specific niche public demands targetted communication and usually technological efforts that the community agents are typically not well prepared to make.

Challenges in the implementation of creative tourism in communities

Communities differ in their levels of development and infrastructure. When talking about tourism in communities in Brazil, perhaps the most familiar image is of the tours in Favela da Rocinha, in Rio de Janeiro. The experience related in this chapter takes place in peripheral communities in the northeast of Brazil that, despite being in an urban area, have strong characteristics of low development. The challenges presented here might not be the same as those of communities located in regions with greater economic flow.

The work in vulnerable territories takes time and can be quite challenging. The development of activities in the peripheries does not happen in a linear, constant, and growing way and demands constant attention to the movements and nuances of the group. One of the aspects that most influences engagement is the financial return. It is difficult to join a process where financial return is not certain and can take time, while there is a real and immediate need for subsistence. Thus, the demands of immediate survival and the lack of quick results strongly influences group engagement. These characteristics turn the work into a process that expands and contracts, and it is not uncommon that few people remain at the end.

Another factor related to subsistence and livelihood is that, initially, the tourist activity does not generate enough recurring demand for subsistence, which makes it a complementary income for the entrepreneurs, who usually have the cultural manifestations as their voluntary work. A vicious circle results from this, as it is not possible to have regular experiences because people are not available full time as a consequence of the lack of regular demands. Moreover, there are no financial conditions for entrepreneurs to invest in the activity to expect it to establish itself and generate a constant return, and a guaranteed livelihood is more urgent. Brazil has no subsidy policies for this type of enterprise, and there is no confidence on the part of entrepreneurs to access any credit line to invest in the business. Activities happen, but not in a sustainable way. The investment return time is also an issue for both public and private sponsors. Even those who understand the difficulties of working with vulnerable groups expect great results in an often unrealistic time period. This issue impacts the result of transformation because changes in the cultural structure of a community take time to happen and demand repetition. After carrying out all the training, it is necessary to maintain some guidance to support the new way of doing things that was presented during the qualification sessions.

In my experience as an agency entrepreneur, I provided that maintenance to entrepreneurs of the products I sold and, later, continued to play this role with the members of the Recria network (see below). My friends and I keep volunteering so as not to lose what has already been developed. I cannot define an ideal time to determine that there is autonomy, but I can say that autonomy happens more easily when public policies are acting in line with interventions. If public policies offer structure, people become more confident to invest in the process. Public policies are necessary for the creation of infrastructure. When people see the ongoing construction of structures, they feel more confident about investing in the development process because they realize that they are not alone and there is a greater chance that their efforts will be rewarded.

As the intervention process starts to present positive results and people engage in making activities happen in the community, public policies perceive the movement and tend to join forces, providing the necessary support. Unfortunately, this is the usual cycle: we do our work first, proving that it is worth investing in that territory, and only then do public policies begin to appear. Being in a network amplifies the voices and accelerates the process because it allows people to occupy governance spaces and claim for their rights, making peripheral places more visible and therefore more visited.

Selling is likely to be the most challenging activity since it is difficult to guarantee the regularity of the experiences. As mentioned previously, most people have other paid jobs and are unable to dedicate themselves full time to this activity, which complicates the creation of a long-term calendar. Thus, we can only manage to establish calendars for up to 3 months, but no more than that, which contrasts with tourism plans that are made with greater advance planning. Another nuance is to tackle the resistance from buyers, as there is still a lot of fear of entering a poor neighbourhood due to misconceptions and prejudice, even if a local agent accompanies them.

Network articulation: Recria – Rede Nacional de Experiências e Turismo Criativo (Brazilian Experiences and Creative Tourism Network)

As an evolution of this work and an attempt to respond to some concerns regarding the promotion and commercialization of this type of product, in 2017 I was involved in co-founding, with a group of creative entrepreneurs, the organization Recria – Brazilian Experiences and Creative Tourism Network (Fig. 24.1). This is a new level in learning how to use creative tourism and collaboration as a tool for social development.

The members of Recria strongly believe the full potential of local knowledge and experiences will only be harnessed when there is an exchange of knowledge between territories. Communication technologies fulfil that function partially, but the periphery is disadvantaged even when it has access to technology, due to structural issues. Acting according to new social dynamics can balance this challenge. Their organization in groups enables the strengthening of local competitiveness based on the unique identity elements of that location. At the same time, being connected to the world makes it possible for them to communicate their identity and production, as well as to learn and exchange experiences so that this feedback cycle benefits the communities involved and generates development. Growth is only possible in diversity because those exchanges enrich their repertoire, boosting creativity and the potential for generating solutions.

The network emerges as an intelligence capable of connecting initiatives that were previously isolated and informing the construction of a collective strategy to strengthen each segment. This organization strategy creates a repertoire from the diversity of participants; we become empowered through articulation, commercialization volume, and expanded service capacity. In this system, entrepreneurs feel strengthened and can communicate with travel agents, who act as partners in the marketing of products; with the government, which supplies the necessary infrastructure to stimulate a creative environment; and with tourists, who establish a

Fig. 24.1. Recria's launch – 'The fabulous world of creative tourism' seminar was an event produced to present Recria to broader society. Photo by Letiene Ferreira, used with permission.

different relation with that place, understanding and valuing the experience offered.

In Recife, there are two strong communities in the modality of community-based creative tourism: Ilha de Deus and Bomba do Hemetério. Endowed with great potential for creative tourism and products already formatted and in operation, they face similar challenges so they decided to work together to support each other. Bomba do Hemetério developed the first community-based tourism experience in an urban area in north-eastern Brazil and features an infinite cultural diversity (Fig. 24.2), while visitors are enchanted by the way of life in the small fishing community of Ilha de Deus (Fig. 24.3), placed within the largest urban mangrove in Brazil. Despite the differences between the locations, they share similar desires and ambitions, as well as problems and concerns, and they both have the desire to radically change the traditional way of doing tourism.

The 'Bombando na Ilha' movement emerged from the union of those two communities to discuss

Fig. 24.2. Bomba do Hemetério contains a large number of carnival groups. One of the creative tourism products developed is 'Recife é sempre carnaval' (Recife is always carnival), which promotes a true carnival anytime. Right: An *adereços* (accessories) workshop. Left: A *maracatu* percussion workshop. Photos by Maurício Ferry, used with permission.

Fig. 24.3. 'Cooking with Negralinda', the creative gastronomy experience run by Negralinda, a former shellfish collector who, through creative tourism, became a chef. Photo by Maurício Ferry, used with permission.

initiatives and add new players, experiences, organizations, and bits of intelligence that could strengthen the movement and increase its capabilities to boost creative tourism. Following some meetings after this movement launched, we concluded that many other communities in Brazil were experiencing similar difficulties and that if they were in a network, they could learn from one another and improve their results. We then decided to create Recria – Rede Nacional de Experiências e Turismo Criativo (Brazilian Experiences and Creative Tourism Network).

Recria presents itself as an integrative movement of creative tourism experiences in Brazil that aims to boost the development of people, environment, and local culture. It also seeks to promote and support actions based on creativity and enchantment as a driving tool for tourism and, consequently, territorial development. Recria works to support members and articulate partnerships to strengthen the creative tourism segment.[3] Since its founding in 2017 it has achieved significant results: the Creative Tourism Network nominated it as 2018's best strategy for creative tourism development; it received the Pernambuco Tourism Award as best creative tourism initiative; it connected stakeholders to book events at associated restaurants (this is important because these are events that traditionally would not happen in peripheral spaces); and it influenced public policies by encouraging and working for Recife to participatively develop *Recife: Plano do Turismo Criativo, 2019–2021*, the municipal plan for creative tourism (City of Recife, 2018), and to establish the Recife Creative Tourism Forum. The municipal plan for creative tourism guides the government to invest in and develop activities that stimulate and strengthen creative tourism activities. In order to monitor the implementation of the municipal plan, the Recife Creative Tourism Forum was set up as a governance organ, which acts independently from the city government.

Reflections on results

Despite the challenge, working with tourism as a transformation tool brings consistent results. I have not systematized the information that would allow me to define accurate improvement rates. However, it is evident that the projects accomplished positive results, which I observe not only from the interventions in which I took part actively but also from the knowledge of other territories that use creative tourism today as a platform for social transformation.

The creative tourism movement attracts visitors in search of the best the community has, and the perception of a locale as a place of wealth production changes the community's self-image. More confident, people seek their rights more actively and start to resist forms of violence perpetrated against them.

Their improved self-esteem is also reflected in their appreciation of their own cultural manifestations. Generally, folkloric expressions depend on government subsidies to be performed and, consequently, to continue to exist. Creative tourism experiences show that there are people interested in local cultures who are willing to pay for them. This creates a space for the practice and maintenance of culture as well as financial returns to subsidize activities, reducing government dependence and expanding the vision for other monetization possibilities.

The financial return is real, albeit slow. Although I do not have a historical series to analyse, the financial return can be illustrated as follows: In the community of Ilha de Deus, a former shellfish collector is now a chef and owns a restaurant that opens regularly, where lunch costs R$40.00. Additionally, there are 12 local people working there and the restaurant ingredients are bought in the community. The community hosted approximately 7000 visitors in 2019, each paying a minimum of R$20.00 (the average is R$40.00), which totals at least R$140,000.

As initiatives stand out, other segments of society become interested in the stories, and the enchantment of the periphery emerges. This is demonstrated by the number of journalists interested in telling stories of power and richness in these places, by stakeholders from the traditional market willing to engage in such projects, and by the interest of government offices to display these initiatives as a tourism strategy. Little by little, communities can see a rise in the amount of public investment in infrastructure.

The consequent visibility helps to mitigate prejudice against those territories and, especially, the people who live there. It is a virtuous cycle: as the narratives about the place change, so does the imagery about that space and those people. As the outlook changes, people feel more confident to access the community, and creative tourism encourages a warmer look at that exchange, in which the host is not an attraction on display but the provider of a learning opportunity. This perspective is a game-changer in bonding and in the elimination of prejudice.

Final considerations

My experience with tourism and territorial development has always included working with communities in situations of social vulnerability, which poses great challenges. Many obstacles must be overcome until people are available for the process, and many others may appear until the process generates outcomes for those involved. However, the need to build a more equal society is an irrefutable issue, and I see tourism as a powerful tool to raise awareness and create integration and transformation. Creative tourism in particular favours connections in gentle contacts with different realities. When people visit a territory within a creative tourism experience, they are open to making contact with that reality. My observations of the encounters and learning experiences provided by creative tourism have shown me that it is a valuable tool for human development that brings people together and reduces barriers.

In creative tourism, people are the central element and you must care about them. They are unique, original, and irreplaceable, so it is imperative to work with them to develop their fullest potential. It is paramount to develop social capital that allows for dreaming and planning a common future. Moreover, education-driven work is of great importance. Any time and opportunity should be taken as a learning experience for all people involved in the process, which strengthens community autonomy, encourages creativity and innovation, and keeps things moving. Education can be the difference between enthusiasm and continued actions, mistrust, or indifference.

Culture is a strategic advantage because it is a powerful mobilization agent. It allows people to operate in different ways, using creativity and boldness and building collective meanings and values, dreams, and significant experiences. It can inspire people who never intended to work together to cooperate in order to develop a collective project.

In order to develop creative tourism initiatives capable of developing a territory, it is essential to establish a trust relationship not only within the community, but also with partners who can help to support initiatives in that space. Tourism needs to be integrated within many areas and its development requires joint efforts from both public and private sectors. Development agents such as governments, promotion agents, and big enterprises need to play the role of supporting creative entrepreneurs and connecting stakeholders who can catalyse the activities. I have witnessed this in Recife, my city, where the private sector mobilized the city council to establish a public policy on creative tourism (City of Recife, 2018). Many creative entrepreneurs joined the movement and other stakeholders coalesced in an upward movement that has improved the work possibilities and the relationship between entrepreneurs, citizens, and the city council/government.

The concept of integral development excludes the idea of isolated initiatives: we must understand that we are part of a system and we need each other to exist. As for the learning power in meaningful encounters, being in a network is inspiring because it gathers people from different socio-cultural backgrounds in the same group, and this diversity improves collective intelligence and action power. For example, we may see a well-known visual artist creating products with a carnival crafter from the periphery, in meetings that enable mutual learning and result in extraordinary solutions. Intentionally mingling people and cultures from various socio-cultural realities in order to design products not labelled as peripheral may be a strategy to minimize curiosity about the exotic and stimulate curiosity about the learning potential in such meetings.

Using creative tourism as a tool for development is not simple or easy, but it is feasible. Its worth and potential are closely connected to the kind of tourism we want, and the kind of development we support. It is necessary to think about our position regarding vulnerable people. Are we prepared to make room for them to become protagonists of their lives? Are we ready to drop the saviour role and acknowledge that not everybody must live the way we believe is 'correct' or 'normal'? We might visit places because they are exotic and hear people saying that they have to adapt to receive a larger public, but it is crucial to understand the kind of development they really want.

Believing in people's capacity for creating lies at the core of my perspective and approach. With inputs from the territories and self-confidence, we can create new models of local development that use tools and strategies aligned with contemporary society's needs for collective work and knowledge through organizing networks. If successful, this process may be followed by other transformations in the way of thinking, designing, implementing, and evaluating development initiatives so that they are more sustainable, promote people's autonomy, and minimize inequalities.

Notes

1. *Borogodo* is a Brazilian expression that refers to the quality of what is charming or magical. It is equivalent to the French phrase *je ne sais quoi* but with a shine that only Brazil has. Its general meaning is similar to 'hype', but more enchanting.

2. The principles and methodologies of training I use are those of Participative Management for Local Development (Gespar), designed by the Instituto de Assessoria para o Desenvolvimento Humano (Advisory Institute for Human Development) (IADH). For more information, visit: www.iadh.org.br/metodologia/ (accessed 20 September 2020).

3. For more information about Recria and creative tourism in Recife, visit the websites of: Recria Brasil (www.recriabrasil.com; accessed 20 September 2020) and Turismo Criativo Recife (http://turismocriativo.visit.recife.br/ accessed 20 September 2020). See also the report from an international seminar on creative tourism (*Plano de Turismo Criativo do Recife,* 2018).

References

City of Recife (2018). *Recife: Plano do Turismo Criativo, 2019-2021 (Recife: Creative Tourism Plan, 2019-2021).* Available at: http://turismocriativo.visit.recife.br/wp-content/uploads/2019/05/plano_turismo_criativo.pdf (accessed 1 April 2020).

Molina, S. (2011) *Turismo Creativo: El fin de La competitividad.* Escritores, Santiago do Chile, Chile.

Parente, S. (2014) Abordagem do desenvolvimento local/territorial. In: Arturo, J. (ed.) *IADH Atua: Referências para uma Prática em Desenvolvimento Local/territorial.* Instituto de Assessoria para o Desenvolvimento Humano (IADH), Recife, Brazil, pp. 74–89.

Plano de turismo criativo do Recife / Seminário internacional de turismo creativo: Relatório Geral. (2018). Available at: http://turismocriativo.visit.recife.br/wp-content/themes/turismo-criativo/download/relatorio_geral_seminario_internacional_turismo_criativo-ilovepdf-compressed.pdf (accessed 20 September 2020).

Richards, G. (2009) Turismo cultural: padrões e implicações. In: Camargo, P. and Cruz, G. (eds.) *Turismo Cultural: Estratégias, Sustentabilidade e Tendências.* Universidade Estadual de Santa Cruz (UESC), Bahia, Brazil, pp. 25–48.

Sen, A. (2000) *Desenvolvimento como Liberdade.* Companhia das Letras, São Paulo, Brazil.

Zapata, K. (2014) Economia criativa na perspectiva do desenvolvimento local. In: Arturo, J. (ed.) *IADH Atua: Referências para uma Prática em Desenvolvimento Local/territorial.* Instituto de Assessoria para o Desenvolvimento Humano (IADH), Recife, Brazil, pp. 196–206.

25 Why creative tourism won't work if residents are not involved: A tale of two cities

Kathleen Scherf

Thompson Rivers University, Canada

Every definition of creative tourism cites, in some form, the necessity of a positive and productive relationship between residents and tourists, whether it is expressed as visitors engaging in the everyday life of the community, or as visitors learning from residents a creative skill unique to that location. Tourism agencies and providers are well aware of the desire of tourists to be embedded within a community; the World Tourism Organization cites experiencing the everyday life of the destination as the top tourism trend in 2019 (WTO, 2019, p. 5).

Barcelona

In cities where over-tourism occurs, for example, Barcelona, it is very difficult to convince residents that tourism is beneficial, as they find their independent stores bought up, their rents rising, and their infrastructure crumbling. Their own sense of their community, their cultural heritage, is being erased. They are not happy. Anti-tourist graffiti is rampant, even in the elegant Gracia neighbourhood; some merchants at the popular Mercado de la Boqueria actively and vocally forbid pictures; and locals enjoying afternoon walks on Las Ramblas is a thing of the past – it is now a location to be avoided at all costs. In Barcelona, it seems an almost impossible task to create an atmosphere where visitors can engage with locals to experience the everyday life of the community.

The municipal government of Barcelona is in a difficult place: it is responsible for speaking up for its citizens and properly maintaining the city's infrastructure, yet it cannot lose the immense economic benefits of Barcelona's massive touristic appeal. In 2018, there were approximately 32 million tourist arrivals in Barcelona, whose population is 1.6 million. Unsurprisingly, the city's population has decreased 11% since 2015 (Giner, 2019). Barcelona tourism is perhaps too much of a good thing.

How does Barcelona deal with this situation, when the number one trend in tourism is to live like a local?

Barcelona has officially adopted the practice of un-identifying visitors as 'tourists' or 'outsiders', declaring in its *Strategic Tourism Plan 2020*:

> Far from being an outside phenomenon, tourism produces the city and, at the same time, the city shapes tourism's possibilities. Tourism is an inherent and constituent part of the current urban phenomenon. Tourism activities must not be seen as something foreign to the city, they are not 'out there', but part of its day-to-day activities, intrinsic dynamics, and even daily life. So tourists do not have to be considered passive players 'in the city' but rather as visitors with rights and duties 'of the city'. (Turisme de Barcelona, 2017, p. 7)

Like Copenhagen, which is examined further on in this chapter Barcelona realizes that 'the quality of the tourist experience depends on guaranteeing the well-being of the people who live in the city' (Turisme de Barcelona, 2017, p. 16). In order to relieve the pressure on the central touristic district, Barcelona plans to encourage neighbourhoods outside the centre to develop sustainable tourism plans (Turisme de Barcelona, 2017).

Corresponding author: kath@tru.ca

 Creative Tourism: Activating Cultural Resources and Engaging Creative Travellers (eds. N. Duxbury, S. Albino, and C. Pato de Carvalho)
DOI: 10.1079/9781789243536.0025

Neighbourhoods rather than entire cities thus become the site of experiences co-created by residents and tourists. Or not. Barcelona's Barrio El Gotico, for example, is a neighbourhood abandoned by locals; residents have fled or have used their residences as short-term rentals for visitors and the neighbourhood is unashamedly and completely given over to the business of tourism. To the west of Las Ramblas, however, in the equally ancient Barrio El Raval, many people still reside, despite climbing housing costs, inadequate infrastructure, and hordes of tourists. So how do residents and tourists engage with one another to create shared social capital in El Raval? In other words, is tourism embedded in the community of local residents? If not, they will work to keep the distinction between tourists and residents as sharp as possible. If that is the case, creative tourism activities will not be sustainable, will not be genuinely place-based, and simply will not work.

In the first three months of 2018, I conducted a survey in El Raval. The survey was distributed online in three languages; the survey branched at the point where respondents identified themselves as either a visitor or resident. Both responded to statements on a five-point Likert-like scale. Figures 25.1, 25.2, 25.3, and 25.4 summarize my findings regarding the relationship of residents and visitors in El Raval.

With this study, we see the complex reality of the El Raval neighbourhood of Barcelona. With its history of marginalized status and degradation, and being a major recipient of Barcelona's over-tourism, residents are sceptical about the Turisme de Barcelona's idea that an integration of tourist and resident interest will result in improved social capital in El Raval. But this survey suggests also that although residents are less willing to engage in the creative tourism enterprise than are visitors (Fig. 25.1), about an equal number of residents enjoy interacting with visitors as do not (Fig. 25.4). Since over half of visitors enjoy engaging with residents, perhaps there is room to work on some creative tourism projects, especially given that 55% of residents enjoy sharing experiences with visitors (Fig. 25.2) and telling visitors about everyday life in El Raval (Fig. 25.3).

Residents underestimate the interest of visitors in connecting with them; visitors overestimate the interest of residents in connecting with them. These results suggest that communication is an issue.

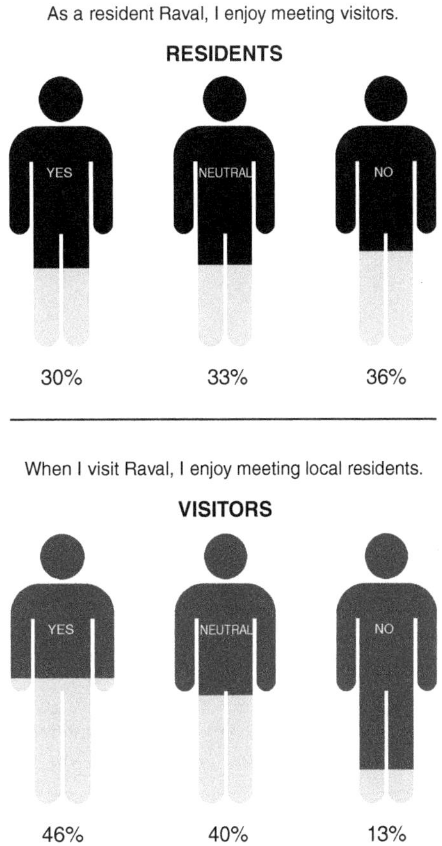

Fig. 25.1. Do residents and visitors enjoy meeting?

Could social media assist in bringing together the narratives of residents and visitors to create shared value? Figure 25.5 shows that about 20% of residents and 55% of visitors currently connect through social media.

Does social media have the potential to connect residents and visitors in El Raval? Figure 25.6 shows that both residents and visitors believe in that possibility: 31% of residents and 61% of visitors, representing an 11-point and 6-point increase over what they currently do, respectively. Given these statistics, I think a collaborative paradigm for locals and visitors could begin by implementing a social media programme, and it is my first recommendation.

If creative tourism is to work in El Raval, some embedding of tourist activities within the local resident population must occur. With this preliminary survey to guide practitioners and local agencies, a

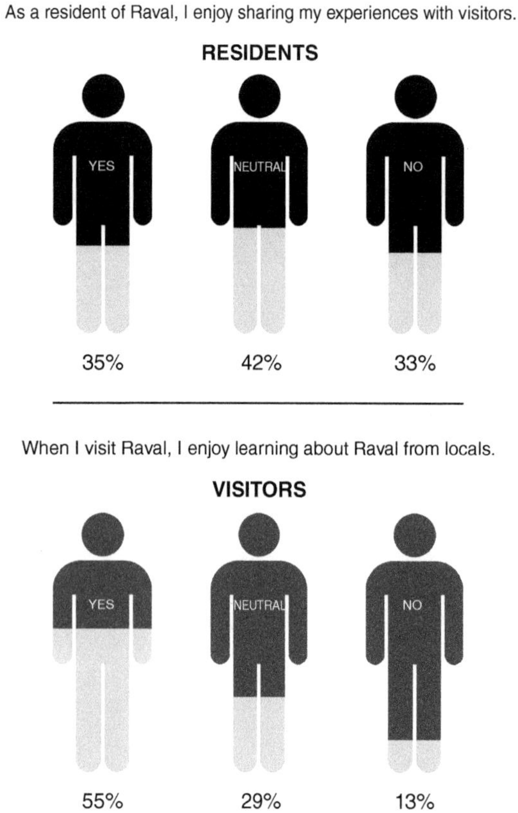

Fig. 25.2. Do residents and visitors enjoy sharing experiences about El Raval?

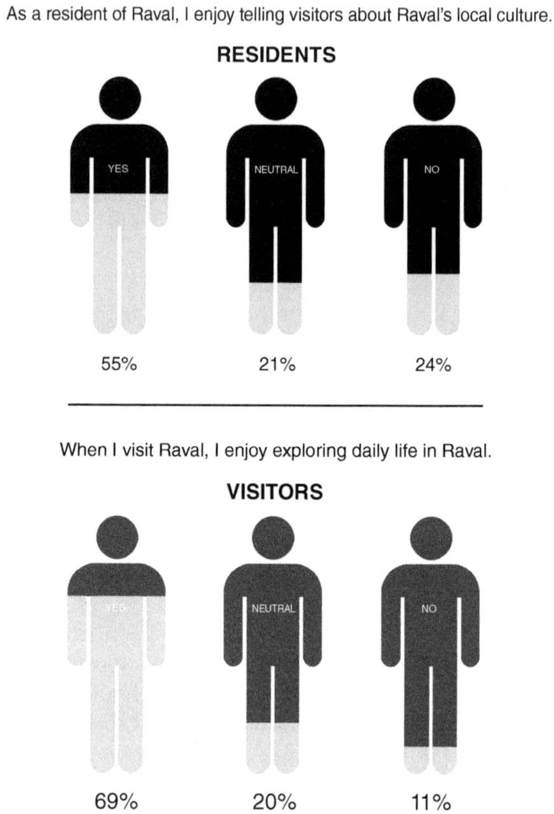

Fig. 25.3. Do residents and visitors learn about and share everyday experience in El Raval?

cultural mapping process could consult and build plans on some of these ideas:

- create a social media forum for residents and visitors to share their stories;
- develop and offer programmes of education for the residents of El Raval about the benefits to them of connecting with visitors through creative tourism (e.g. local community place-making; sustainable development);
- provide ways for visitors to meet residents and engage with the everyday life of El Raval;
- create a real-time forum for residents and visitors to meet;
- offer cross-cultural creative tourism events; and
- develop a community planning process that includes the provision of authentic local touristic experiences. Residents could become the touristic interpreters.

Considering the attraction for visitors of gastronomy, shopping, local culture, arts, and similar topics, combined with the demonstrated interest level of residents in sharing their local culture and the storied history of the neighbourhood, a resident-based initiative to develop and programme a system for sharing their knowledge and experience would give them agency in how to share their own place, and involve them in the activities they develop for visitors. Embedding resident-led creative tourism in El Raval could be a first step to creating a relational tourism experience that brings together residents and visitors in the city of Barcelona if this destination wants to become a more sustainable tourism destination.

But a problem in initiating this kind of project in Barcelona is penetrating the opaque bureaucracy to determine which agencies might take the lead. The official tourism strategy, developed in a consultative process, is overseen and published by the Tourism Division of Barcelona City Council (2017). The tourism plan, available on the city website, indicates that two meetings were held with the

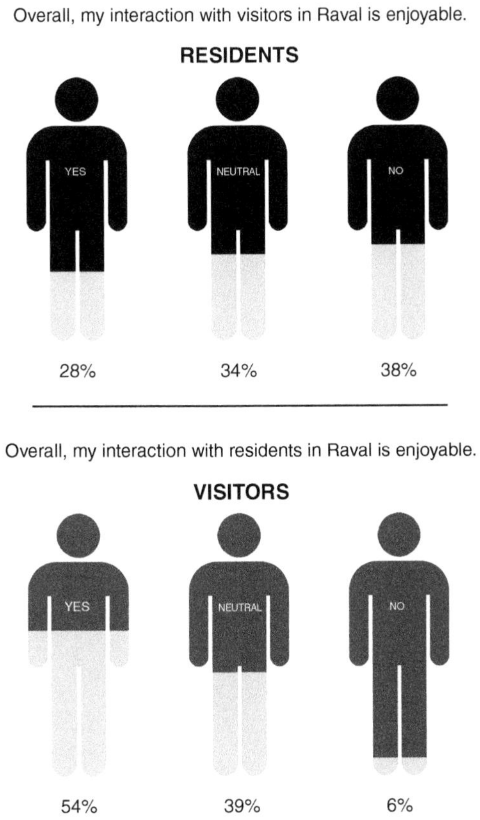

Fig. 25.4. What is the overall feeling of residents and visitors about each other?

city's destination management organization (DMO), Turisme de Barcelona. When I met with the DMO in early 2018, its representatives were quite clear that it was a completely separate entity from the Tourism Division of the city and was unable to assist with any collaborative paradigm regarding the opportunities for creative tourism in El Raval revealed by my initial survey.

There is another organization in El Raval, called Tot Raval, which represents business and the general community. Its staff refused to be involved in exploring the opportunities, as tourism is a prickly subject in El Raval, and Tot Raval did not want to upset or alienate any of their stakeholders. Finally, the head office of the Creative Tourism Network is located in Barcelona; although interested in my project, and happy to send out links to the survey, it is an association for creative tourism providers, and hence has no mandate to help plan creative tourism initiatives.

It seems that implementing a social media programme to better connect residents and visitors will require a sponsor, yet I have been unable to find a likely one. The path to finding local agencies who could collaborate in exploring creative tourism activities is as confusing as the alleys and passages of the old city itself.

Over-tourism is a massive component in Barcelona's tourism strategy; officials, planners, and tourism providers are constantly in a situation where the past lack of planning affects what they can do in the future. They are constantly struggling in a destination where sustainability has not been a planning concern, and where it is now urgent. The situation is duplicated in places like Rome, Venice, and Amsterdam. It seems clear that employing social media as a tourism tool can aid sustainable tourism in a number of specific ways, including building relationships between residents and visitors. Are there destination cities, I wondered, that are not yet quite over-touristed, that have heeded the possibilities of resisting the problems of cities like Barcelona by employing social media? I found there is some interesting activity going on in Copenhagen, particularly with regard to using social media as a shared meeting site for visitors and locals.

Copenhagen

Copenhagen's bold 'Localhood' campaign is a strong example of the collaborative place-making strategy. Proclaiming 'The End of Tourism as We Know It: Towards a New Beginning of Localhood', Copenhagen's tourism plan to 2020 speaks to 'the experience of temporary localhood':

> Today, fewer and fewer [people] want to be identified as tourists. Instead, new generations of travellers seek out experiences that not only provide a photo opportunity, but also get their hands 'dirty' and immerse them in the destination. The travellers seek out a sense of localhood, looking to experience the true and authentic destination – that which makes a destination unique. With the increasing number of providers and businesses that tap into the sharing and collaborative business potential, travellers gain increasing access to the local travel experience. (Aarø-Hansen, 2017, p. 5)

In this way, Copenhagen's destination marketing organization hopes to avoid the unsustainable

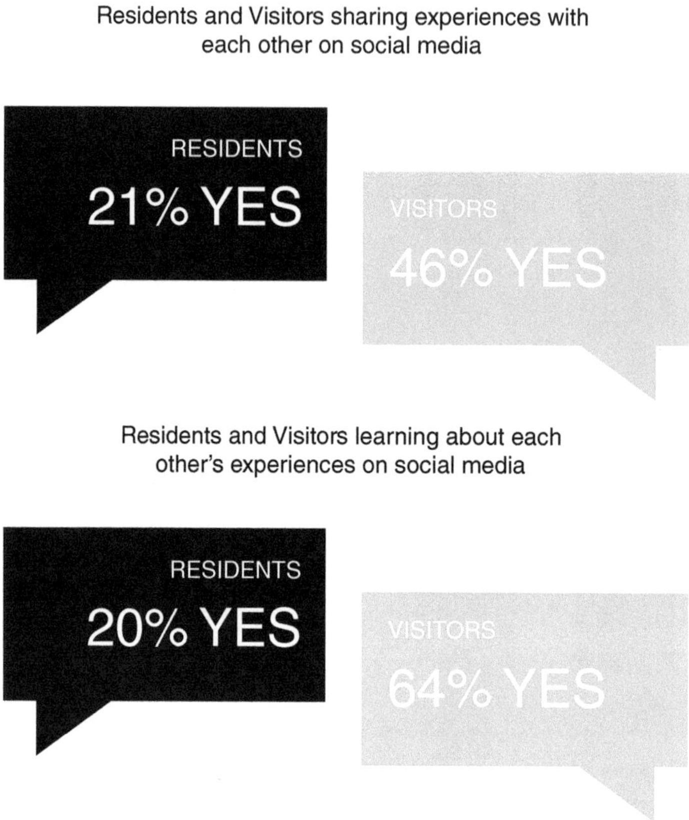

Fig. 25.5. Current social media connection between residents and visitors.

over-tourism of cities such as Amsterdam and Barcelona, where touristic activity is frequently viewed by locals with approbation, if not outright hostility. Instead, Localhood asserts:

> The delivery of an authentic destination experience depends upon the support of locals, whereas the liveability and appeal of our destination – and thereby the advocacy of locals – depends on our ability to ensure a harmonious interaction between visitors and locals. (Aarø-Hansen, 2017, p. 5)

The Localhood for Everyone tourism strategy can be accessed here: http://localhood.wonderfulcopenhagen.dk/wonderful-copenhagen-strategy-2020.pdf (accessed 24 February 2020).

One of the pillars of the Localhood tourism strategy is that 'Shareability is King' (p. 13), and one of the activities to support this pillar is to 'work to enable shareable moments between locals and visitors' (p. 21). More than 80% of locals support this concept of growing tourism on connections between visitors and residents (p. 24). This demonstration of local support is higher than what we see in El Raval; while it is impossible to completely undo the over-tourism El Raval has experienced, the strategies detailed in the Copenhagen tourism strategy could well be applied to up-and-coming neighbourhoods in Barcelona as it attempts to distribute its tourists throughout the city. For example, the hipster beach neighbourhood of Poblenou, seven stops on the Metro from Plaça Catalunya (with a transfer), might be at just the right stage to experiment with Localhood.

Copenhagen has used social media to establish localhood for all. On Instagram, the hashtag #visitcopenhagen is very busy with images from all over the city, with international posts and likes. The Facebook page 'VisitCopenhagen' is administered by the DMO Wonderful Copenhagen but anyone can post, and lots of people do, and people also comment on other posts with questions, advice, suggestions, and reactions. El Raval could really

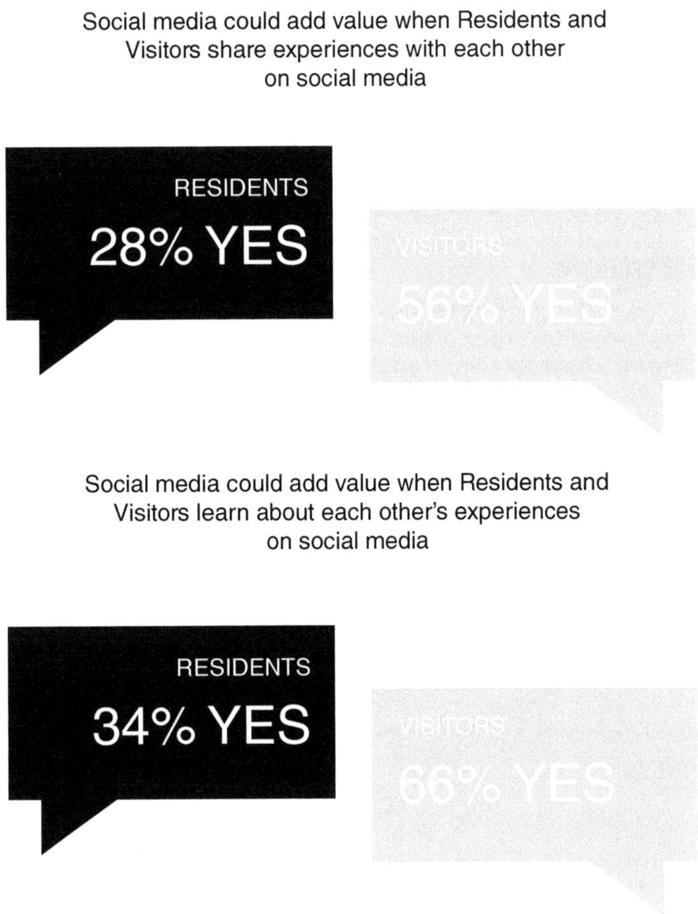

Fig. 25.6. Potential for residents and visitors to connect using social media.

take a note from Copenhagen on using social media to engage visitors with locals.

Furthermore, if you go to the VisitCopenhagen website (www.visitcopenhagen.com; accessed 19 February 2020), you will see that the city has

> asked some of Copenhagen's most well-rounded locals to share their knowledge on where to go, what to eat, drink, and what to do in their home city. This is your gateway to dig a little deeper in wonderful Copenhagen.

The same site offers videos of neighbourhood tours by locals.

While Copenhagen is not, like Barcelona, a city currently confronting over-tourism, it is doing precisely the kind of work involving locals that creates a productive and enjoyable relationship between residents and visitors. The tourism plan also aims to distribute visitors from the central area into neighbourhoods by highlighting these smaller communities within the city. Could neighbourhoods in Barcelona that, like Poblenou, are not quite as over-touristed as El Raval, apply Localhood strategies, so that when over-tourism hits it, it will, like Copenhagen, be ready? Not only that, but having included local stakeholders in the development of the tourism strategy, locals are embedded in the idea of Localhood for all. On this strong foundation, creative tourism – the positive and productive relationship of residents and visitors – can be built. As I write this, all of us are in the midst of sheltering in place, locked down as a result of the COVID-19 coronavirus outbreak. Many of us have hardly been out of our homes for weeks; tourism will no doubt be forever changed. Now is the time for officials, planners, and providers to make use of social media as a tourism tool.

Acknowledgement

All figures were developed by Thomas Sandhoff, Thompson Rivers University, Canada.

References

Aarø-Hansen, M. (2017) *The End of Tourism as We Know It: Towards a New Beginning of Localhood*. Available at: http://localhood.wonderfulcopenhagen.dk/wonderful-copenhagen-strategy-2020.pdf (accessed 24 February 2020).

Giner, C. (2019) Barcelona's tourist headache. *Euronews*, 23 August 2019. Available at: www.euronews.com/2019/07/23/barcelona-s-tourist-headache-as-residential-population-declines-11-since-2015.

Tourism Division of Barcelona City Council (2017) *Barcelona Tourism for 2020: A Collective Strategy for Sustainable Tourism*. Available at: https://ajuntament.barcelona.cat/turisme/sites/default/files/barcelona_tourism_for_2020_0.pdf (accessed 23 April 2020).

Turisme de Barcelona (2017) *Strategic Tourism Plan 2020*. Turisme Barcelona, Barcelona, Spain.

World Tourism Organization (WTO) (2019) *International Tourism Highlights, 2019 Edition*. UNWTO, Madrid, Spain. doi: 10.18111/9789284421152.

26 Youth, ways of production, and community

FERNANDO ALMEIDA

Encontrarte Amares, Portugal

> Your Christ is a Jew. Your car is Japanese. Their pizza is Italian. Your Greek democracy. Your Brazilian coffee. Your Turkish holiday. Your Arabic numerals. Your Latin letters. Only your neighbor is a foreigner.
>
> – Poster scattered on the streets of Berlin, in 1994, cited by Mónica Kostera (2002, p. 13, in Bauman, 2005, p. 33)

We are collecting beings. We collect memories, experiences, and knowledge gathered between relationships established in a globalized world. The activities in which we choose to participate contribute effectively to the construction of our individual and collective identity. Creative tourism, in a world today led by youth unrest, promotes dialogue between local communities and visitors, in an ecosystem of creativity and participation, proposing new possible paths, new interactions with each other, new modes of production.

Today we are experiencing a process of globalization and accelerated technological development where 'identity' is a somewhat unstable concept (Bauman, 1998). The development of transport and communication technologies bring the local and the global closer together, contributing to a constant dialogue between multiple layers and identity references. The distances are getting shorter, the trips more accessible, and the information faster. In this context, the definition of identity, part of a construction unrelated to direct experience, is a meta-experience. We can speak of an identity associated with the postmodern subject (Hall, 2005) that is defined in a state of procedural construction, multiple, and, above all, fluid (between the local and the global). This is an identity built from the participatory dialogue between the physical and the virtual.

It is in this context that we can highlight the contribution of young people to the materialization of these assumptions. The youth community seems to address new issues to the world and develop specific communication tools that are far from traditional interpersonal relationship models. For young people, the virtual world is a space for expression and identity discovery. The internet and online games allow for an experimentation of social roles, expanding the range of relationships and access to information. Here, they can choose idols, create their own cultures far from the authority figures of parents and family, and build relationships with distance and freedom. Collectively, they create spaces of autonomy and participation, assuming multiple identities or continuously reconfiguring the same.

Simultaneously, the market for experiential experiences, such as creative tourism, accompanies this globalization movement. More than ever, finding affordable accommodation and travel is within the reach of any internet user. It is possible to explore the specifics of a distant place, share moments of learning or leisure, integrate into a culture or community on a temporary basis, and create and recreate our identity from the places and people we come across – a composition of collected pieces on trips between relationships. Thus, we cannot help but question how new concepts of identity and new participation settings are being tested in online games, chats, social networks, low-cost hostels, couch surfing, woofing, and experiential and creative tourism?

There are many cases that demonstrate a global-scale community's ability to question the status quo, a community based on the fluidity of information

Corresponding author: direccao@encontrarte.pt

 Creative Tourism: Activating Cultural Resources and Engaging Creative Travellers (eds. N. Duxbury, S. Albino, and C. Pato de Carvalho)
DOI: 10.1079/9781789243536.0026

and communication technologies. Examples, among many others no less relevant, are the case of Edward Snowden (who made public the details of various programmes that make up the US National Security Agency's global surveillance system), which problematized the lack of transparency in access to information; the School Strike for Climate led by the young environmentalist Greta Thunberg, who sensitized us to the need for new models of environmental, cultural, economic, political, and social sustainability; couchsurfing.com, founded in 2004 by Casey Fenton (currently with more than 12 million members in 200,000 cities), which reconfigured the concept of travel and accommodation on a global scale; or Facebook founder Mark Zuckerberg, who revolutionized the media, models of interpersonal relationships, and identity. Young people who believe they have something to contribute to altering established structures can, through their proposals, rebalance the ways in which we relate and how we participate politically in the world.

We can then observe that, in different contexts, activities with repercussions in the collective are addressed, bringing together geographically distant communities to foster cross-cutting discussions with society. These provide material and immaterial instruments that, when shared and made accessible, become universal and give rise to a new paradigm – the commons as a way of producing the community.

However, there are also other dimensions that should not be underestimated. Hyper-participation, fuelled by the proliferation of social networks, calls into question the quality of public debate. Traditional institutions of democratic representation (political parties, unions, parliament, and traditional media) have lost their ability to represent the citizen in the face of new aesthetics in contemporary life. In this context, ideas and ideals with different degrees of radicalization multiply, releasing authoritarian phenomena such as xenophobia or digital radicalism. Hyper-participation quickly takes us to a territory of populist discourses where 'everyone has something to say about everything' regardless of the quality or legitimacy of the contribution.

Cross-cutting these issues, a discussion emerges around the contribution of the community in the definition and constitution of a common territory. The words of the philosopher António Negri are insightful here: 'The important thing is to recognize the common as a mode of production in our society and as a fundamental product of everyone's work' (Negri, 2016, p. 8). Underlying this thought, participatory artistic practices emerge and contribute to expanding processes of discussion probing the dialogue between community and civic, political, and cultural participation. On one hand, these practices question an exhausted view of elite art, only for the access of some, and, on the other hand, they experiment with participatory devices that may be relevant, not only for the creation process, but also for urgent discussions about the state of our society.

Since 2013, Encontrarte Amares (http://encontrarte.pt; accessed 17 May 2012) has been exploring these assumptions, enhancing, crossing, and developing experimentation spaces that may potentially generate new approaches to artistic creation and participatory models (see Box 26.1). An example is the project, 'Don't you ever go (Oh baby)', a co-creation between the Fabulous Future collective and Amares artisans. From a workshop run by local artisans and mediated by the Encontrarte Amares team, the Dutch collective investigated traditions and crafts in order to reflect on how they can influence applied arts and design. The project, inspired by valentine's handkerchiefs, resulted in the co-creation of an artistic installation about love, friendship, affection, and collective (Fig. 26.1). Highlighting the transversal dimensions of this process, it fostered a broader reflection on social models, including intergenerational dialogue, the sharing of modes of production, the intersection between popular and contemporary aesthetics, the international scale, and the search for a shared and global language.

With the same objectives, in 2017, Encontrarte Amares challenged a group of local young creators to accompany the creation process of Arco de Goães (a community-based tradition that involves dozens of residents of the parish of Goães, Amares, for the co-creation of a celebratory arch to the village's patron saint). In an artistic residency format, these young people collaborated and participated in all phases of the materialization of the arch while developing a documentary with a focus on community participation. In 2019, the documentary was presented to the local community and a construction workshop created by residents was produced. Through this activity, the visibility of this tradition was leveraged, contributing to the reinforcement of the local identity and establishing a new instrument for sharing the modes of production, one capable of attracting new audiences for this process of heritage creation (Fig. 26.2).

Encontrarte Amares descends from a heritage of participation that had great expression in Portugal in the second half of the 1970s. Over 11 years, it has challenged more than 160 young creators, national and international, to critically reflect and act on alternative models of artistic participation, politics, and civics (see Box 26.1 and Fig. 26.3).

To frame this reality in Portugal, we must briefly revisit the political context in recent decades. Going back to the years that followed the revolution of April 25, 1974, all over the country, numerous actions were developed based on the principles of cultural democratization. They were inspired by

Fig. 26.1. Art installation 'Don't you ever go (Oh baby)', Amares, Portugal. Photo courtesy of Stichting Fabulous Future, used with permission.

Fig. 26.2. Arco de Goães, Amares, Portugal. Photo courtesy of Filipe Braga, used with permission.

Box 26.1. Encontrarte Amares

The meeting of people, artistic experimentation, and the communion of affections consolidate the intentions that have propelled Encontrarte Amares since 2009. This biennial community arts festival has been proposing, through a pluridisciplinary programme, a live dialogue between the radicality of contemporary artistic thought and the traditional heritage of the region of Minho, Portugal. Community-based art projects such as visual arts exhibitions, animated cinema, performing arts, radio, series of discussions and theoretical reflection, artistic residencies, competitions, and educational activities give shape to an intense programme that culminates in the 4 days of the Festival, promoting an environment of experimentation, participation, gathering, active contemplation, and freedom of thought. Encontrarte Amares promotes access to cultural participation by integrating the local community at all levels of decision-making for the festival as well as by promoting dialogues with artists from different geographies.

compelling artists such as Elvira Leite and Augusto Boal, who developed participation methodologies with the aim of guaranteeing people's access to culture. Cultural dynamism campaigns were developed and the Fundo de Apoio aos Organismos Juvenis (FAOJ) contextualized this in youth policies that enhanced a deep connection with communities, namely through the training of professionals. Actions like these, among others, boosted the direct participation of communities in artistic and cultural creation processes, and contributed to forming a new generation of professionals capable of perpetuating this mission.

However, in 1986, when Portugal entered the European Union, we saw the acceleration of a series of processes related to market logic and consumer-centred views. In 1994, the seminar 'Portugal na Internet' took place, where the internet was presented to the public and journalists for the first time (however, only in 1999/2000 did it reach full commercialization and accessibility for the general public). The 1990s brought the implementation of cultural and social policies to stimulate this area of internet-related activity, but which

Fig. 26.3. Co-creation show 'O Homem do Cávado', Amares, Portugal. Photo courtesy of Filipe Braga, used with permission.

simultaneously led to a stagnation in direct participation by communities, privileging instead the hyper-institutionalization of processes. Following this phase, a period of 'participation crisis' occurred in Portugal, from which the country only began to recover in 2013. At that moment, the country was in the context of a bailout request due to the economic crises, following Ireland and Greece, with all the consequent implications. The Portuguese government then implemented a series of unprecedented austerity measures in the country that had significant impacts on the lives of the populace.

It is in this context that Encontrarte Amares has promoted the development of approaches centred on the collective, with participatory practices boosted by non-formal and/or non-governmental entities. Today, we see a strong presence of these participatory artistic practices promoted by NGOs and integrated into the programming of cultural facilities and municipalities throughout the country, in the structuring of programmes promoted by the European Union, and rising interest in this area within scientific research. More recently, but not less importantly, the emergence of new configurations of participation such as online petitions, crowdfunding, social networks, forums, and internet art are defining new communities associated with digital culture, which are widely explored by the younger segments of society.

Following this trajectory, we have observed that investment in processes of cultural democratization, where participatory artistic practices take a central role, have directly contributed to the definition of new communities and to their effective participation in the discussion of the country's cultural and social policies. Simultaneously, these dynamics have been boosted by the emergence of a new paradigm based on new relational communication technologies. This organic and emancipatory process of the concept of community leaves its legacy in all social sectors – including effects on tourism, namely, through creative tourism.

The concept of creative tourism, which is increasingly established in the world tourism scene, is inspired by the sharing of production modes, projecting the visitor to the role of active participant. It focuses on an element that allows local residents and visitors to share space, time, and knowledge for co-creating products or, more than that, for enabling interpersonal experiences. Today we can share time with an artisan, appropriating his techniques, breaks, materials, impressions, and desires – and redesigning the final product to our needs; we can learn traditional dances by temporarily integrating a folkloric ranch; we can fish and cook with a fishing community; or we can share a studio with visual artists to learn how to paint. Despite the multiple languages – from which we expand our potential for expression – the feeling of belonging to a group, community, context, or a more global society emerges in us. Within this sense of collective identity, of which we are a part and contribute to its co-creation, we review and renew our own desire and individual identities.

Creative tourism proposes to deconstruct the boundaries between the transmitter and the observer, between the material and the immaterial, leveraging the creative potential of those involved to start to define a new intervention territory – the commons. It may be from this central concept of commons and the intersection between participatory artistic practices, civic and political participation, and the plurality of definitions associated with 'community' that new concepts such as creative tourism emerge and are articulated.

The physical frontiers of the 21st century have become porous, bringing together geographies, cultures, ethnicities, and generations. Mobility between territories, ubiquitous communication technologies are, more than ever, tangible realities and accessible to the majority. Young people operate in a global, community-generating world, with a wealth of references, languages, and liberating processes in the face of anthropocentric notions in today's society.

It remains to be seen whether the identity of communities becomes a diffuse, globalized, and uprooted concept released from its original context or whether it organically adapts itself through the appropriation of new instruments of survival, communication, and expression. It is in this still diffuse digital age, in which young people try out alternate models of civic and political participation, that communities see the urgency to reconfigure themselves. Perhaps in a reciprocal dialogue where the communities simultaneously serve as a ground for experimentation and are potentially empowered, or weakened, by the proposed participation logic.

More than creating new concepts, products, or economically based activities, today young people demand an unceasing exploration and reinvention of production methods based on participatory models. Practices such as creative tourism are subject to this condition, leading to a future that brings generations closer together and promotes new forms of communication and knowledge transfer.

It remains for us to contribute, the best we can, to these new paths that are being traced, in a space of transgression between the physical and the virtual, and that challenge us to new and promising configurations of democracy and community. Perhaps it will be in the freedom and humanity proposed in the action of creative tourism that we will find an alternative practice to better understand the role of youth and the evolution of communities.

Acknowledgement

This chapter was developed within the CREATOUR® project (no. 16437), which is funded by the Portuguese Foundation for Science and Technology (FCT/MEC) through national funds and co-funded by FEDER through the Joint Activities Programme of COMPETE 2020 and the Regional Operational Programmes of Lisbon and Algarve.

References

Bauman, Z. (1998) Sobre o comunitarismo e a liberdade humana, ou como enquadrar o círculo. In: Bauman, Z. *O Mal-estar da Pós-modernidade*. Jorge Zahar Ed., Rio de Janeiro, Brazil.

Bauman, Z. (2005) *Identidade.* Zahar Editora, Rio de Janeiro, Brazil.

Hall, S. (2005) *A Identidade Cultural na Pós-Modernidade*. DP&A, Rio de Janeiro, Brazil.

Kostera, M. (2002) *Postmodernizmw zarzadzaniu*. Pozman, Poland.

Negri, A. (2016) *O Comum como modo de produção*. Caderno 52. Edições Chão de Feira, Belo Horizonte, Brazil.

27 Creative tourism as a strategic approach for decolonial thinking and doing in Namibian tourism

Melanie Sarantou* and Satu Miettinen

University of Lapland, Finland

This chapter explores how creative tourism can be applied as a strategy for decolonization, inclusion, and participation. Creative tourism is about engaging tourists in activities that stimulate them to use their creative potential (Runco and Acar, 2012) and confidence (Kelley and Kelley, 2013) in a cultural tourism setting. Creative tourism can occur through either total immersion in cultural activities, for example singing, craft production, painting, or through less interactive roles, such as being spectators in a concert, museum, or more informal cultural settings, for example, community theatres or cultural events.

Often the role of learning and developing new skills with a local community or engaging in creative processes where the outcomes may be concrete artefacts, is essential in creative tourism. This kind of creative interaction presents opportunities for storytelling (Segalo *et al.*, 2015) and informal communication that can offer a safe space to share life stories and personal histories (Karppinen *et al.*, 2018) that can work as counter narratives for colonial pasts. Thus, creative tourism can offer new ways for value co-creation, supporting new divisions of roles and question power structures between various stakeholders throughout the tourism experience. Further, creative tourism can enable the creation of new value offerings from several perspectives: improved value for local craft production, value in artistic expression, as well as collaborative management strategies, practices, and marketing (Miettinen, 2007).

Decolonial thinking and doing emerged in the global south. While postcolonial theories are intrinsically linked to European Western philosophy, decolonizing approaches seek to depart from postcolonialism by looking at the peripheries of colonization, thus including earlier and later incidents and results of colonialism and hegemonic power structures (Mignolo, 2009; Battell Lowman and Mayblin, 2011). The foundation of decolonial thinking is to offer a counterpoint against modernity as a result of colonial structures and processes. What postcolonial and decolonial theories have in common is that they are both political projects that seek to disrupt epistemological hierarchies (Battell Lowman and Mayblin, 2011). Decolonial thinking and doing proposes critical thinking towards Eurocentric and other global power systems in order to limit such hegemony (Mignolo, 2011).

This chapter reflects on postcolonial discussions as the case studies are set in the context of Namibia. This society with its current structures continues to face the consequences of colonial regimes and postmodernism (Ashcroft *et al.*, 2006) that prevails in tourism (Munt, 1994; Nuryanti, 1996). Tourism can be viewed to support hegemonic systems that often support implicit epistemological hierarchies, but creative tourism can promote participatory, tolerant, and inclusive collaboration within tourism communities, thus counteracting dominant and implicit narratives (Sarantou, 2014). Decolonial thinking and doing may be considered a strategy to undo or deconstruct

*Corresponding author: Melanie.Sarantou@ulapland.fi

 Creative Tourism: Activating Cultural Resources and Engaging Creative Travellers (eds. N. Duxbury, S. Albino, and C. Pato de Carvalho)
DOI: 10.1079/9781789243536.0027

the dominant narratives that are associated with tourism, such as some postmodern imagery. Through participatory and inclusive approaches, creative tourism could be an active strategy for the kinds of transformations that result from rethinking and redoing tourism products and services. Creative tourism may further enable resistance and/or disobedience (Mignolo, 2009) that is critical for the disruption of prevailing political and social structures.

Respectful design (Sheehan, 2011; Tunstall, 2013) and service design are important frameworks (Reitsma *et al.*, 2019) that question: (i) the role of leadership; (ii) utilization of power; (iii) ownership of process, outcomes, ideas, and material culture; and (iv) the role of Indigenous community and local knowledge when designing creative tourism experiences. It is essential to foster participatory art-based methods and embrace equality in the actions, when and where possible. The complexities in grasping histories and colonial structures that are implicit in societal, educational, and institutional structures, as well as in the context of tourism (d'Hauteserre, 2004), are immense. The tedious and necessary processes of analysing and explicating such structures underpins the colonial political project of decolonizing (Paschen, 2010) practices. It is important to recognize and understand that these complexities are present and real, and as such can impact various processes and practices. This chapter seeks to deliver a critical analysis on how decolonizing strategies and creative tourism can be mutually supportive and contribute to sustainable community development, followed by a discussion on how creative tourism and artistic expression can positively impact on participation. Narratives, arts-based approaches, improvisation, embodied learning, and local knowledge hold the keys to decolonizing practices as they can create and sustain empowerment. These practices give room for creating and expressing personal stories, transformative social practice, and counter narratives for past oppression and/or animosity (Sonn *et al.*, 2013).

Creative craft practices hold the potential for becoming interactive processes in which craft practitioners, with tourists, can express and accept themselves through the artefacts made. Such outcomes can be presented to others and enjoyed, thus resulting in feelings of empowerment by the makers and tourists alike. Such practices can also enable communication and narrative sharing between craft practitioners and tourists, enabling learning and healing. This chapter considers creative tourism activities in terms of craft making. It draws on two Namibia-focused doctoral thesis projects and ongoing research: one investigated creative tourism in Namibia (Miettinen, 2007), while the other mapped postcolonial Namibian craft and design narratives and identities (Sarantou, 2014). Both analyse how craft production can support decolonial thinking and doing from different perspectives. In both cases, the critical role of narrative practices in craft making production is explored.

The methodology is based on ethnography and reflection. The chapter draws on longitudinal research data and creative tourism case studies from Namibia between 2005 and 2019. Two examples are considered: the Grashoek and Penduka communities in Namibia. Ethnographic observations and research diaries were used to collect data from the field in 2011, while more recent research (2007–2019) was based on self-reflection and reflective discussions between the researchers (Fook, 2011). Both PhD processes span over a decade (2002–2014) and created extensive data sets on tourism and creative production in Namibia. Additional observational data from the My Dream World project (2010–2012) will be used. The Namibian craft organization, Penduka, participated in this project. In a follow-up study in 2017, one of the authors collected interview data with women who work for Penduka on topics of women's and personal empowerment. Co-design workshops and arts-based approaches, such as textile art installations, were employed. All the mentioned field studies used several methods for data collection. Apart from observations, interviews, focus group and reflexive discussions, note taking and collaborative note taking were used, but this chapter, more particularly, draws from observations and note taking.

The longitudinal nature of the projects enabled the documenting of the shift of approaches from postcolonial theories to (personal) approaches and discussions. These processes resulted in various collaborative processes with and by Namibian communities that were funded by the Local Cooperation Fund of the Embassy of Finland in Windhoek and the Finnish Ministry of Foreign Affairs. These activities were critically reflected on, evaluated, and reported to the ministry.

Theoretical framework for decolonial thinking and doing on Namibian creative tourism

Theoretically, the paper is grounded in the discourses involved in postcolonial Namibian and South African tourism (Miettinen, 2007; Hottola, 2009; Sarantou, 2014). Additionally, literature is considered for defining decolonial thinking and doing (Mignolo, 2009), socially engaged arts (Bishop, 2005; Badham, 2010), and decolonizing approaches in creative tourism. A theoretical framework for decolonial thinking and doing needs to borrow insights from postcolonial discussions related to Namibian tourism as well as an understanding of the nature of creative tourism, craft production, and other creative processes that can enable more equitable relationships between stakeholders.

Creative tourism can serve as a strategy for initiating encounters with local communities. It engages the senses, such as smell, touch, hearing, and seeing, which can enrich cultural experiences, especially when making crafts, which are tactile and embodied processes. Tourists can be involved in craft making or learning about various techniques with the craft makers and their communities of practice. Such personal encounters between craft makers and tourists may engage local community members, thus cross- and intercultural experiences may come about, including learning about one another, such as historical and social contexts. Such approaches may bring about the development of new kinds of tourism products and services that come forth from the encounters and roles between craft practitioners and tourists in diverse processes of value co-creation (Binkhorst and Den Dekker, 2009). The value for the tourism experience can thus be collectively constructed through learning, the sharing of information, exchanging skills, and immersion into cultural activities.

When attention shifts from objective authenticity in the relationship between craft making and the tourism trade to the process of creative tourism, some of the power structures and static images can be redefined (Miettinen, 2007). In the context of textile heritage and craft making, tourism sites can be places where heritage, such as the use of traditional textile techniques, is lived and performed (Richards *et al.*, 1998). Traditional crafts and their production can be attractive for cultural tourism (Horjan, 2011). At craft or creative tourism sites, living heritage is about people and the artefacts made and used within the spaces. Living heritage sites are, therefore, specific environments, in which individual and group identities are continuously performed and shaped (Sarantou, 2014). Within these environments, social and material meanings are continuously changing (Sarantou, 2014).

Postcolonial discussion on Namibian tourism products and services

Participatory approaches in southern Africa should draw on the inherent strengths of narratives and the roles of stories in African hospitality as media that fabricate connections between people and their material culture (Sarantou, 2014). The importance of imbedded narratives in craft objects is illustrated by their relationship to tourism in southern African craft and design worlds. Exotic and romantic values are connected to crafts from 'developing' countries because these same values and meanings are connected to tourism and the places and environments people seek as escapes from their everyday lives (Sarantou, 2014).

Craft objects and photographs are typical memorabilia that tourists accumulate during their tourist experiences. Such objects hold romantic memories of exotic people and places (Sarantou, 2014). This connection between tourism and crafts continues to fuel exotic and romantic ideas about crafts. In some cases, tourism can have significant potential for ensuring the cultural, social, and economic sustainability of craft people and their communities. Craft purchases can benefit both the tourist and the craftsperson. Tourists gain and construct personal symbolic meanings during their encounters with 'other' cultures and craftspeople benefit economically through craft sales, while their cultural identities and traditions are maintained through craft practices that may raise awareness of their significance (Miettinen, 2007).

Enhancing local creative practices means that resources such as local materials, expertise, and labour are utilized, and the physical transportation of artefacts and goods are minimized. This approach develops human creativity and inclusion because design solutions focus on local challenges and importance of sustainable use of local resources (Fletcher and Grose, 2012). In the process, jobs are created, and resources are made use of at a local level. These locally focused approaches are essential for environmental sustainability not only in

Namibia, but also globally (Sarantou, 2014). Such approaches increase profitability, help in creating local markets, and do not create an overdependence on tourism.

The imagery used in the promotion of tourism products and services has been discussed in the postcolonial tourism discourse as representing 'otherness' through peoples and landscapes (Wels, 2004). The images depend on the static image of tradition and stereotypes of 'African' culture (Miettinen, 2007). The majority of the pictorial representations in the Namibian tourism context carry implicit semiotic representations that are similar to colonial representations, showing images of Indigenous people in an unchanged landscape of the past (Miescher and Rizzo, 2000). Promotion materials strongly depend on the desires of tourists for the exotic, traditional, and authentic, thus these representations construct meanings that are based on a traditional image of the local context (McLaren, 2004) instead of contemporary ways of living. Tourists also travel to experience aspects of past traditions that they cannot personally experience, but (want to) imagine, and so tourists uphold ideas of an immutable 'past', which is often socially constructed (Molz, 1998).

However, some craft communities use a new and challenging kind of visual discourse in tourism promotion; thus, one can argue for the shifting from postcolonial to decolonial thinking and doing through resistance and disobedience (Ndlovu-Gatsheni, 2013; Craggs, 2014). Even though local communities have scarce resources, they often find ways to communicate a different set of values through the visual materials they produce, which can often be used for tourism promotion (Miettinen, 2007). Often local communities operate on the margins of global economic power structures which gives them freedom to communicate their own set of values and ideas. The visualizations of various tourism products differ depending on how they are made. When local makers illustrate their own products or services, they often communicate values that depict them as members of a community. Through self-selected visualizations, the craft communities are able to resist dominant pictorial discourses that are often associated with tourism (Miettinen, 2007). They can represent themselves and their communities as active participants in community development instead of passive objects (see Rodney, 1997) or subjects of a colonial gaze (Miettinen, 2007; see also McLaren, 2004).

In the development of creative experiences, it is essential that tourists can also develop their own narratives and engage their imagination in the activities they undertake. One of the potentials in creative tourism is that direct interaction between tourists and the local communities can undo the usual power relationships of the host–guest relationship (Richards and Wilson, 2006). Additional negatives of tourism in southern African communities are that the host may feel resentment towards visitors who are the descendants of those who once raided their cultures and removed tangible and intangible assets and now seem to expect Indigenous people from these regions to superficially 'perform' their cultures as if essentially 'divorced from historical process' to 'justify their colonial policies' (Doğan, 1989; Sarantou, 2014). In effect, tourism itself is essentially an intervention and performed activity (Sarantou, 2014). It is incumbent upon tourists to adopt an ethical approach when they visit what are still vulnerable communities due to the unequal power relations that exist between the visitors and the visited (Sarantou, 2014). It is evident that tourism influences the viability of craft practitioners in Namibia and other parts of southern Africa. In the light of postcolonial theory, postcolonial tourism requires host communities to be accorded respect and dealt with sensitively and responsibly (Sarantou, 2014).

Decolonizing place and the role of creative practices

While published tourism research considers postcolonial theory, it has little to contribute to decolonial thinking and doing due to the large body of information that is still focused on implicit colonial values (Chambers and Buzinde, 2015). In line with the critical turn in tourism, which proposes the decolonization of tourism research (Aquino, 2019), this chapter suggests that creative tourism, narrative, and creative practices can enable such a turn to be practically implemented through inclusion and participation by local communities. Creative practices and methods have been decolonized through place-based arts, which adopted a deep consideration for 'place ethic' (Lippard, 1997). This ethical consideration demands respect for any place that reaches further than only drawing from aesthetic inspiration that is known to derive from the 'tourist gaze', which often results from superficial and short-term interaction with place. Place ethics consider the cultural value of place, acknowledging that it is deeply rooted in cultural meanings and traditions, which are often rendered invisible or silent due to hegemonic forces at play. The role of creative practices, such as arts, in creative place making (Markussen and Gadwa, 2010) is acknowledged by governments and as a result, embedded in cultural and development policy (Grodach, 2011). Creative place making through creative tourism activities can enable ways to make power shifts or propose new roles and ways to experience and comprehend the narratives by local communities.

Postcolonial and postmodern tourism discourse critically reflects on the dominant narratives that romanticize the exotic 'people' living in Africa and their even more exotic 'things' (Sarantou, 2014). These dominant narratives can be harmful as they propose otherness, which in turn reproduces systematic representations of unrealistic ideals (Miettinen, 2007). These dominant narratives need unmaking for counter narratives to emerge so the political rights, social interests, and ecological concerns of communities can be recognized (Stanley, 2007). Harmful dominant narratives can be connected to unjust political systems such as colonialism and Apartheid, examples that apply to Namibia (Sarantou, 2014; Miettinen, 2007). Such harmful narratives potentially stimulate negative behaviours that increase ever-growing global problems such as overconsumption, depletion of natural resources, and the undermining of human well-being. Negative behaviours harm the well-being of often peripheral communities, the 'natural' environment, and valuable opportunities for the diversification of livelihoods, while they stimulate social and economic inequity and isolation (Sarantou, 2014).

Participatory and collaborative approaches that draw on creative practices, arts, and craft making enables interactive and inclusive ways for designing services for tourism, especially creative tourism services. Participatory and creative practices can enable more innovative and multi-modal ways for counter narratives and resistance that enable decolonial doing within tourism fields. The approaches have in common the democratization of the creative practices through collaboration, with lesser interest in discipline-specific aesthetics, thus making it more approachable and accessible to communities (Bishop, 2005; Badham, 2010).

Attempts in Australia to decolonize community cultural development (CCD) placed an emphasis on creative practices 'with communities', which entailed community-based creative practices that are based on self-determination and cultural growth through community capacity building, narrative practices, and expressions of identity (Badham, 2010). During the early 1990s, the CCD approach evolved to represent creative engagement by communities. This approach proliferates aspects of cultural pluralism, storytelling, self-determination, embedding arts practice within existing working cultures and organizations, and sustaining initiatives over longer periods of time (Badham, 2010). Some examples exist of community-based tourism activities that can lead to successful implementation of creative tourism, with the potential to create new counter narratives that challenge the proposed romantic representations of tourism and promote more equal and independent local agency.

Two creative communities from Namibia: Penduka and the creative community of Orwetoveni

Ethnographic observations and research diary entries were used to collect data during several field studies in Namibia, in which both communities participated. Both communities were regularly engaged in a variety of creative tourism practices, that enabled them to diversify their means for sustained livelihoods.

Orwetoveni

Penduka is located in the capital of Namibia, in the township of Katutura that offers a modern context to urban living and livelihoods in postcolonial Namibia. The township, with its complex Apartheid history, has developed into modern Namibian suburbia. Penduka and its activities are mainly organized by local women, with international volunteers assisting from time to time. The Dutch Penduka organization, which was involved in the establishment of Penduka some 20 years ago, continues to offer support. The Penduka craft groups, who are situated in Katutura and Otjiwarongo, are widely acknowledged for their embroidered textiles that narrate stories about life and activities on Namibian commercial farms where these women had lived at some stage of their lives, usually in their youth. Although many of these women live in townships in Windhoek and Otjiwarongo, others who embroider for this group live on commercial farms in the surrounding regions of Khomas and Otjozondjupa.

Groups of piece workers embroider pieces of cloth that are then sewn and finished into artefacts such as household linens and small accessories. The craft organization manages remuneration for piece work through some of their group leaders' bank accounts, while they use courier services to transport embroidered piece work and raw materials between Windhoek and Otjiwarongo. The artefacts are all sewn up and finished in this organization's factory in Windhoek where between 7 and 12 skilled women use industrial sewing machines to fill orders and manufacture artefacts for their outlets. The embroidered textiles are used in different product collections.

At the Orwetoveni craft producing site in the Otjiwarongo communal farm area, located in the Otjozondjupa region of Namibia, which is part of Penduka, a group of women from various Namibian and Indigenous cultural groups – including the Damara, Ovambo, San, Tswana, and Herero – regularly gather to continue their craft practices. The languages spoken by these women are mainly Damara/Nama (Khoekhoegowab) and Afrikaans. There are between 600 and 800 women in the Otjozondjupa region who work as members of this embroidery group from time to time. Some of the women in the group embroidered regularly, while others managed to fit in craft production when they could, depending on the demands of their domestic responsibilities. The translator explained that many of the group's members had to travel a lot (so that they, their partners, or other family members had access to work) and therefore lived somewhat unstable lives depending on work opportunities for them and their partners or other family members. Orwetoveni embroidery is sold at the Penduka shop and Namibian Craft Centre in Windhoek. Some of their artefacts are also exported to the Netherlands and Norway through volunteer efforts and collaborations between Namibia and these countries.

Embroidery themes vary, but the most common narratives are about farm life. These narratives are mostly expressions of individual experiences and reference themes such as partnerships between people, love, work, domestic activities, living and coping with HIV, domestic violence, and other social issues, such as the dangers of unprotected sex. Illness and poverty are also expressed. There are many gendered activities in the themes, illustrating a traditional lifestyle and stereotypical roles for men and women. Activity such as bickering, gardening, harvesting, collecting and carrying water, and looking after domestic animals are portrayed, while men are depicted as stereotypically masculine. Narratives about men picture violence, hunting, driving cars, and chopping wood. Children are depicted at play or going to school and looking after domestic animals. Illustrations of the women depict them busy with domestic activities.

Orwetoveni provides a safe environment so that the craftswomen are able to bring their young children with them. When I visited, at times there were 14 children playing in the yard, while about 20 women were embroidering. Some children were playing and running around. They were busy and vocal, often demanding attention. Other children were relaxed and quiet, playing close to their mothers with tape measures, mimicking sewing motions, and playing with cut-off yarns. Two mothers were carrying their children in *abba karoses* where they slept peacefully. Most children were preschool age, although one teenage girl was present, playing on her mobile phone while assisting with the care of her siblings and other children. This craft production site exemplifies the mingling of craft and domestic activities. Three young women who were embroidering were described as early school leavers by the translator. They were being taught crafting because they had few other work opportunities. Women in this craft group use mobile phones to communicate with one another about work and related opportunities. There are three group leaders in the Odjozondjupa region who assist the craftswomen in three different locations.

The animals and other 'stories' were embroidered in bright primary colours, thus turning into uniquely identifiable animations. The women sat on chairs, boxes, and empty plastic crates and they worked in their laps, where they kept threads and tape measures, while some had bags in which they carried the necessities for their embroidery. Movements, sounds, smells, strokes, and images all contributed to the sometimes implicit and sometimes physical contact between these embroiderers and their surrounding material realities. That they took this mélange of activities and actions in their stride was demonstrated by the relaxed atmosphere that prevailed throughout my observations. Blanket stitches were favoured by this group, but in the past chain stitch, back stitches, knotted stitches, and satin stitches were used, but are now less common. The women reported that they had learned blanket stitch about 4 years ago and since then preferred to use it because it is less time consuming. Moment-by-moment, the embroiderers were guided by their knowledge of what to do, even as they were entangled in actions, emotions, and previous experiences during the process of making and expressing new stories on cloth (Miettinen, 2007).

Grashoek

The Indigenous Ju/'hoansi San Living Museum was initiated in 2004 with the Ju/'hoansi San community at Grashoek, located close to the C44 road between Grootfontein and Tsumeb (Sarantou, 2014). Grashoek is a positive, pleasant, and open community of Ju/'hoansi. After arriving at Grashoek, tourists are met by Ju/'hoansi in Western clothing but they will notice that the facilitators of the cultural events will soon after change into traditional attire, and to some extent this indicates the performative aspect of the Ju/'hoan Living Museum (Sarantou, 2014). Members of the Grashoek community normally wear Western clothing. Homes, schools, and commercial premises are mostly built using Western materials and methods, although you could find some constructed from local wood and clay. Specific cultural products are created and then sold for profit, after which community performers don their cultural costumes with handcrafted personal adornments to perform traditional rituals. Roles for community members include those of hunters, dancers, craft sellers, storytellers, craftspeople, translators, guides, and many more. Popular activities that tourists participate in include cultural dancing, craft making, and hunting.

In Grashoek, most community members were involved in an activity that sustains one or more of the 'products' offered to tourists. These include traditional hunting, dancing, and game playing along with storytelling, tool making, craft workshops, gathering of bush foods, and accommodation. A wide variety of artefacts produced by craftspeople are also sold to visiting tourists. There is a significant flow of cash to the Grashoek community due to the income they generate from artefacts and services sold to tourists at the Living Museum. They independently produce all products and services sold to tourists, while their private lives are lived at Grashoek and their income-generating activities conducted somewhat separately from Grashoek, in the nearby 'Living Village'. Tourists' intrusion in their private lives may, therefore, be less of a challenge for them. The atmosphere in Grashoek's public spaces was relaxed due to specific 'boundaries', such as a designated area, organized activities, programmes, and preset costs, having been put in place for tourists according to terms agreed upon by the community.

Tourists can meet the artefact makers, see how these craftspeople live and work, and learn how they relate to the materials used in their artefacts. Tourists can even learn how to make these crafts during a workshop offered by the community. 'Hunting' excursions with San bows and arrows are also offered as a cultural activity. Hunting is controlled by the Nyae Nyae Conservancy and therefore these 'hunting' excursions are offered as educational activities in which tourists learn about tracking and the traditional hunting tools of the Ju/'hoansi instead of hunting to kill animals unnecessarily. A large number of men were craft producers of both traditional hunting tools and jewellery. At the craft selling site, no imported resources were used in any of the artefacts made. Even the twine used for stringing necklaces and making arrows and bows were usually made by hand using animal tendons or plant fibres.

At the Grashoek craft selling site women and men interact with tourists to show and sell their artefacts. Women used a wide variety of natural seeds, pods, ostrich eggshell, wood, and leather to make the personal adornments they sold to tourists. Ostrich eggshell, however, was used sparingly as ostriches have become almost extinct in this region, with the result that some NGOs 'import' ostrich eggshell from farms in South Africa to sustain the production of crafts in the nearby Nyae Nyae Conservancy (Sarantou, 2014). The Grashoek community did not have access to such supply in ostrich eggshell, thus they relied on thrift, finding natural seeds and pods from their direct environment to produce the memorabilia sustainably. Although Shostak (1981) argues that craft production, like most other aspects of Ju/'hoan life, is not based on gender or hierarchical structures, this aspect of Ju/'hoan life is changing due to cultural influences and changes to their lifestyles, thus more gendered dimensions to Ju/'hoan craft production were observed.

All artefacts displayed in this craft site were crafted by hand. Women made personal adornments such as necklaces, bracelets, and embroidered leather wraps, while men made personal adornment and weapons such as bows, arrows, arrow vessels, and knives. At Grashoek there were no disadvantaged craftspeople trying to beg buyers or trade high-quality artefacts from them for distressingly small amounts of money. Producers around Grashoek were dependent on materials found within their immediate environment, such as pods, seeds, wood, animal tendons, and recycled cotton, yarn, and paper. These delicately fabricated artefacts sustain the Grashoek community in a particular way, because they are unique memorabilia that are purchased by visitors for the sake of the meanings that are already invested in them at the site where they are purchased.

Discussion

Opportunities for using creative tourism as a strategy for decolonial thinking and doing within the Namibian tourism field are illustrated in the two exemplary cases reflected upon in this chapter. The creative tourism activities that were offered at both sites were different and diversified. Penduka is a creative tourism destination in Windhoek. It offers several activities to tourists, for example, learning about Penduka craft production and the stories depicted in Orwetoveni embroidery, and observing textile printing and sewing activities at workshops. Storytelling and narratives that are shared about the textiles and how Penduka contributes to local income generation offer key cultural experiences to tourists.

Penduka has succeeded in creating a new kind of income generation model that is related to the tourism market, which enables women to work independently. This kind of activity supports the distribution of power, leaving room for decision making at a local level. It also contributes to the (re)shaping of dominant narratives creating a counter-narrative on active women working to provide for their families and actively developing their business models and opportunities in the local tourism context. Penduka women are represented as skillful craft producers and as an innovative group that are actively engaged in developing their collections of artefact offerings rather than stagnating in past histories and power structures coming forth from the colonial era and Apartheid. To visiting tourists, Penduka's story is a fresh representation of Namibian urban and rural communities, multi-cultural contexts, and the living community of Katutura.

The living village at Grashoek represents a wide variety of creative tourist activities that sustain livelihoods, educate their children, and embody their hybrid and changing lifestyles, which can be described as partly traditional and modern, in a remote Namibian community. Despite their remote location, the Grashoek community succeeds in generating a sustainable income from tourism. Supported by mostly international volunteers from Germany, the community also created a unique income generation model that enables the community to work independently. Grashoek represents a stark contrast to the Ju/'hoansi communities living elsewhere in the vicinity where Indigenous Namibian craft practitioners were often found begging tourists to buy their artefacts at unsustainably low prices. These kinds of tensions that are created by dominant power structures were not observed in Grashoek where decision making was at a local level, leaving the community empowered and enabled, at least to some extent, to influence their own futures. The community engaged in performative and narrative practices to familiarize tourists with their cultural knowledge about their traditional craft and hunting techniques.

A significant theme that emerged from observations in both communities was the important role of stories, as embodied and performative practices. The variety of stories that were told, crafts made, and cultural and tourism practices performed illustrated the intrinsic creativity that prevails in Namibian communities. Stories were shared through singing, gossiping, talking, demonstrating, showing, touching, and acting. Stories were received by listening, crying, laughing, giggling, and gesturing (such as nods or shrugs). Different ways of sharing stories in working processes were demonstrating, instructing, and drawing. Varieties of stories were found in working processes, skills, and methods such as embroidering, weaving, carving, cutting, drilling, painting, sewing, and tool making. All these activities and actions eventually formed valuable individual and unique storylines because of the way they were told.

Orwetoveni was a unique site that was based around the home lives of the textile embroiderers. It may be argued that creative tourism in this setting may become too intrusive to sustain as many craftswomen had to balance child rearing with craft and other income-generating activities. The Penduka premises, where the crafts are sold at a business premises for creative tourism, presents a more suitable and less intrusive site. At Grashoek tourists arrive at an Indigenous village after many hours of driving, but the unique setup enables the community to separate, to some extent, their private lives from tourism activities. The setup of the craft shop almost melts into its natural environment and the arrangement of tourist and private spaces in Grashoek illustrates creative placemaking on the terms of the Indigenous community. From the two examples presented from Orwetoveni and Grashoek, emerging themes were used to create a framework for decolonial thinking and doing by using creative tourism as a strategy.

Reshaping narratives

In the examples discussed, the two creative tourism communities, Penduka and Grashoek, are communicating and expressing their skills, traditional knowledge, and identities, but also demonstrate innovation and novelty in their product and service offerings. The communities have reshaped and created alternative narratives that challenge the traditional 'African and exotic' dominant narratives that are connected to tourism developments.

New means for sustainable livelihoods

In her study of tourism and craft communities Miettinen (2007) shows that crafts sold to tourists were one of the main sources of income for many of her participants. New models for sustaining livelihoods are coming forth from Penduka and Grashoek, including business development and the redistribution of power structures in favour of the local communities. Both communities of Penduka and Grashoek have ownership of the services and products which are offered. They are involved in creative tourism activities that include craft making and skills transfer, drawing on local traditions, but they have established boundaries that determine their personal spaces and activities, as well as the roles that visiting tourists will play in the interactions that are offered (Simpson, 2008).

Narratives of care and interpersonal relationships

Women used stories and explanations to illustrate the effort, time, and skill required to make crafts. Individuals were demonstrating, explaining, and breaking down processes in steps and sequences so that storylines were able to emerge alongside the sequences emerging from craft processes (Sarantou, 2014). These groups' interactions with one another, with the children whom they were looking after while producing crafts, their working processes and interactions with their raw materials all revealed narratives of care, support, and social cohesion (Sarantou, 2014). Apart from the more obvious roles that narratives play in artefact production, such as providing themes and inspiration for craft making, narratives related to craft production assist craftspeople to make sense of their lives and to translate their life situations and their life narratives into the artefacts they produce (Sarantou, 2014). The Orwetoveni embroidery groups and their collaborations with, and inputs in, a collaborative business model provides creative spaces for developing interpersonal relationships and giving care in environments where work overlaps with other care activities towards families or the environment. This kind of creative and personal space is unique in a postmodern society and responds well to the idea of decolonial doing and dismantling traditional structures and hierarchies.

Creative approaches

Namibian artefact makers get various kinds of emotional sustenance from their creative practices that support their identity expressions. These benefits include: (i) gaining a means of self-expression through making processes in which shaping forms and materials 'converse' with makers, hence reinforcing their sense of purpose and of self; (ii) the therapeutic qualities creative, improvised, and repetitive practices such as artefact making offer; (iii) pride in what they are doing and accomplishing; (iv) pleasure and creative expression found in the tactile contact with shaping materials into new forms; and (v) recognition received from clients and customers, the general public, the media, and smaller social groups such as friends and families (Sarantou, 2014).

Local knowledge and identities

The embroidered expressions from Orwetoveni narrate both reality and fiction. These animated narratives are for the narrator experienced realities that are informed by local knowledge, selected from complex identity processes and expressed in identity statements (Sarantou, 2014). They are narrative transformations and a means of making sense of life for these craftspeople. That means the transformative power of fiction assists these women to express their realities in their embroidered narratives. The Grashoek community uses their local and traditional knowledge of collecting, selecting, refining, and applying local materials from the environment to produce delicate crafts sustainably. Their identity expressions are used for wearing, adornment, and hunting, and so they serve to express identities within their local community and in performed tourism interactions. In these hybrid spaces, Indigenous, traditional, and modern identities overlap, enabling powerful expressions that are closely related to local realities instead of implicit 'exotic' expectations by tourists.

Embodied learning

Learning is broadly understood as the acquisition of skills or knowledge through experience, iteration, study, or being taught (Sarantou *et al.*, 2019b). In both craft communities, bodily spaces and surfaces, such as laps, were used in creative tourism practices. Thus, these more private bodily spaces, that were flexible and expandable to accommodate dependent children, tools, materials, and making processes, supported embodied practices and learning. Changes in bodily positioning also offers solutions to resource limitations. Lap work not only supports embodied learning through the work of hands, but it provides space for storytelling, thinking, and reflecting while doing and making (Sarantou *et al.*, 2019b). Makers are often more willing to experiment, improvise, make mistakes, and redo processes to find best solutions and new means of expression when they are engaged in such private and embodied practices and spaces (Sarantou *et al.*, 2019b). Such practices offer new windows to cross-cultural learning and communication. Some forthcoming advantages may be the undoing and reshaping of dominant narratives that are implicit in tourism fields.

Improvisation

Improvisation as an approach towards decolonizing practices in Namibian craft and design worlds has been illuminated (see Sarantou *et al.*, 2019a). Namibian artefact making is holistic and makers are not distanced from their making processes and environments, or the textilities of their raw materials. Rather, they are deeply involved in their procedures, materials, and social realities during making (Sarantou, 2014). Namibian craftspeople and designer–makers regularly draw on improvised processes because they are regularly confronted with design problems and the limitations that they face within their particular environments of making, which often lead to the making of unique artefacts in small quantities (Sarantou, 2014).

Conclusion

Miettinen (2007) confirms that many of her research participants were reluctant to speak about tourism's negative effects because these women felt they were dependent on income they generate from tourism. While the financial gains craftspeople get from tourism are obvious, its negative impacts on their worlds also need to be considered. The interaction with the Penduka and Grashoek individuals and their broader communities enables tourists to experience a different culture as part of modern life and modern times. Hence exotic notions of the Indigenous Ju/'hoan, or 'Africans' are minimized (at least for some visitors and tourists) while cultural awareness about these groups and way of life is raised. These are all reasons why the creative tourism practices of the communities are socially, economically, and most importantly, environmentally sustainable. Although their practices are cultural and place-specific, this does not mean that such communities should be excluded from mass market activities. Their approaches to tourism could be strategically redesigned with a critical awareness of sustainable outcomes.

Lessons learned

- Creative tourism can enable resistance for the shaping of new and alternative dominant narratives related to place and practices that can promote decolonial thinking and doing.
- Creative tourism can enable new innovation in product and service development that is based on local knowledge transferal, which can contribute to identity expressions, such as a variety of creative forms of making, including place making.
- Creative tourism can enable and promote sustaining livelihoods in local communities and in rural contexts.

References

Aquino, R.S. (2019) Towards decolonising tourism and hospitality research in the Philippines. *Tourism Management Perspectives* 31, 72–84.

Ashcroft, B., Griffiths, G., and Tiffin, H. (eds.) (2006) *The Post-colonial Studies Reader*. Routledge, London, UK.

Badham, M. (2010) Legitimation: The case for 'socially engaged arts'—navigating art history, cultural development and arts funding narratives. *Local-Global: Identity, Security, Community* 7, 84–99.

Battell Lowman, E. and Mayblin, L. (2011) Introduction from the editors: Theorising the postcolonial, decolonising theory. In: Battell Lowman, E. and Mayblin, L. (eds.) *Studies in Social & Political Thought* 19, 3–8.

Binkhorst, E. and Den Dekker, T. (2009) Agenda for co-creation tourism experience research. *Journal of Hospitality Marketing & Management* 18(2–3), 311–327.

Bishop, C. (2005) The social turn: Collaboration and its discontents. *Artforum* 44(6), 178–186.

Chambers, D. and Buzinde, C. (2015) Tourism and decolonisation: Locating research and self. *Annals of Tourism Research* 51, 1–16.

Craggs, R. (2014) Postcolonial geographies, decolonization, and the performance of geopolitics at Commonwealth conferences. *Singapore Journal of Tropical Geography* 35(1), 39–55.

d'Hauteserre, A.M. (2004) Postcolonialism, colonialism, and tourism. In: Lew, A.A., Hall, C.M., and Williams A.M. (eds.) *A Companion to Tourism*. Blackwell Publishing, Oxford, UK, pp. 235–245.

Doğan, H.Z. (1989) Forms of adjustment: Sociocultural impacts of tourism. *Annals of Tourism Research* 16(2), 216–236.

Fletcher, K. and Grose, L. (2012) *Fashion and Sustainability: Design for Change*. Laurence King, London, UK.

Fook, J. (2011) Developing critical reflection as a research method. In: Higgs, J., Titchen, A., Horsfall, D., and Bridges, D. (eds.) *Creative Spaces for Qualitative Researching: Living Research*. Practice, Education, Work and Society, Vol. 5. Sense Publishers, Rotterdam, The Netherlands, pp. 55–64.

Grodach, C. (2011) Art spaces in community and economic development: Connections to neighborhoods, artists, and the cultural economy. *Journal of Planning Education and Research* 31(1), 74–85. doi: 10.1177/0739456X10391668.

Horjan, G. (2011) Traditional crafts as a new attraction for cultural tourism. *International Journal of Intangible Heritage*, 6(7), 46–56.

Hottola, P. (ed.) (2009) *Tourism Strategies and Local Responses in Southern Africa*. CABI, Wallingford, UK, and Cambridge, Massachusetts, USA.

Kelley, T. and Kelley, D. (2013) *Creative Confidence: Unleashing the Creative Potential within us All*. Currency, New York, New York, USA.

Karppinen, S., Poutiainen, A., Kairavuori, S., Rusanen, S., and Komulainen, K. (2018) ImproStory: Social improvisation and storytelling in arts and skills subjects in teacher education. *International Journal of Education & the Arts* 19(9), 1–24. doi: 10.18113/P8ijea1909.

Lippard, L.R. (1997) *The Lure of the Local: Senses of Place in a Multicentered Society*. New Press, New York, New York, USA.

Markusen, A. and Gadwa, A. (2010) *Creative Placemaking*. National Endowment for the Arts, Washington, District of Columbia, USA.

McLaren, B. (2004) From Tripoli to Ghadames: Architecture and the tourist experience of local culture in Italian Colonial Libya. In: Medina Lasansky, D. and McLaren, B. (eds.) *Architecture and Tourism: Perception, Performance and Place*. Berg, Oxford, UK, and New York, USA, pp. 75–92.

Miescher, G. and Rizzo, L. (2000) Popular pictorial constructions of Kaoko in the 20th century. In: Miescher, G. and Henrichsen, D. (eds.) *New Notes on Kaoko*. Basler Afrika Bibliographien, Basel, Switzerland, pp. 10–47.

Miettinen, S. (2007) Designing the creative tourism experience a service design process with Namibian craftspeople. Doctoral dissertation. University of Art and Design Helsinki, Finland.

Mignolo, W.D. (2009) Epistemic disobedience, independent thought and decolonial freedom. *Theory, Culture & Society* 26(7–8), 159–181.

Mignolo, W.D. (2011) Epistemic disobedience and the decolonial option: A manifesto. *Transmodernity* 1(2), 3–23.

Munt, I. (1994) The 'other' postmodern tourism: Culture, travel and the new middle classes. *Theory, Culture & Society* 11(3), 101–123.

Molz, J.G. (1998) Tasting an imagined Thailand. *Culinary Tourism* 55(3), 53.

Ndlovu-Gatsheni, S.J. (2013) *Coloniality of Power in Postcolonial Africa*. African Books Collective, CODESRIA, Dakar, Senegal.

Nuryanti, W. (1996) Heritage and postmodern tourism. *Annals of Tourism Research* 23(2), 249–260.

Paschen, J.A. (2010) Decolonizing the gaze at Uluru (Ayers Rock). In: Burns, P., Palmer, C., and Lester, J. (eds.) *Tourism and Visual Culture, Vol.* 1. CABI, Cambridge, UK, pp. 64–75.

Reitsma, L., Light, A., Zaman, T., and Rodgers, P. (2019) A respectful design framework: Incorporating Indigenous knowledge in the design process. *The Design Journal* 22(sup1), 1555–1570.

Richards, G. and Wilson, J. (2006) Developing creativity in tourist experiences: A solution to the serial reproduction of culture? *Tourism Management* 27(6), 1209–1223.

Richards, G., van Oss, F., Fernandes, C., Sousa, L., Tzanakaki, K., and Miettinen, S. (1998) *Developing and Marketing Crafts Tourism: The EUROTEX Project*. ATLAS, Arnhem, The Netherlands.

Rodney, W. (1997) How Europe underdeveloped Africa. In: Grinker, R.R. and Steiner, C.B. (eds.) *Perspectives on Africa: A Reader in Culture, History and Representation*. Blackwell Publishers, Oxford, UK, pp. 585–596.

Runco, M.A. and Acar, S. (2012) Divergent thinking as an indicator of creative potential. *Creativity Research Journal* 24(1), 66–75.

Sarantou, M.A.C. (2014) Namibian narratives: Postcolonial identities in craft and design. Doctoral dissertation, University of South Australia, Australia.

Sarantou, M.A., Beaulé, C.I., and Miettinen, S.A. (2019a) Decolonising Namibian arts and design through improvisation. In: Bohemia, E., Gemser, G., Fain, N., de Bont, C., and Almendra, R.A. (eds.) *Conference Proceedings of the Academy for Design Innovation Management*, 2(1), pp. 174–186, doi: 10.33114/adim.2019.02_364.

Sarantou, M.A., Sillgren, S.K., and Pokela, L.M. (2019b) In her lap: Embodied learning through making. In: Jokela, T. and Coutts, G. (eds.) *Relate North: Collaborative Art, Design and Education*. InSEA Publications, Viseu, Portugal, pp. 84–107.

Segalo, P., Manoff, E., and Fine, M. (2015) Working with embroideries and counter-maps: Engaging memory and imagination within decolonizing frameworks. *Journal of Social and Political Psychology* 3(1), 342–364.

Sheehan, N.W. (2011) Indigenous knowledge and respectful design: An evidence-based approach. *Design Issues* 27(4), 68–80.

Shostak, M. (1981) *Nisa: The Life and Words of a Kung Woman.* Earthscan, London, UK.

Simpson, M.C. (2008) Community benefit tourism initiatives—a conceptual oxymoron? *Tourism Management* 29(1), 1–18.

Sonn, C.C., Stevens, G., and Duncan, N. (2013) Decolonisation, critical methodologies and why stories matter. In: Stevens, G., Duncan, N., and Hook D. (eds.) *Race, Memory and the Apartheid Archive*. Palgrave Macmillan, London, UK, pp. 295–314.

Stanley, C.A. (2007) When counter narratives meet master narratives in the journal editorial-review process. *Educational Researcher* 36(1), 14–24.

Tunstall, E. (2013) Decolonizing design innovation: Design anthropology, critical anthropology, and Indigenous knowledge. In: Gunn, W., Otto, T., and Smith, R.C. (eds.) *Design Anthropology: Theory and Practice*. Bloomsbury Academic, London, UK, pp. 232–250.

Wels, H. (2004) About romance and reality-popular European imagery in postcolonial tourism in southern Africa. In: Hall, C.M. and Tucker, H. (eds.) *Tourism and Postcolonialism: Contested Discourses, Identities and Representations.* Routledge, London, UK, pp. 76–94.

28 Connecting local artisans to tourism in extra-metropolitan areas: The importance of social embeddedness

Fiona Eva Bakas,^*

Lusófona University, Lisbon, Portugal

This chapter presents highlights from recent research into the emergence of artisan entrepreneur-mediators who link artisans to creative tourism initiatives in extra-metropolitan areas (i.e. rural areas and small cities) in Portugal (Bakas *et al.*, 2018). Creative tourism is aligned with contemporary trends to revive local crafts and traditions in rural areas as it stimulates artisan entrepreneurs to co-create and co-preserve local traditions (Duxbury and Richards, 2019) while engaging with the local community (Landry, 2010). In extra-metropolitan regions, economies are often fragile and small businesses often find it hard to stay solvent. 'The countryside' or the 'rural' must be considered as a place where the creative economy is differently manifested and articulated from the now standard 'creative script' based on cities (Bell and Jayne, 2010). It is imperative to understand the critical elements within a small business's operating environment or 'ecosystem' that support or thwart entrepreneurial activity (Kline *et al.*, 2014). With artists and artisans increasingly used to represent, market, and enhance the visual image of destinations, looking more closely at the link between artisan activity and tourism is timely (Morpeth and Long, 2016).

The artisans who were investigated in this research are part of a national project on creative tourism, CREATOUR®, in Portugal. 'CREATOUR®: Creative Tourism Destination Development in Small Cities and Rural Areas' is a three-and-a-half-year (2016–2020) interdisciplinary research-and-application project that was developed for the first time in Portugal, a network of creative tourism packages, many offered by artisans (see www.creatour.pt; accessed 21 April 2020). The CREATOUR® approach to creative tourism defines it as a tourist experience that includes four dimensions: active participation, creative self-expression, connection to place, and community engagement (Bakas and Duxbury, 2018). Creative tourism is fuelled by the new paradigm of the twenty-first century according to which tourists strive to find 'transformative' experiences (WTO, 2016), focus on self-development, and want to be active participants in tourism experiences of the 'everyday' (Duxbury and Richards, 2019).

Linking artisans, social embeddedness, and entrepreneurship

An artisan is viewed as a skilled person who creates objects of aesthetic and/or functional value, mainly by manual labour, using traditional craft techniques and/or materials (UNESCO, 1997). Artisans as businesses and residents in rural areas can serve to diversify rural areas' economic base and thus can be highly valuable in rural development programmes (Duxbury *et al.*, 2011). This is partly because when handicraft/artisanal production is paired with another industry, such as tourism, the entrepreneurial venture has a better chance of being profitable and contributing to rural development (Teixeira and Ferreira, 2018).

*Corresponding author: fiona.bakas@ulusofona.pt; creatour@ces.uc.pt

^This chapter was written while Fiona Bakas was a post-doctoral research fellow at the CREATOUR project at the Centre for Social Studies, University of Coimbra, Portugal

 Creative Tourism: Activating Cultural Resources and Engaging Creative Travellers (eds. N. Duxbury, S. Albino, and C. Pato de Carvalho)
DOI: 10.1079/9781789243536.0028

There is growing recognition that economic behaviour can be better understood within its social context, and that entrepreneurship is also embedded in social relations (Granovetter, 1985). The concept of social embeddedness can be summarized as the nature, depth, and extent of an individual's ties into a community and these links configure economic process. In the research presented here, social embeddedness is used as an analytical tool to illuminate the dynamism and complexity of entrepreneurial situations and to investigate the influence of social networks on entrepreneurial action (McKeever *et al.*, 2014).

In terms of entrepreneurship, a loose definition of an entrepreneur is 'someone who perceives an opportunity and creates an organisation to pursue it' (Bygrave, 1997, p. 2). Given the diverse cultural contexts within which entrepreneurship takes place, it becomes increasingly important to view entrepreneurship as an economic activity derived from sociocultural processes (Tucker, 2010). According to the entrepreneurial ecosystem approach, entrepreneurs are important players themselves in creating their ecosystem and keeping it healthy (Stam, 2015). While the concept of entrepreneurial ecosystems is largely used for high-growth entrepreneurship (Spigel, 2017), it is used here as a concept to investigate some of the social, political, economic, and cultural elements within a locale that support the development of innovative start-ups within the context of creative tourism in extra-metropolitan areas.

Findings

Table 28.1 presents an overview of the creative tourism activities of five artisan entrepreneur-mediators within the CREATOUR® project. Joining the CREATOUR® network seems to have acted as a catalyst for the emergence of the artisan entrepreneur-mediator activities to foster connections between artisans and tourism in these initiatives. Previously, the main contact artisans had with tourism was through fairs, where they exhibited their products and tried to sell them. Creative tourism enabled them to connect with tourists in a different way, and this was often achieved via a person who played the role of an intermediator.

Artisan entrepreneur-mediators take on leading roles in connecting artisans to creative tourism, including:

- advertising the creative tourism events;
- registering participants;
- collecting registration fees;
- booking the physical space where the creative tourism workshops take place;
- acting as guides;
- translating;
- transporting elderly artisans to the creative tourism workshops; and
- accessing funding programmes to develop the creative tourism workshops.

In short, they present their vision to the artisans and are responsible for advertising the events, registering participants and collecting registration fees, booking the physical space where the workshops take place, and acting as guides. As the proprietor of Quico Turismo says, 'I deal with the bureaucracy so the artists can be creative' (interview, March 13, 2018). These roles are taken up by the artisan entrepreneur-mediators since the artisans involved may have little experience with giving workshops and limited knowledge about contacting tourists and the various funding programmes that are available to develop tourism products.

A recurring theme among artisan entrepreneur-mediators is that since the creative tourism workshops often feature ancient techniques and the know-how lies mainly in the hands and minds of the elderly, many of the artisans are at an advanced age. As a new type of activity, the artisans may find delivering the workshops challenging and outside their 'comfort zone'. For example, both the proprietor of LOOM New.Tradition and Cerdeira Village comment on how the elderly craftspeople they invite to be part of their workshops often resist the role of teacher and workshop leader because it is not a role they have taken up before. The interviews indicate there may be a gap between the expectations of the entrepreneur-mediators and the willingness and the capacity of the artisans, and the need for time, training, and support to design and implement these new activities.

Regarding artisan entrepreneur-mediators' success in terms of social embeddedness, two aspects appear to be critical: connection to place and type of community engagement.

- *Connection to place.* While the majority of artisan entrepreneur-mediators have a strong connection to their place of operation in terms of them or their partner feeling 'local to the place' where they operate their creative tourism project, levels of social embeddedness are also influenced by the ways in which they express engagement to

Table 28.1. The CREATOUR® artisan entrepreneur-mediators examined

Artisan entrepreneur-mediator (and location)	Creative tourism project	Size of artisan network	Community engagement activities
VERde NOVO (Ribeira da Pena, Norte region)	*Linho de Cerva e Limões*: Offers workshops and guided tours on the cycle of linen – from sowing to weaving	Medium – 6 weavers	VERde NOVO has created a network of the community's weavers, bringing them together and helping them re-design some products and modify production techniques in order to meet contemporary market needs. Interested in running a community entrepreneurship programme. Local partners include linen field owner, tourism animation companies, and restaurants. Difficulties collaborating with local authorities
Cerdeira Village/ ADXTUR (Cerdeira, Lousa, Centro region)	Offers artist residencies and artisan workshops to tourists on a variety of arts, including woodcarving and pottery	Large and international	Cerdeira Village employs eight people from the locale in the business. For the fourth year, it will organize workshops for children on International Children's Day. Cerdeira Village exhibits items of each artist in residence at the local museum
Quico Turismo (Nazaré, Centro region)	*Nazaré Creativo*: To date, offers photography expeditions with a cultural-historic focus and workshops sewing key-rings in the shape of 'carapau' (horse-mackerel) fish, which are traditionally associated with Nazaré	Small – 2 artisans	Quico Turismo has been involved in numerous cultural activities in Nazaré, including publishing a book for children about Nazaré and making a documentary on the Nazaré fishermen. She aims to create an image of Nazaré that is not just about the 'big wave' (and surfing), showcasing its history and art instead
LOOM New.Tradition (Évora, Alentejo region)	Offers workshops and guided tours related to fabric-weaving, plant stamping, and dyeing and develops creative fabric-art related trails	Small – 4 female weavers/ artisans	In a context of 'slow fashion', LOOM New. Tradition is focused on helping to revive the Alentejo blanket tradition and preserving local customs by engaging local weavers and other creators in its activities
Proactivetur (Loulé, Algarve region)	Offers various half-day or full-day workshops on ancestral craft techniques such as cane basket weaving, palm weaving, and making traditional floor tiles	Large – more than 40 artisans linked through the TASA project (https://projectotasa.com)	As a travel agency, Proactivetur engages with the local community mainly through business connections with accommodation providers, restaurants, taxis, resorts, and tourism providers

'their' community(ies) when conceptualizing their entrepreneurial roles. Some entrepreneur-mediators view their community engagement mainly in business terms, such as co-operating with local restaurants and accommodation providers. For example, the proprietor of Proactivetur says: 'As a travel agent we are totally engaged with the local providers of inland communities, such as accommodation, restaurants, and taxis' (interview, March 16, 2018). For others, built into their conceptualization of entrepreneurship is a concern for helping the local community. For example, VERde NOVO has a deep sense of responsibility towards the artisans they work with, saying, 'It's not in our DNA to just do tourism animation and leave! We want to be part of the community and get public institutions to work with us' (interview, March 12, 2018).

- *Community engagement.* The importance of social embeddedness for entrepreneurial success is also expressed in terms of strengthening the links amongst the actors within the creative tourism entrepreneur ecosystem. Artisan entrepreneur-mediators make efforts to get private and public institutions to work with them in facilitating and hosting the creative tourism workshops. For example, a local hotel has provided space for a photography exhibition for VERde NOVO. The proprietor of Quico Turismo encourages the artisans in her micro-network to interact with each other by inviting them to each other's workshops. She feels that this interaction and bonding will 'help create a community of creative people' and that it also makes good tourism entrepreneurial sense. As she says, 'through reinforced bonds and increased collaboration, you can also attract more people to Nazaré' (interview, March 13, 2018).

Some of the artisan entrepreneur-mediators are increasing their social embeddedness by offering services to the rural communities that are otherwise missing. For example, artisans in rural areas often provide early training grounds for local young people, in the absence of the kinds of infrastructure available in the city (Luckman, 2012; Bakas, 2014). The proprietor of Quico Turismo created a book for children on the local history of Nazaré and the proprietor of Cerdeira Village is planning artisan workshops and an art exhibit for children in the fourth grade of a local school to coincide with International Children's Day. As the proprietor of Cerdeira Village says, 'Children are part of the culture that is forgotten' (interview, February 5, 2018).

Conclusion

The research discussed here was the first time that a critical analysis of artisan entrepreneur-mediators who facilitate the link between artisans and tourism is carried out in terms of social embeddedness, their roles and connections to creative tourism, and types of community engagement. We found that artisan entrepreneur-mediators in rural areas or small cities take on multiple roles as networking agents who organize and offer creative tourism experiences, providing the missing link between artisans and tourists. While social embeddedness is influenced by the level of connection to place and community engagement, we find that artisan entrepreneur-mediators are also able to increase their social embeddedness in various ways.

For artisans, this research helps illuminate and clarify the roles of entrepreneur-mediators who can link artisans to wider market opportunities. Some issues involved in creative tourism initiatives appear to be associated with 'start-up' phases when the entrepreneurs are not yet deeply embedded in a community. This suggests that these entrepreneurs should consider various avenues through which to develop their social embeddedness, such as engaging in community-centric activities, for example offering free classes for children in the community, which help expand and deepen social ties. As active agents living out their chosen lives and lifestyles in rural environments, the artisans and the artisan entrepreneur-mediators are gradually finding ways to work together to revitalize cultural traditions and develop new revenue streams. However, as new initiatives, this process of mediation and trust-building takes time, ongoing communication, and mutual understanding and respect.

Take-aways

- Artisan entrepreneur-mediators can connect artisans to tourism by finding the physical space for the workshop, advertising, registering participants, managing payments, translating, and acting as guides.
- Social embeddedness can improve opportunities for entrepreneurs, especially in small cities and rural areas.
- Artisan entrepreneur-mediators can increase their social embeddedness by engaging in community-centric activities such as offering free classes for children and training for the youth.

Acknowledgements

This chapter was developed within the CREATOUR® project (no. 16437), which is funded by the Portuguese Foundation for Science and Technology (FCT/MEC) through national funds and cofunded by FEDER through the Joint Activities Programme of COMPETE 2020 and the Regional Operational Programmes of Lisbon and Algarve.

References

Bakas, F. (2014) Tourism, female entrepreneurship and gender: Crafting new economic realities in rural Greece. PhD dissertation. University of Otago, Dunedin, NZ. Available at: http://hdl.handle.net/10523/5381 (accessed 21 April 2020).

Bakas, F.E. and Duxbury, N. (2018) Development of rural areas and small cities through creative tourism: The CREATOUR project. *Anais Brasileiros de Estudos Turísticos (ABET)* 8(3), 74–84.

Bakas, F.E., Duxbury, N., and Vinagre de Castro, T. (2018) Creative tourism: Catalysing artisan entrepreneur networks in rural Portugal. *International Journal of Entrepreneurial Behavior & Research* 25(4), 731–752. doi: 10.1108/IJEBR-03-2018-0177.

Bell, D. and Jayne, M. (2010) The creative countryside: Policy and practice in the UK rural cultural economy. *Journal of Rural Studies* 26(3), 209–218. doi: 10.1016/j.jrurstud.2010.01.001.

Bygrave, W.D. (1997) *The Portable MBA in Entrepreneurship*. J. Wiley & Sons, New York, NY, USA.

Duxbury, N., Campbell, H., and Keurvorst, E. (2011) Developing and revitalizing rural communities through arts and culture. *Small Cities Imprint* 3(1), 111–122.

Duxbury, N. and Richards, G. (2019) Towards a research agenda for creative tourism: Developments, diversity, and dynamics. In: Duxbury, N. and Richards, G. (eds.) *A Research Agenda for Creative Tourism*. Edward Elgar Publishing, Cheltenham, UK, pp. 1–14.

Granovetter, M. (1985) Economic action and social structure: The problem of embeddedness. *American Journal of Sociology* 91(3), 481–510. doi: 10.1086/228311.

Kline, C., Hao, H., Alderman, D., Kleckley, J.W., and Gray, S. (2014) A spatial analysis of tourism, entrepreneurship and the entrepreneurial ecosystem in North Carolina, USA. *Tourism Planning & Development* 11(3), 305–316. doi: 10.1080/21568316.2014.890127.

Landry, C. (2010) Experiencing imagination: Travel as a creative trigger. In: Wurzburger, R., Aageson, T., Pattakos, A., and Pratt, S. (eds.) *A Global Conversation: How to Provide Unique Creative Experiences for Travelers Worldwide*. Sunstone Press, Santa Fe, New Mexico, USA, pp. 33–42.

Luckman, S. (2012) *Locating Cultural Work: The Politics and Poetics of Rural, Regional and Remote Creativity*. Palgrave Macmillan, London, UK.

McKeever, E., Anderson, A., and Jack, S. (2014) Social embeddedness in entrepreneurship research: the importance of context and community. In: Chell, E. and Karataş-Özkan, M. (eds.) *Handbook of Research on Small Business and Entrepreneurship*. Edward Elgar Publishing, Cheltenham, UK, pp. 222–236.

Morpeth, N.D., Long, P. (2016) Creativity, the visual arts and tourism. In: Long, P. and Morpeth, N.D. (eds.) *Tourism and the Creative Industries: Theories, Policies and Practice*. Routledge, London, UK.

Spigel, B. (2017) The relational organization of entrepreneurial ecosystems. *Entrepreneurship Theory and Practice* 41(1), 49–72.

Stam, E. (2015) Entrepreneurial ecosystems and regional policy: A sympathetic critique. *European Planning Studies* 23(9), 1759–1769. doi: 10.1080/09654313.2015.1061484.

Teixeira, S.J. and Ferreira, J.J.M. (2018) Entrepreneurial artisan products as regional tourism competitiveness. *International Journal of Entrepreneurial Behavior & Research* 25(4), 652–673. doi: 10.1108/IJEBR-01-2018-0023.

Tucker, H. (2010) Peasant-entrepreneurs: A longitudinal ethnography. *Annals of Tourism Research* 37, 927–946.

UNESCO (1997) *Final Report: International Symposium on "Crafts and the International Market."* UNESCO, Manila, Republic of the Philippines.

World Tourism Organization (WTO) (2016) *Global Report on the Transformative Power of Tourism: A Paradigm Shift towards a More Responsible Traveller*. Available at: www.e-unwto.org/doi/book/10.18111/9789284417834 (accessed 21 April 2020).

Part V: Addressing challenges of impact and sustainability

Part V addresses the potential greater impacts of creative tourism initiatives, from the well-being of the local community in which they are embedded, to addressing society's existential need to address our relationship with the planet and each other in more sustainable manners. The chapters present different pathways to think about the longer horizon of the impacts of creative tourism initiatives, in a holistic manner incorporating social, economic, environmental, and cultural dimensions. Part V closes with a series of policy recommendations for the development of a locally embedded and networked creative tourism sector that brings together culture, tourism, and local/regional development.

In Chapter 29, 'An impact self-assessment tool for creative tourism with insights from its application to the CREATOUR® project', Maria Assunção Gato, Elisabete Tomaz, Pedro Costa, Ana Rita Cruz, and Margarida Perestrelo present an integrated and systematic impact self-assessment process to accompany the development and implementation of creative tourism activities. This process can support strategic planning and decision making and help to envision and document a wide range of impacts resulting from a creative tourism initiative. In turn, this process can provide important arguments and evidence to support a project's sustainability. The chapter is organized into two parts. In the first part of the chapter, the authors present two related impact assessment tools they developed within the CREATOUR® project to support practitioners' planning and decision making in developing creative tourism projects, and to evaluate local impacts over time. The self-assessment process and tools were informed by a multidimensional analysis that took into account economic, social, cultural, and environmental factors and the quality of life of local communities. The first diagnostics tool provides a framework to guide a strategic analysis process that defines objectives, gaps, challenges, and opportunities. The chapter then presents an impact self-assessment matrix to review and monitor the results of creative tourism initiatives. In part, this matrix helps practitioners become more aware of the multifaceted array of possible benefits of artistic, cultural, and creative tourism experiences for territorial development. The second part of the chapter is dedicated to an overview of the application and summary of results from the diagnosis phase with the 40 pilot initiatives in CREATOUR® and reflections on lessons learned from these processes. The chapter ends with a set of 'take-away' points and advice for practitioners relating to assessing the impact of their creative tourism activities.

In Chapter 30, 'A sustainable future for creative tourism in Lapland', Maria Huhmarniemi, Outi Kugapi, Satu Miettinen, and Laura Laivamaa present creative tourism as a channel for pursuing and realizing the roles of culture in the broad context of sustainable development. Connecting micro/individual experiences with macro-scale considerations of global climate change and environmental sustainability, the chapter reflects on how using arts-based methods in creative tourism can enable learning and, possibly, transformation of a traveller's worldview and values towards environmentalism. The authors' analyses and reflections are rooted in their experiences in Finnish Lapland, a context where culture is bond to nature, and where cultural and linguistic diversity derives from the Indigenous Sámi population and other cultural groups living in the region. In this context, arts-based methods are considered as a strategy to share the specificity of local knowledge systems and experiences of Arctic nature and culture. Culture and creative tourism in Lapland are explored through seven discourses related to cultural sustainability: heritage, vitality, economic viability, diversity, locality, eco-cultural resilience, and eco-cultural civilization (cf. Soini and Birkeland, 2014). Linking analysis with pragmatic points of advice for practitioners, the authors conclude that creative tourism can offer outstanding conditions for promoting connectedness with nature as it allows the tourist an immersion in creative activities that promote direct contact with nature. Transformative experiences such as those enabled by creative tourism can also have a positive impact on healing and promoting personal and social balance between nature and heritage.

The final chapter of Part V brings a policy development perspective that considers how to design an enabling milieu for the development and sustainability of a creative tourism sector in small cities and rural areas. In Chapter 31, 'Policy recommendations on creative tourism development in small cities and rural areas: Insights from the CREATOUR® project in Portugal', Alexandra Gonçalves, Maria do Rosário Borges, Cláudia Pato de Carvalho, Nancy Duxbury, and Pedro Costa share a set of policy recommendations for authorities and agencies at all levels (national, regional, and local) relating to the development of creative tourism initiatives in small cities and rural areas. These recommendations were informed by the needs of the CREATOUR® pilot organizations, and discussions that occurred throughout the project. Specific needs include: involvement in partnership approaches, networks, clusters, and territorially embedded creative milieus at both regional and national levels; development of inter-sectoral mediation and communication skills to build bridges among different sectors; qualified and motivated human resources involved in all cycles of the creative tourism value chain; improved, targeted communication and marketing approaches; and strategies for community involvement in creative tourism initiatives. What follows is a set of recommendations for tourism and cultural/creative sector policymakers that are organized, at the macro level, in five interlinked domains: product and enterprise development; inter-sectoral co-operation, networking, and soft infrastructure; capacity-building and training; communications and marketing; and research, monitoring, and industry intelligence. This set of recommendations, reinforced by the need to develop new structures to foster ongoing interconnections between the cultural and tourism sectors, is presented as the basis of a public policy framework for short and medium-term strategic planning measures.

Reference

Soini, K. and Birkeland, I. (2014) Exploring the scientific discourse on cultural sustainability. *Geoforum* 51, 213–223. doi: 10.1016/j.geoforum.2013.12.001.

— C.P.C. and N.D.

29 An impact self-assessment tool for creative tourism with insights from its application to the CREATOUR® project

Maria Assunção Gato,* Elisabete Tomaz, Pedro Costa, Ana Rita Cruz, and Margarida Perestrelo

ISCTE – Instituto Universitário de Lisboa, DINÂMIA'CET – ISCTE (Centro de estudos sobre a mudança socioeconómica e o território), Portugal

Given the enormous potential of sustainable creative tourism activities in low-density territories and local communities, creative tourism organizers/promoters should include a creative tourism impact assessment as an integrated and systematic process accompanying these activities. It can be used to identify the effects generated by such activities and support decision making for their sustainability. It can also examine the diversity of causal links between motivations for action, networking synergies, desired results, and identified impacts for the organizers/promoters, stakeholders, and the wider community.

Considering the small scale of creative tourism, the limited resources in any specific case, and the high engagement level of stakeholders, a *self-assessment exercise* (i.e. carried out by the organizers/promoters of creative tourism activities) is most appropriate in order to monitor activities and results. Thus, within the CREATOUR® project, a self-assessment process for creative tourism initiatives was developed, beginning with an initial diagnosis and proceeding to an assessment of outcomes and impacts (see Fig. 29.1). The process and tools were co-developed between researchers and practitioners to help various types of creative tourism organizers to measure, monitor, and evaluate the course of their activities in order to improve management and decision-making processes. The main objective was to assist practitioners and stakeholder organizations to become more aware of the intentional and unintentional effects generated by creative tourism activities, highlighting the potential benefits of artistic, cultural, and creative experiences for territorial development.

This article is divided into three parts: part I outlines the development of the self-assessment process and tools; part II presents a synthesis of overall findings from the CREATOUR® project resulting from this self-assessment process; and part III lists take-aways for practitioners.

Part I: Self-assessment process and tools

Searching for a multidimensional assessment for creative tourism

As background to the self-assessment exercise, we conducted a multidimensional analysis that took into account economic, social, cultural, and environmental factors as well as the quality of life of local communities, their cultural involvement, civic participation, and the conservation and preservation of both tangible and intangible cultural forms. Looking at the existing research, we found that the few studies that try to assess the impacts of creative tourism activities feature challenging conceptual

*Corresponding author: maria.gato@iscte-iul.pt

 Creative Tourism: Activating Cultural Resources and Engaging Creative Travellers (eds. N. Duxbury, S. Albino, and C. Pato de Carvalho)
DOI: 10.1079/9781789243536.0029

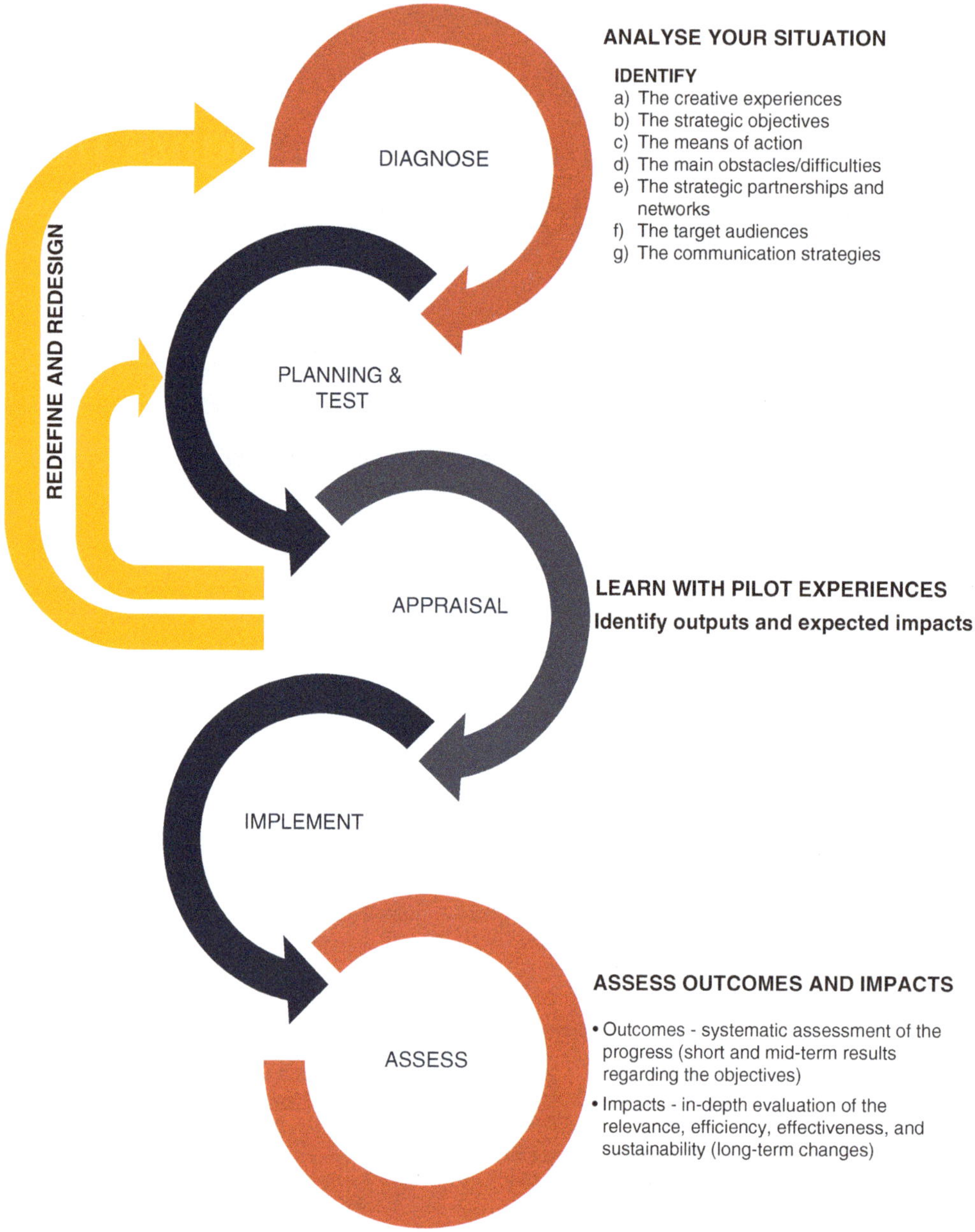

Fig. 29.1. Creative tourism initiatives assessment in five steps. Image developed by the authors.

and operational issues. These concern the nature of the cultural and creative activities, the fundamentals of the creative tourism experience itself, and the specific circumstances of each location, scale, and project. Moreover, there is a lack of systematic evidence of the multiple effects and the sustainability of these initiatives in the territories and on the local community's engagement (Richards and Wilson, 2007; Korez-Vide, 2013; Buaban, 2016; Qiu *et al.*, 2019). They also lack information on the networks that support creative processes, the context in which these processes take place, and their spill-over effects (Jeffcutt and Pratt, 2002; Comunian, 2010). In addition, none of the studies examined are addressed to practitioners and their needs for conducting a self-assessment of their creative tourism activities.

Given this situation, we found that one of the most relevant approaches to framing the analysis and impact assessment of creative tourism is the integration of culture as the fourth dimension of sustainable development (Hawkes, 2001; UNEP and WTO, 2005; Nurse, 2006), along with the economic, social, and environmental ones. Thus, a four-dimensional structure was used as the basis to think about the operationalization of an impact self-assessment tool, as illustrated in Fig. 29.2.

Inspired by this four-dimensional structure, the question then became: from what perspective(s) should impacts be measured? To be relevant to the individual case and context, the impact assessment will vary depending on the perspective of interest (and what can be accomplished). For instance, one must consider questions such as:

- What kind of effects do we want to assess (i.e. direct, indirect, induced)?
- From whose perspective do we want to assess those effects (e.g. the promoters, the local community, society as a whole)?
- What is the time frame (i.e. immediate effects, short-term, long-term)?
- How can these effects be measured (i.e. quantitative/qualitative effects, tangible/intangible effects)?
- What purpose contextualizes these impacts (e.g. quality of life and well-being, sustainable development, cultural value, etc.)?

With these questions in mind, many arguments can be raised for the relevance and usefulness of a multifaceted impact self-assessment in the context of sustainable creative tourism.[1] This relevance gains even more strength given the diversity of organizers involved in creative tourism (such as municipal entities, private entities, museums, cultural associations,

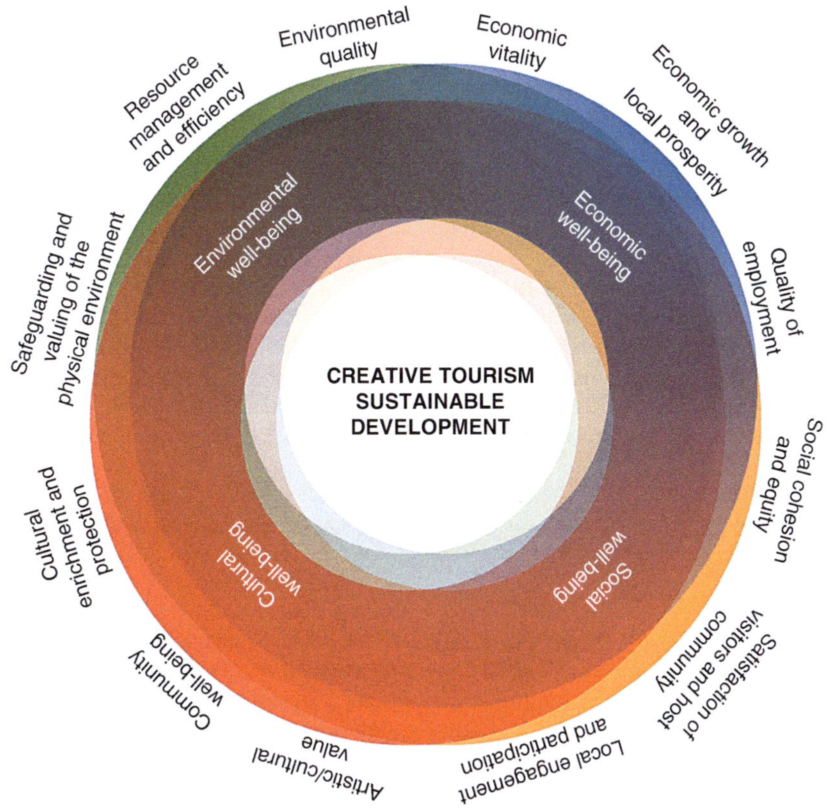

Fig. 29.2. Impact assessment framework for sustainable creative tourism. Image developed by the authors (Tomaz *et al.*, 2020), drawing upon the work of UNEP and WTO (2005, p. 20).

local development associations, craftsmen, artists, and tour operators, among others), each with different action foci, implementing strategies, objectives, and a wide range of creative tourism experiences as driving forces behind all these elements.

Co-constructing a self-assessment tool for creative tourism

A self-diagnosis exercise allows creative tourism practitioners to gain a clearer perspective on the main topics that will guide the development of their activities, as well as any necessary readjustments. Notwithstanding the nature of the creative tourism activities which are carried out, or the type of organizer involved, an initial diagnosis addressing fundamental questions is essential. This diagnosis not only facilitates the planning and implementation of initiatives, but it can also be fundamental for carrying out an impact self-assessment exercise that is relevant to the project organizer and their context.

Questions involved in a self-diagnosis should include:

- What are the major objectives you want to achieve with the creative tourism initiatives?
- What are the available means of action to implement these initiatives (e.g. economic, infrastructural, material, human, communication, technological, etc.)?
- What partnerships do you have at the local, regional, and national levels and with what regularity of connection? In these networks, do you recognize interesting partners for the development of your creative tourism activities?
- What opportunities do you expect to achieve with (or arising from) the activities?
- What obstacles, difficulties, and challenges do you identify to implementing the activities?
- Which target audiences do you want to reach and how?

Thoughtfully responding to these questions can improve a creative tourism organizer's ability to perceive and reflect on activity planning processes and help to define objectives, gaps, challenges, and opportunities. The responses also set a foundation for considering the multiple effects (economic, social, cultural, and environmental) that can arise from these activities at the territorial level and involving local communities.

The questions outlined above were identified through an extensive process of monitoring and knowledge co-production during which they were developed and tested with the organizers of the pilot project initiatives of the CREATOUR® project (see Box 1.1 in Chapter 1 for a brief description of the overall project). The diversity of creative experiences proposed within this project, the different types of organizers involved, and the variety of objectives implicit in their action strategies required the development of methodological procedures that would allow for both co-monitoring the activities implemented and the subsequent evaluation of their effects in the four sustainable development dimensions. For this purpose, and to address the questions raised in the two key phases of this process, we developed an evaluation tool consisting of two distinct exercises: a strategic analysis diagram for the diagnosis phase (Fig. 29.3), and an impact self-assessment matrix for the assessment phase (Table 29.1).

Based on the diagram presented in Fig. 29.3, the diagnosis phase developed in the CREATOUR® project produced 40 strategic analysis diagrams based on individual interviews with project organizers at different stages of the implementation process (100 interviews approx.). This information was initially collected and returned to the individuals interviewed, allowing for a regular follow-up on the evolution of their activities. In addition to the possibility given to each organizer to monitor their activities and make necessary readjustments, sharing the results of this analysis exercise at a regional level also allowed for a process of knowledge production and interaction among the various participants in the CREATOUR® project, thus fostering the exchange of experiences and new partnership networks.

Once creative tourism activities were implemented, it was necessary to assess their effects on the project organizers and the network of partners involved, as well as local communities and tourists. Structured by four dimensions and 12 sub-dimensions,[2] an impact self-assessment matrix was built with a view to being accessible and flexible for different types of organizers and entities operating in the field of cultural and creative tourism (see Table 29.1).

The co-research work carried out between CREATOUR® researchers and pilot project organizers demonstrated the need to create a qualitative self-assessment tool that could respond to the diversity of initiatives and entities operating in this

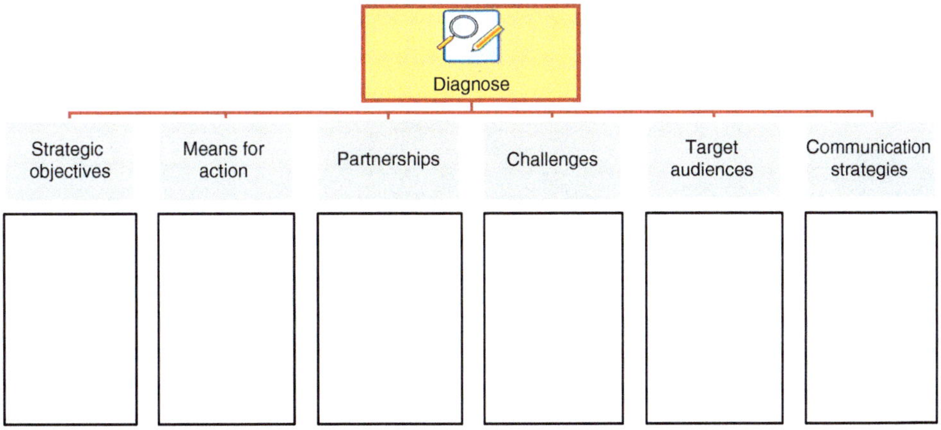

Fig. 29.3. Strategic analysis diagram for creative tourism initiatives. Developed by the authors.

sector. In addition, it also highlighted a diversity of issues linked to the sub-dimensions that should be included in this matrix in order to make it more comprehensive and flexible. As such, in using the matrix, a variable and non-exhaustive completion of the various rows and columns is allowed, in accordance with the nature of the creative tourism initiatives, the degree of maturity they have reached, and the organizers' objectives and intervention focus.

One of the main purposes of this matrix is to help creative tourism organizers assess their activities from several perspectives and to then make complementary analyses based on the initial diagnosis. This linked process increases their capacity to understand and monitor the multiplicity of effects generated, both within their business project and in interactions with partners and local communities. It also takes into account tourists' or visitors' reactions to the experiences offered, in order to better meet the expectations of both parties.

Part II: Applying the tool to CREATOUR®

The diagnostics phase was methodologically supported by a strategic analysis focused on the creative tourism activities that the 40 CREATOUR® pilot project organizers intended to develop. As mentioned above, this exercise resulted in the completion of 40 diagrams, one per pilot project, in order to allow the project organizers and the research team to monitor the activities at different stages of the process. In addition to the issues covered by the diagrams, their joint analysis also made it possible to classify each organizer in terms of typologies, the main focus of action, and type of engagement with the local community. This characterization enabled us to systematize some essential information about the project organizers and to find common ground and patterns among them, while providing an interesting comparative analysis among the pilot initiatives and the regions in the project. A summary of information about the 40 pilots is provided in Tables 29.2 and 29.3.

This analytical exercise was essential to design the multidimensional matrix and to better frame the 40 outcomes extracted from it. The particular characteristics of the creative tourism projects involved in CREATOUR® – small-scale activities in low-density areas – also brought additional challenges to this impact assessment exercise, such as the need to capture tangible and intangible elements, directly or indirectly. Additionally, the quantitative data available at the scale of these territories are rarely consistent and robust enough to support economic, social, cultural, and environmental impact assessments. Thus, qualitative information must be collected, and strong interpretation skills are determinant for making accurate associations and relational links between creative tourism experiences and outcomes.

Accordingly, and following the initial exercise, the evaluation phase resulted in the production of 40 multidimensional impact self-assessment matrices. (In the Appendix to this chapter, the example of ADERE Peneda-Gerês, a local development association located in the Norte region of Portugal, is presented to illustrate the exercises carried out.) The completion of these matrices was also based on

Table 29.1. Multidimensional impact self-assessment matrix. Developed by the authors

Dimensions	Subdimensions	My Creative Tourism Activity	Specify the effects (How / In what way / Direct or induced effects / Time frame)			
			For me/my business/my entity	For stakeholders	For local community	For visitors/tourists
ECONOMIC	1. Economic vitality	Generated new sources of income? Fostered new investments in the creative sector and tourism? Contributed to expand markets/audiences? Promoted Economic sustainability?				
	2. Economic growth and local prosperity	Increased the value of traditional local products and sectors? Included creative products and services in tourism development? Fostered synergies between sectors/other actors? Increased the visibility of the territory as a tourism destination?				
	3. Quality of employment	Created new job opportunities? Supported the developed of business skills? Overcame the seasonality of tourism employment?				
Social	4. Social cohesion and equity	Promoted the inclusion of socially excluded or/and economically disadvantaged people? Increased the access to knowledge and creative experiences? Supported social innovation projects in creative tourism?				
	5. Satisfaction of visitors and host community	Promoted community well-being? Strengthened local cultural identity? Encouraged social innovation in tourism and community development? Encouraged sustainable and open behaviours in the community?				
	6. Local engagement and participation	Fostered creative opportunities to meet the needs of the local population and visitor expectations? Increased the involvement of all stakeholders in the implementation stages? Raised public awareness of the nature and benefits of promoting creative tourism?				
Cultural	7. Artistic and cultural value	Promoted the intrinsic value of the arts? Contributed to the training and empowerment of artists? Contributed to the recognition and integration of different cultural expressions, both formal and informal?				
	8. Participant satisfaction	Developed high-quality creative experiences that increased participant satisfaction? Improved the participants' cultural capital? Encouraged change in mentalities and behaviours? Promoted personal development?				
	9. Cultural enrichment and protection	Ensured the management and conservation of local tangible/intangible heritage? Valued and promoted local knowledge? Integrated local heritage and distinctive qualities to create unique experiences?				
ENVIRONMENTAL	10. Safeguarding and valuing the physical environment	Included biodiversity values in the planning and implementation of creative tourism? Contributed to reducing the physical impact of tourism? Enhanced the landscape for creative tourism?				
	11. Resource management and efficiency	Motivated the efficient use of local resources? Created or reused local physical infrastructure for creative tourism?				
	12. Environmental quality	Explored the small-scale benefits of creative tourism experiences? Reduced potential negative externalities resulting from tourism? Raised awareness of the need to reduce the ecological footprint?				

individual interviews with the project's creative tourism developers/organizers. A content analysis subsequently carried out on the strategic analysis diagrams and self-assessment matrices produced 40 short reports. These reports sought not only to summarize the results achieved by the activities undertaken, but above all to present some guidelines to the creative tourism pilot project organizers.

One of the starting points for the analysis of the results obtained from these assessment exercises was the dominant intervention focus of the CREATOUR® pilot initiatives, given their relevance in guiding the implementation of the creative tourism experiences. Overall, the territory is the dominant focus of intervention for 50% of the creative tourism pilot initiatives, followed by culture (30%), and then tourism (20%). This lower emphasis on tourism in comparison to the territory (although regionally diverse) confirms the prevailing perception that creative tourism is a significant activity to leverage new territorial dynamics in low-density contexts.

Table 29.2. Characterization of the first 20 creative tourism pilot initiatives. Developed by the authors

Region (NUTS II)	Pilot Initiatives	Type of Entity: Municipality or Other Public Body	Type of Entity: Company	Type of Entity: Non-profit Organization	Dominant Intervention Focus: Culture	Dominant Intervention Focus: Tourism	Dominant Intervention Focus: Territory	Local Engagement Type: Predominantly Endogenous	Local Engagement Type: Predominantly Exogenous
Norte	VERde NOVO		●				■		●
	CM São João da Madeira	●					■	●	
	LRB		●				■		●
	CM Amares & ENcontrARTE	●			■			●	
	ADERE PG – Peneda Gerês			●			■	●	
Centro	ADXTUR		●			■		●	●
	Quico – Turismo		●			■		●	
	Museu Conímbriga	●			■			●	
	Associação Luzlinar			●	■			●	
	Domínio Vale Mondego			●	■				●
Alentejo	Loom New Tradition		●		■				●
	CM Mértola	●					■	●	
	CM Beja	●					■	●	
	CECHAP			●	■				●
	CM Reguengos Monsaraz	●					■	●	
Algarve	Tertúlia Algarvia		●			■		●	
	PROACTIVETUR		●			■			●
	Associação Odiana			●			■	●	
	CM Loulé	●			■			●	
	Barroca		●			■		●	

Table 29.3. Characterization of the second 20 creative tourism pilot initiatives. Developed by the authors

Region (NUTS II)	Pilot Initiatives	Type of Entity: Municipality or Other Public Body	Type of Entity: Company	Type of Entity: Non-profit Organization	Dominant Intervention Focus: Culture	Dominant Intervention Focus: Tourism	Dominant Intervention Focus: Territory	Local Engagement Type: Predominantly Endogenous	Local Engagement Type: Predominantly Exogenous
Norte	Galandum Galundaina			●			■		●
	ADRAT – Revitalizar Vilar			●			■	●	
	CM Bragança – Cuscus	●					■		●
	CM Esposende	●			■			●	
	Desteque – Caretos			●			■	●	
Centro	Ruralidades e Memórias – 3B			●			■	●	
	New Hand Lab+Museu Lanifícios	●					■	●	
	VIC//Aveiro Arts House		●		■			●	●
	CM Abrantes + Canal 180	●			■			●	●
	Associação Destino Caldas			●	■				●
Alentejo	Genuine Alentejo		●			■		●	
	Marca – ADL			●			■	●	
	CACO – Artesãos Odemira			●			■	●	
	João Cutileiro – Pedra+			●	■				●
	Vagar Walking Tours		●			■		●	
Algarve	Espírito de Terra-Boliqueime			●			■	●	
	Associação In Loco			●			■		●
	C. Ciência Viva + AP Matemática			●	■				●
	Eating Algarve Food Tours		●			■		●	
	Associação BackUp			●			■	●	

As shown in Fig. 29.4, there is a visible regional prevalence in the distribution of the three foci of intervention among the 40 pilots. Except for the Alentejo region, which has a relatively balanced distribution between the three foci, the other regions reflect a more specific positioning both around the territory, in the case of the northern region (Norte), and around culture, in the central region (Centro). Although they had not been predetermined, these guidelines can be strategically relevant to increase the capacity for regional co-ordination of the pilot initiatives. As for the southern Algarve region, the

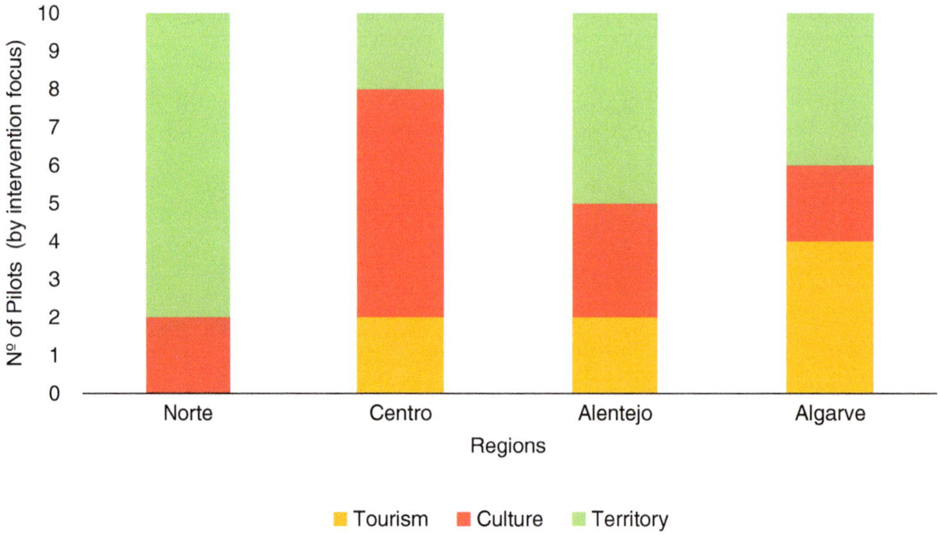

Fig. 29.4. Distribution of dominant intervention focus of pilot projects by region. Developed by the authors.

balanced territory–tourism results can be attributed not only to the specific characteristics of this region (the smallest in geographic terms), but above all to the strong predominance of tourism as the main economic activity of the region and a long and robust investment in products seeking to complement the traditional 'sun and sand'. Nevertheless, the Algarve is the region where the cultural dimension of the proposed creative tourism activities is less obvious when compared to the other three regions.

The internal coherence of the dominant intervention focus by type of organizer is also notable. Fig. 29.5 illustrates the obvious weight of the territory in the actions of *non-profit organizations*, based on their concern to attract benefits to the population and territories in which they intervene, where creative tourism is envisaged as a means to achieve that goal. The territory development is also the main focus for *municipalities and other public bodies* that have incorporated the 'creative narrative' into their public policies in recent years. This is also an expected result, given the purposes that led these entities to focus on organizing and promoting creative tourism. According to these organizations, this differentiated tourism offer is seen as an opportunity to leverage economic, social, and cultural vitality in small towns and rural areas, due to its strong connection with the local communities and their tangible and intangible assets. Furthermore, investing in creative tourism can be a way to retain the population in these territories and promote their empowerment to play a more active role in unique and tailored interactive experiences. In the case of *companies*, their intervention is related to their survival being dependent on the economic benefits generated by tourism, so their main focus is inescapably the development of their tourism activity and client satisfaction. Most of these companies have been developing activities in the tourism sector for some time and creative tourism has appeared in this context as a complementary offer and, in some cases, as a means to counteract seasonality.

In terms of the project organizers' expectations regarding the impacts generated by the creative tourism activities they have implemented, the analysis revealed that a concern with the economic dimension prevails regardless of the type of organizer. However, the distribution between the four dimensions is different in each case.

In an overview, it is observable that *municipalities and other public bodies* expect more economic impacts from creative tourism activities, followed by those expected at the cultural and social levels. The strategies of intervention are focused on endogenous resources seeking to boost the local economy and culture through the enhancement of tangible and intangible heritage assets, in addition to reputation-building mechanisms targeted at tourism. Strategically, some municipalities have

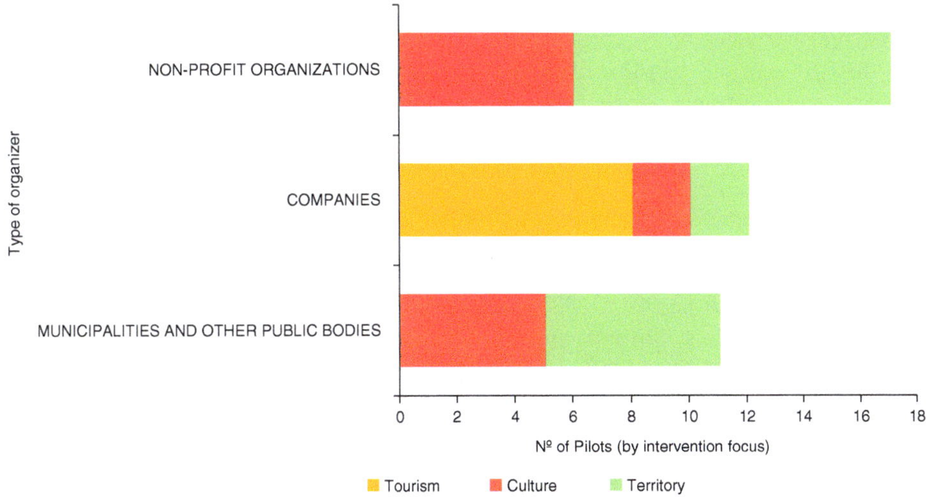

Fig. 29.5. Distribution of dominant intervention focus of pilot projects by type of organizer. Developed by the authors.

assumed a leading role in promoting and developing creative tourism initiatives, while others are seeking partnerships to lead and expand creative tourism offers to enhance entrepreneurship and empower small local businesses to provide activities that are able to complement tourism. The proposed initiatives are not directly focused on natural resources, which explains a relatively lower emphasis on environmental impact expectations.

Non-profit organizations show a more uniform distribution of impacts in the four dimensions of development. Impacts in the economic and social dimensions are expected in the same proportion, with little difference from the social ones. And although the environmental dimension is the one where these organizers expect to generate the least impact, they are the ones that value this dimension the most. In parallel, non-profit organizations seem to be the best positioned to achieve better results through creative tourism in low-density territories, not only due to the greater institutional flexibility that frames these local development organizations, but also because of their partnerships and privileged knowledge of their fields of action. In general, they mobilize around territorial and local development commitments, looking for innovative opportunities to articulate culture and tourism. However, their maintenance often implies a diffusion in relation to the focus of their multiple interventions (or a certain fluctuation according to the priorities of the main funding programmes), which can result in more superficial investment in certain dimensions or a change regarding their priorities for action.

Companies also express the perception of important impacts at the economic level, followed by those at the social and cultural levels. That this expectation is mainly linked to economic development is not surprising since it is the sustainability of these small companies that is at stake. However, this implies establishing partnerships in the territory in which they operate and empowering and valuing their partners to respond spontaneously and creatively to new business opportunities. Thus, the social dimension derives from the strategic priorities of developing local partnerships and tourism networks and, in a certain way, the cultural dimension acknowledges the added value conferred by their socio-economic incorporation. As such, some of them easily recognize the value of partnerships and seem to be in a better position to strengthen a national creative tourism network in the future. The environmental dimension does not take a prominent position for companies, although local endogenous resources and public spaces are valued within the scope of the tourism activities they develop.

As a result of these exercises, we can point to some challenges that require attention from creative tourism project organizers, such as the relevance of involving local communities and establishing collaborative partnerships in creative tourism experiences. We also wish to highlight that the pilot project organizers considered and incorporated a great diversity of endogenous resources that can be valued and should be protected in terms of sustainability, and they all generally acknowledged the diversity of assessment dimensions during the assessment exercises. The diverse array of organizers and pilot projects enabled us to identify some challenges, opportunities, and constraints to creative tourism initiatives as varied as those contemplated by CREATOUR®, and to take them into account in the design of the impact self-assessment matrix.

The diversity of drivers for action (i.e. motivations and rationales) animating different types of project organizers, as well as the variety of geographic areas, experiences, and social contexts under consideration unmistakeably demonstrated the difficulty of making generalizations about the role and impacts of creative tourism. Moreover, it has improved our knowledge about each particular situation, and underlined the need for a closer look at the specifics of individual initiatives and the geographic and organizational settings in which they are embedded.

Other challenging lessons that emerged include:

- The difficulties of converting and exploring certain endogenous resources within creative tourism experiences.
- The constraints caused by an irregular demand in the planning and sustainability of some projects.
- The informal and voluntary link that characterizes some established partnerships has implications for the predictability of a constant supply of experiences.
- The different projects' maturity levels presented by the participating organizations limited their performances and expected results.
- The socio-cultural and political contexts in which the pilot project organizers are immersed also determined their achievements.
- The organizers' varying knowledge regarding this type of assessment tool required additional investment in the clarification of some concepts.
- The relatively long timeframe between the performance of the activity and the observation of its results.

Concluding remarks

A multidimensional impact assessment is essential to identify, monitor, and evaluate the various tangible and intangible effects of creative tourism initiatives and increase knowledge about them to create rewarding and sustainable experiences for tourists, residents, and project organizers. The identification of positive and negative repercussions through this evaluation can allow local organizers to take preventive or corrective measures and build innovative solutions that improve the competitiveness of an initiative or, more broadly, the destination, making it more viable and attractive.

The objective of the initiative presented in this chapter was to demonstrate these advantages through an assessment tool aimed at practitioners in the creative and cultural tourism sector, trying not only to meet their needs, but also to fill a methodological and operative gap. The proposed tool comprises two methodological exercises aimed at two distinct phases in the process of creative tourism activity development, but complementary in terms of the resulting analysis. The example taken from the CREATOUR® project as an inspirational case study (see Appendix) demonstrates this complementarity and highlights the decisive role that the creative tourism project organizers have had in the whole process of co-creation and co-production of knowledge within CREATOUR®.

In general terms, we can conclude that this co-creation and impact self-assessment exercise has contributed to:

- Collect and analyse relevant information, often forgotten or undervalued.
- Create unique opportunities for dialogue, participation, and collaboration around the priorities, objectives, and circumstances of different stakeholders.
- Provide a greater understanding of essential issues, such as the potential of these activities in low-density territories.
- Assist informed and reasoned decision-making to improve the effectiveness, benefits, and sustainability of future activities.
- Realize the relevance of different players in the process and the potential of their complementary roles for the results.
- Make the best use of the different motivations of those involved in creative tourism initiatives to maximize the benefits of their actions.
- Recognize the multidimensionality of sustainable development and the advantages of looking and communicating beyond economic issues.
- Discuss how creative tourism can contribute to the well-being of communities by improving socio-economic and environmental conditions.

- Empower local communities through the cultural involvement of their members and the enhancement of their tangible and intangible cultural heritage.
- Make creative tourism a strong offer to revitalize and promote a better quality of life in low-density territories.

The work developed has also brought to light some vulnerabilities of impact assessment procedures, particularly in the specific field of small-scale creative tourism activities, and the need to continue further developing the conceptual and analytical tools that are used. The tool that we have proposed is certainly part of this and are open to improvements and further developments. This tool is able to identify positive and negative repercussions to take preventive or corrective measures, and to support the development of innovative solutions that improve the competitiveness of destinations, making them more viable and attractive. However, making it operational implies an integrative and sustainable approach that goes beyond looking at the traditional economic effects. There is a need to continue to challenge conceptions, practices, routines, and instruments on this field, and a network of creative tourism agents such as the one that is being launched by this project can be a promising path, working together in the development of these models and tools.

In parallel, this exercise confirmed the difficulties of providing (and communicating) evidence on some qualitative effects, despite the acknowledgement of their importance in the planning and design of the projects and the creation of synergies on these innovative forms of tourism. But it also confirmed that following that path is essential if we want to effectively achieve some insights into the contributions that these projects bring to the effective sustainable development of those territories and communities.

Unsurprisingly, the process and tools discussed in this chapter require a broader and longer application to better understand more deeply the various impacts of these experiences and to improve the methods to measure them on a given territory with its specificities, but also in different sectors of activity and with different types of participants. That is the work that will follow and, for that, the contribution of the academic community as well as the creative tourism project organizers in the field is essential. Toward this objective, we need to continue working together within in-depth collaboration projects that enable us to co-create knowledge in this field.

Part III: Take-aways for practitioners

- Given the potential benefits for territorial development that may arise from artistic, cultural, and creative experiences, an impact self-assessment exercise should not be limited to economic criteria to justify the relevance of such activities.
- A multidimensional analysis can obtain a more comprehensive look at particularities, both of the practitioners and the local contexts.
- Qualitative criteria are essential to better understand direct, indirect, and induced effects; to distinguish the tangible and intangible; and to identify spill over effects that can be generated by creative tourism experiences.
- An impact self-assessment exercise should start from an initial diagnosis and contemplate the whole process through to impact assessment in order to achieve a broad analysis of the results reached and to be pursued, taking into account the various time frames that may be involved.
- A self-awareness of project intentions and priorities as well as the diversity of impacts that are possible, combined with a self-monitoring of effects, can be very useful to guide and (re)direct experiences to target audiences as well as stakeholders' objectives.
- The option for a more flexible assessment model may not allow direct comparisons between several projects but can provide useful insights that can be fundamental for each particular case and transferred and applied to other projects and situations.

Acknowledgements

This work was developed within the CREATOUR® project (No. 16437), which is funded by the Portuguese Foundation for Science and Technology (FCT/MEC) through national funds and co-funded by FEDER through the Joint Activities Programme of COMPETE 2020 and the Regional Operational Programmes of Lisbon and Algarve.

Notes

1. The analysis is more complex in the specific case of creative tourism compared to that for creative activities

or tourism activities in general. For instance, challenges will be faced when these activities involve the co-creation of tourism experiences between provider and user; when the objectives to be achieved relate to sustainable development and are therefore multidimensional; when the objectives are related to community involvement; when these activities are performed in low population density areas or for very small-scale projects; or when the actions of a diversity of project organizers have very different objectives and rationales.

2. For each of the 12 sub-dimensions, a set of criteria was presented reflecting the sustainable objectives, collectively contributing to the design of the creative tourism impact assessment framework.

References

Buaban, M. (2016) Community-based creative tourism management to enhance local sustainable development in Kanchanaburi Province, Thailand. Doctoral dissertation, University of Exeter, UK. Available at: https://ore.exeter.ac.uk/repository/handle/10871/24246 (accessed 15 December 2020).

Comunian, R. (2010) Networks of knowledge and support: Mapping relations between public, private and not for profit sector in the creative economy. Paper presented at 50th Congress of the European Regional Science Association (ERSA), "Sustainable Regional Growth and Development in the Creative Knowledge Economy," 19–23 August 2010, Jönköping, Sweden.

Hawkes, J. (2001) *The Fourth Pillar of Sustainability: Culture's Essential Role in Public Planning*. Common Ground Publishing Pty Ltd in association with the Cultural Development Network, Melbourne, Victoria, Australia.

Jeffcutt, P. and Pratt, A.C. (2002) Managing creativity in the cultural industries. *Creativity and Innovation Management* 11(4), 225–233.

Korez-Vide, R. (2013) Promoting sustainability of tourism by creative tourism development: How far is Slovenia. *Innovative Issues and Approaches in Social Sciences* 6(1), 77–102.

Nurse, K. (2006) Culture as the fourth pillar of sustainable development [prepared for Commonwealth Secretariat]. *Small States: Economic Review and Basic Statistics* 11, 28–40. doi: 10.14217/smalst-2007-3-en. Available at: www.thecommonwealth-ilibrary.org/commonwealth/economics/small-states/culture-as-the-fourth-pillar-of-sustainable-development_smalst-2007-3-en (accessed 15 December 2020).

Qiu, H., Fan, D.X.F., Lyu, J., Lin, P.M.C., and Jenkins, C.L. (2019) Analysing the economic sustainability of tourism development: Evidence from Hong Kong. *Journal of Hospitality & Tourism Research* 43(2), 226–248. Available at: http://journals.sagepub.com/doi/10.1177/1096348018777046 (accessed 12 December 2019).

Richards, G. and Wilson, J. (2007) Creativities in tourism development. In: Richards, G. and Wilson, J. (eds.) *Tourism, Creativity and Development*. Routledge, London and New York, pp. 255–288.

Tomaz, E., Costa, P., Gato, M.A., Cruz, A.R., and Perestrelo, M. (2020) *An Analytical Model for Impact Assessment on Creative Tourism: Creative Tourism Sustainable Development Impacts Matrix.* DINAMIA'CET Working paper series, DINAMIA'CET-ISCTE, Lisbon, Portugal.

UN Environment Programme (UNEP) and World Trade Organization (WTO). (2005) *Making Tourism More Sustainable – A Guide for Policy Makers*. UNEP, Division of Technology, Industry and Economics, Paris, France.

Appendix

To illustrate the exercises carried out, the example of ADERE Peneda-Gerês, a local development association located in the Norte region of Portugal, is presented here.

As the findings in Fig. 29.6 illustrate, this creative tourism initiative focuses on the diversity of endogenous resources of a relatively large territory as well as on the cultural and heritage values of the local communities located within the territory. The objective of counteracting seasonality through creative tourism experiences presents itself as an important added value and with successful expectations. The development of local partnerships in terms of creativity and tourism is viewed as a resource to be enhanced and developed in order to achieve a more constant attractiveness to the territories and to promote them with greater visibility.

As illustrated in Table 29.4, the cells in the matrix were not filled in comprehensively. On the one hand, the characteristics of each creative tourism initiative meets the dimensions and sub-dimensions in a different way. On the other hand, there are effects that require a longer time frame to be properly evaluated, which does not invalidate the results already attained in this illustrative exercise.

Organization: ADERE (Associação de Desenvolvimento das Regiões do Parque Nacional da Peneda-Gerês) **local development association** (Norte Region)
Creative Tourism Initiative: Creative experiences with 5 senses

Diagnosis

Strategic objectives	Means for action	Partnerships	Challenges	Target audiences	Communication strategies
1. Increase the economic profitability of tourism for the local population 2. Provide tools for tourism companies to develop creative tourism with local populations 3. Counteract the tourism seasonality in the region 4. Increase the length of visitors' stays 5. Involve other local/regional entities in creative tourism	1. Local communities (cultural heritage, traditions, knowledge) 2. Existing network of partnerships and contacts 3. Human resources 4. Other local development associations	1. Joint activities with 3 CREATOUR regional partners 2. Extended regional and local partnerships in the tourism sector	1. Communication and promotion of initiatives 2. Getting local communities involved	1. Foreign visitors and tourists 2. Portuguese visitors and tourists 3. Regional visitors	1. Partnership with international creative tourism network 2. Getting publicity through national and regional tourism authorities 3. Project promotion in national and international tourism events

Fig. 29.6. ADERE Peneda-Gerês self-assessment diagnosis. Developed by the authors.

Table 29.4. ADERE Peneda-Gerês self-assessment matrix. Developed by the authors

Dimensions	Subdimensions	**My Creative Tourism Activity** ADERE Peneda-Gerês (local development association)	Specify the effects (How/ In What way/ Direct or induced effects/ Time frame)			
			For me/my business/my entity	For stakeholders	For local community	For visitors/tourists
ECONOMIC	1. Economic vitality	Generated new sources of income? Fostered new investments in the creative sector and tourism? Contributed to expand markets/audiences? Promoted Economic sustainability?	An opportunity to diversify the supply and expand markets towards economic sustainability	An incentive for new investments and activities	Contributes to promote local heritage and products and greater visibility for the region as a creative tourism destination	
	2. Economic growth and local prosperity	Increased the value of traditional local products and sectors? Included creative products and services in tourism development? Fostered synergies between sectors/other actors? Increased the visibility of the territory as a tourism destination?	An opportunity to explore new sources of income in the tourism sector	Have fostered synergies and more partnerships between sectors	Engaged some residents in new activities and learning	
	3. Quality of employment	Created new job opportunities? Supported the developed of business skills? Overcame the seasonality of tourism employment?	A support to the development of new skills, activities and products in the region	Still residual contribution to mitigating seasonality		
Social	4. Social cohesion and equity	Promoted the inclusion of socially excluded or/and economically disadvantaged people? Increased the access to knowledge and creative experiences? Supported social innovation projects in creative tourism?	Indirect benefits arising from the engagement of local communities and promotion of local heritage and products		Recognition and valorisation of informal skills and know-how	Personal and cultural enhancement
	5. Satisfaction of visitors and host community	Promoted community well-being? Strengthened local cultural identity? Encouraged social innovation in tourism and community development? Encouraged sustainable and open behaviours on community?		Some economic benefits in local trade resulting from the presence of visitors	Indirect benefits arising from the presence of visitors	
	6. Local engagement and participation	Fostered creative opportunities to meet the needs of the local population and visitor expectations? Increased the involvement of all stakeholders in the implementation stages? Raised public awareness of the nature and benefits of promoting creative tourism?	Reinforcement of partnerships and conjunction of local interests with visitor expectations	Awareness of local partners for new activities	Raising local community awareness of creative tourism	
Cultural	7. Artistic and cultural value	Promoted the intrinsic value of the arts? Contributed to the training and empowerment of artists? Contributed to the recognition and integration of different cultural expressions, both formal and informal?	Promotion of traditional arts and stimulus for innovation skills	Empowerment of local artists to develop innovative products based on the traditional ones		
	8. Participant satisfaction	Developed high-quality creative experiences that increased participant satisfaction? Improved the participants' cultural capital? Encouraged change in mentalities and behaviours? Promoted personal development?	Satisfaction coming from the visitors as an incentive for the development of more activities			
	9. Cultural enrichment and protection	Ensured the management and conservation of local tangible/intangible heritage? Valued and promoted local knowledge? Integrated local heritage and distinctive qualities to create unique experiences?	Encouragement for social innovation	Promotion of local heritage and cultural dynamism	Recognition for the enhancement of local heritage	Knowledge sharing
ENVIRONMENTAL	10. Safeguarding and valuing the physical environment	Included biodiversity values in the planning and implementation of creative tourism? Contributed to reduce the physical impact of tourism? Enhanced the landscape for creative tourism?	Promoting the fight against seasonality effects	Possibility to benefit from the break in seasonality	Possibility to benefit from the break in seasonality	Possibility to benefit from the break in seasonality
	11. Resource management and efficiency	Motivated the efficient use of local resources? Created or reused local physical infrastructure for creative tourism?	Enhancement of natural resources	Valorisation of endogenous services and products	Raising awareness on the preservation and promotion of natural ecosystems	Raising awareness on the preservation and promotion of natural ecosystems
	12. Environmental quality	Explored the small-scale benefits of creative tourism experiences? Reduced potential negative externalities resulting from tourism? Raised awareness of the need to reduce the ecological footprint?	No negative environmental effects			

30 A sustainable future for creative tourism in Lapland

Maria Huhmarniemi,* Outi Kugapi, Satu Miettinen, and Laura Laivamaa

University of Lapland, Finland

Creativity, arts, and craft-based services to support sustainability

In this chapter we present practical ideas and potential strategies for developing the use of arts-based methods in creative tourism. Arts-based methods have come to be widely discussed and promoted in various disciplines such as the social sciences, education, and tourism research (Leavy, 2009, 2018; Barone and Eisner, 2012) and their strength is seen in their ability to engage participants' creative expression and to obtain participation and knowledge that is not verbal. In the fields of art education and community art, the impact of art-based methods is seen on human growth and well-being, the sense of belonging, the enhancement of kinship, empowerment, and the ability to create meaningful symbols (see, for example, Hiltunen and Rantala, 2015). Inspired by this, in this chapter we consider the potential of arts-based methods and creative tourism experiences for enabling learning and transformation.

Our interest in creative tourism lies in place-related aspects of culture in which local knowledge and some aspects of traditions, as well as environmental debates, are shared with visitors. Therefore, in this chapter, we discuss cultural sustainability through arts-based creative tourism, which can be seen as time- and place-specific so that each community needs to evaluate their own needs, rights, and responsibilities for more sustainable living (Härkönen *et al.,* 2018). As researcher Hanna Lempinen (2019) has described the emphasis of Arctic cultures and their sustainability on change: the ongoing societal and environmental changes are experienced as rapid and unforeseen, and local cultures are seen as important for development of the region.

In Finnish Lapland, where the case studies of this chapter are found, there are many issues – especially in tourism development – concerning the ways nature is utilized without deeper respect, cultural heritage is exploited, and the visual elements of Indigenous people's culture are used without cultural sensitivity and true connection to local residents. Often, the services are built on imaginary stories instead of authentic traditions, such as folktales and poetry (see, for example, Lehtola, 1999; Niskala and Ridanpää, 2016; Sámi Parliament, 2018; Olsen *et al.,* 2019), and the winter tourism can re-produce false representations of the Arctic (Rantala *et al.,* 2011, 2019). As Daniel Chartier (2018) has described it, the Arctic has been historically marginalized by Western research, writings, and artistic presentation as an 'Imaginary North', as an empty and cold landscape (see also Visit Finland, 2019) and not as a multi-ethnic, -cultural, and -lingual place with a rich cultural history and living traditions (Chartier, 2018). The cultural and linguistic diversity in Lapland derives from the Indigenous Sámi population and other cultural minorities inhabiting the area with their own cultures, languages, and, to some extent, their own livelihoods. From the viewpoint of cultural heritage, Indigenous knowledge systems and Arctic eco-social environmental culture include aspects of social and ecological sustainability. In the Arctic, culture is bound to nature. Arts-based methods implemented for creative tourism can be a way to

*Corresponding author: maria.huhmarniemi@ulapland.fi

 Creative Tourism: Activating Cultural Resources and Engaging Creative Travellers (eds. N. Duxbury, S. Albino, and C. Pato de Carvalho)
DOI: 10.1079/9781789243536.0030

express and share the richness of culture, knowledge systems, heritage, and nature connectedness.

Some interdisciplinary studies on the potential and practices of creative tourism in Lapland have been conducted before (e.g. Veijola *et al.*, 2018; Miettinen *et al.*, 2019; Kugapi *et al.*, 2020). For example, environmental art and snow sculpting studies have been organized for tourism companies and public art has been produced to create environments for experiencing culture (Jokela *et al.*, 2014; Huhmarniemi and Jokela, 2019). The ongoing development projects *Handmade in Lapland* (2018–2021 ESF) and *Art-based Services for Tourism Project* (2019–2021 ERDF) concentrate on developing new tourism products and services for the creative sector in Lapland, and a forthcoming project, *Smart Cultural Tourism* (2020–2023 Horizon 2020), will provide a multi-country platform for exploring further the potential of cultural tourism for regional development (research programme). The authors of this chapter are involved with these initiatives as researchers as well as service design, art education, and tourism experts.

Since culture includes all aspects of human life, consists of everyday life and interactions with non-human nature, and it is constantly changing, it is impossible to define coherently. Therefore, in this chapter, we explore culture and creative tourism in Lapland through seven discourses of cultural sustainability: heritage, vitality, economic viability, diversity, locality, eco-cultural resilience, and eco-cultural civilisation (Soini and Birkeland, 2014). We also seek inspiration from Valene L. Smith's (1996) 4H concept – heritage, habitat, history, and handicrafts – in Indigenous tourism due to its significance in Arctic areas. Although Indigenous communities are not the only focus of this chapter, their importance to the local cultural heritage and history cannot be discarded. Moreover, Smith's ideas correlate well with the issues that arise from the place – Lapland – where visual arts, crafts, and design are commonly interwoven (see Huhmarniemi and Jokela, 2020). The chapter is supported with presentations of various art productions and pilots of service productions as well as insights of artists, designers, and producers who work in creative tourism. Moreover, we illustrate creative tourism services that are based on art and contribute to holistic sustainability. As an outcome, we ponder whether creative Arctic tourism could be transformative. In other words, could an experience of Arctic nature and culture change a traveller's worldview and values towards environmentalism?

Creative tourism experiences through seven discourses of sustainability

There are many views about, and definitions of, sustainability as well as sustainable development (see Buckley, 2012; Swarbrooke, 2002). While the demand for social and economic sustainability generally rules the development of society and business together with ecological sustainability, cultural sustainability has received less attention (see Tervo-Kankare, 2017). Cultural sustainability must be seen as an important fourth pillar or even interwoven into all other aspects of sustainability: culture is both an enabler and a driver of the economic, social, and environmental dimensions of sustainable development. Moreover, sustainability stands in opposition to consumerism, consumption, and exploitation as well as ideologies of eternal economic growth. Creative tourism that is focused on dialogues and maker cultures such as crafting moves us from consumers of goods to gathers of experiences and wisdom.

Tourism is frequently seen as a more sustainable livelihood than the industrial use of natural resources in which all four aspects of sustainability are important. Nevertheless, various challenges of sustainability through growing tourism – such as uncontrolled construction of tourism sights damaging the fragile Arctic nature and causes of climate change – are currently under discussion (see, for example, Swarbrooke, 2002; de Bernardi *et al.*, 2018; Similä and Jokinen, 2018). In addition, the cultural and social sustainability of tourism services are under critical consideration and many tourism researchers are calling for enhancing sustainability through participatory research (see, for example, Hakkarainen, 2017; Haanpää *et al.*, 2018; Höckert, 2018; Höckert *et al.*, 2021). With environmental crises challenging both tourism and education, the focus has now been placed on transformative education for sustainability (Jickling and Sterling, 2017). Overall, there is a need for transformative experiences that change our values to being holistic and respectful towards non-human nature or, in other words, to encourage our lifestyles to be humbler (Lin and Oxford, 2011). Thus, we are enhancing overall sustainability through creative tourism.

Creative tourism development offers a place for cultural encounter, dialogue, and knowledge exchange, which can be achieved in several ways, evolving from visiting exhibitions and creative spaces to direct participation in creative activities (see, for example, Richards and Wilson, 2006, p. 131; Miettinen, 2007; Duxbury and Richards, 2019), and shifting the focus from artefacts to their production process, as shown in Fig. 30.1. For tourists, the creative tourism experience is a multisensory experience in which the feel of materials, smells, the relationship with local craftspeople, and the working context become important parts of the experience. Moreover, arts-based practices offer *prosumption*, meaning 'production by consumers', which is an important part of the experience and is shaped by the characteristics of the user (personality, skills, background, cultural values, and motives) as well as those of the product (shape, texture, colour, and behaviour) (see Desmet and Hekkert, 2007) and the producer (creative practitioners). Physical actions and perceptual and cognitive processes (perceiving, exploring, using, remembering, comparing, and understanding) will contribute to the experience, which is further influenced by the context of interaction (physical, social, and economic). In a creative tourism experience, the social interaction between the local artisans, other members of the creative sector, and tourists plays a key role whereby a flow is constructed around the engagement in the creative activity (Dewey, 1934). This may involve learning to knit, weave, or sing a local song, baking a local dish, or engaging in other cultural activities. The creative tourism activities engage both the host and the guest in a mutual creative process (Miettinen, 2007).

B. Joseph Pine and James H. Gilmore (1999) discuss how companies offer experiences that intentionally use services as 'a stage' and goods as 'props' to engage an individual. Experiences are derived from the interaction between the staged event and the individual's prior state of mind and being (see MacCannell, 1973). The more effectively an experience engages the senses, the more memorable it will be. Furthermore, Pine and Gilmore (1999) propose that in an experience, the employees of the staging company are actors performing parts, creating roles, and building characters to engage guests in entertaining, educational, escapist, and/or aesthetic ways. In the context of Lapland, one can experience a multitude of staged activities around Christmas. In this context, the role of creative tourism might be to create more dialogue between the stage and the audience or even re-distribute the power relations between the host and the guest through collaborative engagement in the act of cultural doing.

Researchers Katriina Soini and Inger Birkeland (2014) have studied scientific discourses on cultural

Passive creative tourism	Tourists consuming creative places, such as exhibitions and/or art events and passively taking part of them.	In Lapland, creative places include museums and galleries with exhibitions from Finnish and local artists as well as visitor centers in national parks that have exhibitions on nature and culture. This is the most common creative tourism service in Lapland, but unfourtunately, it does not bring much income for local artists.
Interactive creative tourism	Visits to creative spaces with an artistic, architectural and/or design focus. Tourists learning processes and enjoying the atmosphere interactively.	In Lapland, some artists have opened up their studios in order to give interactive experiences for visitors. The field has not yet built up to its full potential and only a few artists get income from these visits.
Participative creative tourism	Participating actively in creative activities and experiencing skills development.	These activities are often located in rural or peripheral areas of Lapland. Activities include street art painting with graffiti artists, craft workshops or jewellery making in artisans' own studios.

Fig. 30.1. Creative tourism in Lapland. Note: The first two columns are based on categories of creative tourism, presented by Richard and Wilson (2006, p. 131).

sustainability and its diverse meanings. Their analysis shows that the discourse on cultural sustainability includes seven storylines: heritage, vitality, economic viability, diversity, locality, eco-cultural resilience, and eco-cultural civilization. Similarly, Valene L. Smith (1996) introduces the idea of 4H (habitat, heritage, history, and handicrafts) for Indigenous tourism with the aim of developing tourism from the local perspective. As Smith argues, these four aspects describe the phenomenon 'as a culture-bounded visitor experience which, quite literally, is a micro-story of man-land relationships' (p. 287). Moreover, the locally based development will benefit the local communities offering new and innovative activities for tourism (Smith, 1996). Both of these concepts are relevant to creative tourism, where the experience typically enables tourists to participate in local activities such as crafts production, photographing, singing, and other similar practices. In addition, these concepts can help create innovations as well as shape existing practices to be more sustainable. These concepts are discussed with practical examples in the following section.

Seven storylines of cultural sustainability in creative tourism from Lapland

Storyline 1: Cultural heritage

As Thomas Heyd (2007) has described it, the idea of heritage involves reference to something enjoyed and appreciated nowadays, coming from the past and shared in common by all belonging to a nation, ideological affiliation, or another group. In a similar vein, Smith (1996) discusses cultural heritage and its relation to knowledge and skills by arguing that heritage is, in many cases, the factor that attracts tourists. Soini and Birkeland (2014) explain that when discussing cultural heritage in relation to cultural sustainability, some challenges relating to the preservation of heritage stem from globalization flows of people (immigration and outmigration) and trade and the impact of these on languages, ethnic heritage, identity, and the distinctiveness of landscapes. This is also the reality in the Arctic where peripheries are impacted by the centralization of national policymaking and globalization and immigration to and from Northern villages lead to notable changes in the communities (see Olsen *et al.*, 2019). Auclair and Fairclough (2015) describe heritage as a continual process of remaking that is rooted in social construction and emphasize its vitality. Heritage as an intergenerational process in which histories and traditions, for example, are passed on from one generation to another and reshaped in contemporary and spatial contexts of society, has also been inspiring contemporary artists (Härkönen *et al.*, 2018; Guttorm, 2015).

Heritage is frequently defined as cultural heritage and natural heritage. According to Heyd (2007), cultural heritage refers to, for example, architectural monuments and sites with historical and aesthetic value, while natural heritage refers to biological and geological formations with specific conservation or aesthetic value. In discussing these issues from the point of view of the Arctic, it is important to acknowledge that the distinction between cultural and natural heritage is blurry: human and non-human nature are in continuous interaction, with some of the recognized natural heritage also intertwined with human culture. In Sámi culture, nature and culture are not separate concepts (Sámi Parliament, 2006).

Following the guidelines of the Sámi Parliament, we are not separating these elements in this chapter; instead, they walk hand in hand. In Lapland, there are few architectural monuments and heritage consists of nature: Sámi culture, and the history of the Second World War, just to name a few dimensions. Culture in the heritage storyline is constantly changing and it also accumulates as time passes (Soini and Birkeland, 2014). This means that the history of the culture and heritage is important to acknowledge in any future actions done, for example, in tourism. One good example of this is the recent decision made by the Visit Inari company (offering accommodations and tours) in Finnish Lapland not to take their customers to Ukonkivi, the sacred site for the Sámi community (Niemelä, 2019). This accords well with Soini and Birkeland's (2014) idea of choosing the objects for conservation according to both cultural and economic reasoning, considering their importance for the continuation and vitality of culture when everything cannot be preserved. Instead of site visits, Visit Inari takes their cruise around the island and tells the story and its sacred history to their visitors in order to assure the continuation of this cultural knowledge.

Another practical example from Lapland is the winter art, as illustrated in Fig. 30.2. Winter art is one way of reforming Arctic culture but also respecting the core of it: not making permanent changes to

Fig. 30.2. Producing cultural signs on snow with guests. Design of the activity and photo by Juho Hiilivirta, 2017, used with permission.

nature. The art vanishes every spring or even sooner when new snow falls on snowy tracks. Similar aesthetically oriented and ethically sound activities could be done with tourists interested in the concepts of cultural heritage and arts-based methods. If Arctic heritage is understood as a way of thinking – respecting nature and not making permanent signs – this worldview could be shared between generations as well as hosts and guests through arts-based methods.

Take-aways

- Respect cultural heritage and consider its importance for the continuation and vitality when everything cannot be preserved.
- Think of new ways of making art in sustainable ways, together with tourists. For example, making art with snow as a touristic activity and not leaving permanent signs in nature.

Storyline 2: Cultural vitality

In the second storyline, cultural vitality, artists are seen as having the potential for transferring heritage and other cultural capital in a way that is respectful of cultural continuity and cultural identities and promotes social inclusion (Soini and Birkeland, 2014). Creative tourism includes art- and craft-based activities that have the potential to increase the appreciation of traditions and crafts as well as the knowledge of the cultural heritage of Lapland. The aura of tourism art should be shifted from exploitation of the visual language of local (Indigenous and non-Indigenous) crafts to souvenirs and craft-based services that can act as mediators of traditions, cultural values, and discussions (Kugapi *et al.*, 2020). The revitalization of traditions is often discussed in relation to creative tourism whereby traditions are continued and reformed in a cross-generational manner. Revitalization is a complex aim; it is impossible to restore a culture to a certain time or certain experience, and the issues of protecting and renewing cultural heritage are multi-layered and often involve politicized questions of ownership and power (Härkönen *et al.*, 2018). However, creative tourism can breathe new life into active practices and foster appreciation for traditional skills and designs.

Social and cultural aspects of sustainability have, to some degree, an iterative and reciprocal relationship in which culture constructs society and society shapes culture. The cultural dimension adds to sustainability values, creativity, and temporal and spatial dimensions of development that are place-based (see Daniels *et al.*, 2015).

A practical example of revitalization in creative tourism, in Fig. 30.3, comes from Maria Huhmarniemi, who began creating himmeli sculptures by extending traditional decorations into free-form sculptures (Härkönen *et al.*, 2018). Himmeli is a traditional Finnish decorative mobile made together with one's family from short strips of rye straw tied together with string to form a symmetrical three-dimensional structure. Maria has created several series of himmeli sculptures that also carry a message of interculturalism: The himmeli sculptures have Swedish and German names, underlying the fact that art develops in interaction with other cultures,[1]

Fig. 30.3. Revitalizing traditional forms by making decorative crafts from straw. Photos by Maria Huhmarniemi, 2017, used with permission.

and quite surprisingly, it originates from a canopy decoration used in Medieval and Renaissance festivities in Central Europe. The organized himmeli workshops have been cross-generational and intercultural, attended by participants from all ages and diverse cultural backgrounds. As a creative tourism activity, the himmeli workshop includes a dialogue on Finnish culture as well as on cultural vitality and interaction.

Take-aways

- Creative tourism workshops, at their best, foster dialogues of cultures, promote cultural vitality, and offer a beautiful place for interaction.
- Traditions can be used in modern ways for modern needs, but sensitivity is demanded, especially if the visual language is from other cultures.

Storyline 3: Economic viability

The third storyline of cultural sustainability adopts culture (such as heritage, traditional foods, and crafts) for economic vitality in tourism (Soini and Birkeland, 2014). For example, cultural centres both nurture, develop, and sustain local culture and also act as tourist destinations (Soini and Birkeland, 2014; see also Smith, 1996). A growing focus on the economic impact of culture can be noticed in the Arctic Human Development Report (AHDR, 2015) but, unfortunately, when combining cultural and economic activities there can be many clashes, interconnections, and paradoxes (Kramvig, 2017). In any case, the challenges related to the applied use of culture are seen necessary in the current global economy: the instrumental use of culture for sustainability is emphasized and legitimized in many fora (Soini and Birkeland, 2014).

This storyline focuses on the possibilities for creating sustainable small-scale production–consumption relationships (Soini and Birkeland, 2014). In Finnish Lapland, artists have shown an interest in the applied use of their skills and expertise since new ways of earning income are needed (Huhmarniemi and Jokela, 2019). Moreover, crafters experience tourism as necessary to their livelihood, but until now the income has come mostly from souvenirs, while growing expectations are placed on services (Kugapi *et al.*, 2020). Often crafters are lifestyle- and micro-entrepreneurs who give priority to creative work instead of producing large numbers of products and hesitate to get involved in the commercial tourism business (Kugapi *et al.*, 2020). Furthermore, souvenir production has some challenges, such as price competition, quality, and cultural sensitivity (Kugapi *et al.*, 2020; see also Smith, 1996). Similarly, research from India shows how tourists expect the items to be locally made from durable material, fitted to their luggage, and cheap (Smith, 1996). This causes issues for the economic viability of crafters: Are their products suitable for tourism and should they even be?

In practice, to be economically viable and do creative tourism, crafters need to be present in different places and contexts and be able to tell their stories to possible buyers. As shown in Fig. 30.4 (on the left), the crafter makes jewellery and at the same time invites visitors to make earrings with her. Moreover, she is selling her products, talking with her customers, and telling stories, like the crafter on the right. All these activities are happening at the same time. Thus, the role of the crafter is diverse, and the crafter needs to be prepared for surprises.

Take-aways

- Many crafters need to have different skills in order to be economically viable.
- Crafters also need an understanding of how to be involved and work in tourism. This might need new education and training to learn new skills.

Storyline 4: Cultural diversity

Central to the fourth storyline, cultural diversity, is that contemporary society is increasingly diverse and different values are in constant negotiation (Soini and Birkeland, 2014). This is also true in Lapland, as well as the whole Arctic, which is a multi-ethnic and cultural region with various minorities. The dominant narratives in tourism marketing, as well as in international visual arts and literature, do not honour this richness of cultures, and the presentations of Indigenous people also fall many times into stereotypes (see, for example, Olsen *et al.*, 2019). The national strategies in Arctic countries deal with cultures mainly in the context of the regions' Indigenous populations. In other words, when culture is addressed in the strategy texts, it is most often done in the specific context of the regions' Indigenous peoples (Lempinen, 2019). For non-Indigenous residents, who comprise the vast majority of the inhabitants of the circumpolar North, hardly any personal culture is seen as worth sustaining and maintaining from this strategy-level perspective (Lempinen, 2019). At the same time, in general tourism marketing, Indigenous people's culture is strongly presented and celebrated, and many representatives of Indigenous cultures blame tourism for the exploitation of their cultures (Smith, 1996; Lüthje, 1998; Müller and Viken, 2017; Sámi Parliament, 2018). What is needed in the Arctic is a revitalization of various local cultures and traditions, supporting cultural identities (plural), and a celebration of cultural diversity. As Schilar and Keskitalo (2018) note, crafts can be considered as markers and practices of culture, identity, and ethnicity while they have also become touristic representations and idealized imagery. In Lapland, the land of mass-produced souvenirs without reference to cultural heritage, it becomes

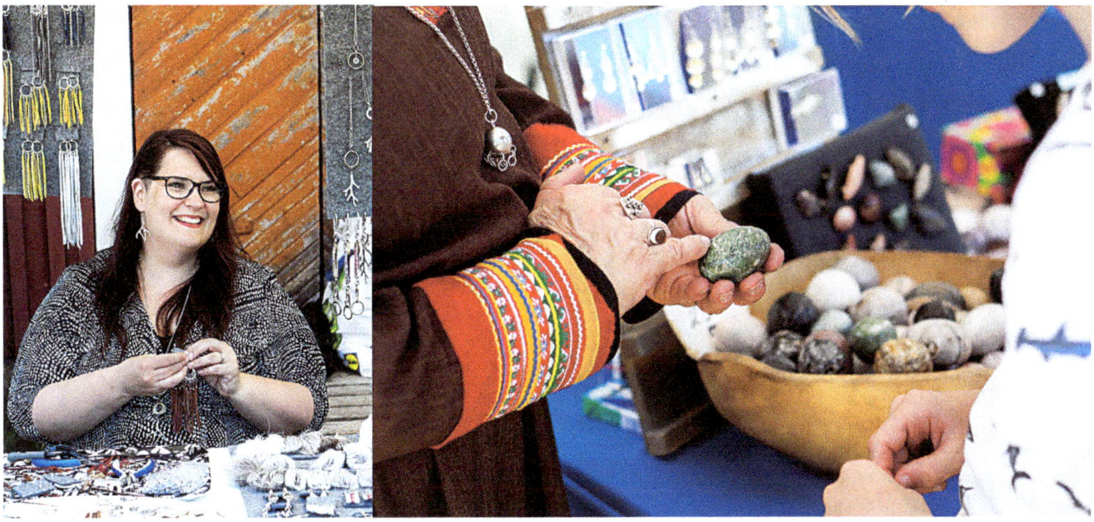

Fig. 30.4. Crafters selling their products, telling stories, and giving workshops at a small-scale market event in a music festival in Lapland. Photos by Mari Parpala, 2019, used with permission.

relevant to ponder whether tourism has destroyed or is destroying some of the original culture.

Härkönen *et al.* (2018) argue that craft-based creative tourism activities with place-specific intercultural approaches could create an open space for dialogue in which the values and the perceptions of cultural heritage can be negotiated. Practical examples of this idea can be told via two artworks. The first, the *Shared Woollen Sceneries,* was a dialogical art project in which Elina Härkönen, Maria Huhmarniemi, and Annamari Manninen invited participants to knit images of their inner landscapes as well as memories of (outer) landscapes. The second was a knitted piece by Maria Huhmarniemi, *Imaginary Identities I* (2019), in which the knitted mitten combines traditional patterns from the Enontekiö and Alahärmä regions in Finland and Viljandi in Estonia (Fig. 30.5). These artworks shared the same values: the knitted crafts reflect cultural, ethnic, and regional identities with their patterns from all around the world. Moreover, the mitten artwork highlights the cultural right to express identities in between cultures (such as Indigenous and non-Indigenous cultures) as well as cultural influences in handicrafts as one very rich way of learning. In creative tourism, it is essential that tourists can develop their own narratives and engage their imagination in the process (Richards and Wilson, 2006).

Take-aways

- Crafted objects can reflect cultural, ethnic, and regional identities.
- Travellers can make their own narrative via crafts.
- Creative tourism deepens the experience of cultural richness and diversity.
- Our cultural identities have many layers: creative tourism can enhance understanding the value of one's own identity.

Storyline 5: Locality

Tourism products and services in Lapland are not always based on local culture(s) due to the fact that most of the tourists are seeking nature-based services in Lapland (Rantala *et al.,* 2011; House of Lapland, 2018) and there has been a strong focus on developing Santa Claus tourism. In addition, there has been a tendency to please tourists from abroad by producing sites and services familiar to them, for example, alp-style houses at skiing centres. This brings challenges to the fifth storyline, locality, which calls for local involvement in (tourism) planning and decision-making processes (Soini and Birkeland, 2014). Unfortunately, in discussing this in the Arctic, and especially in Indigenous contexts, many local residents are left out of the tourism decision-making processes (see, for example, Müller and Viken, 2017). This pattern is also visible in souvenir production.

The commoditization of cultures in tourism occurs when local cultural symbols, rituals, and heritage become part of tourism services and products (Cohen, 1988). Mass market tourism trinkets, staged authenticity (MacCannell, 1973), and kitsch 'airport art' (Graburn, 1967, cited in Cohen, 1988)

Fig. 30.5. Combining patterns from home regions and tourism destinations in workshops. Photos by Annamari Manninen, 2018 (left) and Maria Huhmarniemi, 2019 (right), used with permission.

are good examples of this. In contrast, handicrafts as souvenirs have traditionally represented those locales that have not been mass tourism destinations (Hitchcock, 2005), like Finnish Lapland, where Sámi handicrafts have been available to tourists for decades. These souvenirs were locally made and brought extra income to Sámi to supplement other activities such as, for example, reindeer herding. In Finnish Lapland, this was true before mass tourism started in the 1970s. After that, imported products started filling the shelves of souvenir shops and Sámi handicrafts were sold less and less as a souvenir. These imported goods exploit the local culture without bringing advantages to the locals (Hitchcock, 2005).

According to Timothy (2012), the souvenir industry violates the locals by providing incorrect images and impressions of the culture itself. Nevertheless, it must be emphasized here that the local culture is made visible not only in local products but also through other elements around the destination, like is done in The Nordic Knitting Symposium, a week-long craft workshop organized every summer in different Nordic locations by the non-profit organization Gavstrik (Fig. 30.6).

In the symposium, locality is emphasized in different ways. Firstly, many of the workshops originate from the area in which the symposium is held and have local teachers. Secondly, the side activities allow participants to gain knowledge of the local culture, draw inspiration from the scenery, and learn new ways of doing and working. Thirdly, locality is discussed in lectures held by local crafters. Fourthly, the knowledge of locality and beyond is discussed and learned from other participants. Usually the symposia have participants from all Nordic countries and therefore the event allows the locality to be shared from many perspectives. This example of a creative tourism experience has demonstrated that cultural workshop encounters with local craftspeople engage tourists in a deeper understanding of the local cultures. The narratives work as vehicles for both sharing empathy and creating understanding about the visited place and community. The creative tourism process involves not only consuming cultural tourist activities but also becoming more aware of the society and local ways of living. It shifts the role of the tourist from being an outsider into an active process of learning and becoming aware of it (Miettinen, 2007).

Take-aways

- Locality and/or regionality is a key to sustainability.
- Collaboration with community members enhances cultural and social sustainability.
- Also, other factors, such as scenery and local lifestyle, affect the creative tourism experience and present locality in different ways.

Storyline 6: Eco-cultural resilience

The sixth storyline, eco-cultural resilience, interlinks the ecological and the cultural (Soini and Birkeland, 2014). Troubles, disturbances, and crisis can be seen as opportunities for renewal, and innovative thinking and resilience help individuals to deal with these changes. The concept of resilience refers to the capacity of an individual, community,

Fig. 30.6. Workshops and side-activities in the Nordic Knitting Symposium, Kiruna. Photos by Outi Kugapi, 2016, used with permission.

or even a natural entity such as a forest to deal with changes. Resilience thinking embraces learning to stay within planetary boundaries. The concept of resilience is based on perceiving humans and non-human nature as coupled and in need of each other. As Soini and Birkeland (2014) describe it, the storyline of eco-cultural resilience highlights that ecological and cultural processes are teamed and that community management of natural resources fosters a sense of responsibility and care for nature.

While urban centralization and city lifestyles have led to a disconnection from nature, a strong and interactive relationship with nature is typical for Arctic communities. The Arctic worldview and human–nature relationship emphasize the experience of nature's liveliness, will, and agency. This so-called animistic view of nature is typical in traditional Sámi cultures as well as more widely in the belief traditions of the North. Here, creative tourism can lead to understanding the interdependence of human and non-human nature as well as support participants' capacities to face environmental crises, stay resilient, and act for environmentalism.

In fact, engaging with the need to address the environmental crises is one of the new possible innovations for creative tourism. Nowadays, the media vividly discuss wide-spread eco-anxiety, which refers to severe or mild fears, restlessness, melancholy, and stress caused by ecological concerns such as climate change or the sixth mass extinction of species (Searle and Gow, 2010). As the Arctic is warming more quickly than the rest of the globe and the bio-diversity is weakening rapidly, the Arctic is also rife with anxiety. In designing creative tourism services for guests suffering eco-anxiety, the services can support hope and build the capacities and capabilities to act in an environmentally friendly manner.

This is visible, for example, in the socially and environmentally engaged art workshop 'Bark Boat of Hope' by art educators Mari Parpala and Emilia Laatikainen, where participants make boats of barks with expressed hopes for the future (Fig. 30.7). The workshops are held in a small touristic village, Äkäslompolo, in Finnish Lapland, where locals are battling against iron mining plans that involve many risks, such as the destruction of the clean nature, and worries are especially focused on the purity of small rivers. The arts-based methods used in the workshops have an ability to create an existential and spiritual impact that affects the body, mind, and feelings and thus can relieve eco-anxiety (see Pihkala, 2017). Moreover, artistic expressions can also support the resilience of hosts and guests.

Take-aways

- Creative tourism in nature can contribute to eco-cultural resilience and battle against eco-anxiety.
- Incorporating innovative thinking, arts-based methods, and resilience in creative tourism activities can help individuals to deal with troubles, disturbances, and crises in life.

Storyline 7: Eco-cultural civilization

Soini and Birkeland (2014) explain that the seventh storyline, eco-cultural civilization, refers to an

Fig. 30.7. Creating hope for people suffering eco-anxiety and discussing local environmental conflicts in workshops such as the 'Bark Boat of Hope'. Photos by Mari Parpala, 2019, used with permission.

ecological turn of the values and behaviour of people: the importance of cultural aspects in achieving the overall aims of sustainability is emphasized, and the integration of culture in sustainability is seen as crucial for coping with change. At the same time, transformative experiences and deep changes in values are seen as necessary for the transition to sustainability; educational theorists highlight the need for increasing empathy, kinship, and a sense of community and belonging (Lin and Oxford, 2011; Jickling and Sterling, 2017; Foster *et al.,* 2019). The concept of eco-cultural civilization refers to these transformative experiences that lead to humbler consumerism and sense of fullness of life in relation to nature. In Finnish Lapland, the landscape itself has become a venue for creative tourism: it allows the tourist an immersion in creative activity through walking, sensing the silence and soundscape of nature, and experiencing, for example, winter art such as snow and ice sculptures (Miettinen *et al.,* 2019; see also Fig. 30.8). As a space for learning, these activities could lead to instilling values that underpin an eco-cultural civilization.

A practical example of fostering eco-cultural civilization was the *Wooden Spoons* art project where artist and researcher Antti Stöckell invited various groups of people to participate in walks in forests, make their own spoons by blowing into glowing red charcoal, and share stories of craft-based cultural activities and heritage. This method of using fire in processing wood was innovative, as it is not common nowadays although it is part of the Finnish cultural heritage. After the workshop, the participants received either a finished wooden spoon or a yet incomplete spoon as a homecoming gift, as well as the good feelings brought by the walking and encounters with the forest (Stöckell, 2018). Altogether, this art project included aspects of personal revitalization as well as transformational experiences and explored ways of interweaving practices of crafting, dialogical art, and hiking (Stöckell, 2018).

Take-aways

- Eco-cultural civilization is an important element of sustainability and also in coping with change. Transformative experiences are needed, and creative tourism offers a place for this.
- Creative activities could enhance nature connectedness as it allows the tourist an immersion in creative activity through walking, sensing the silence and soundscape of nature, and experiencing, for example, winter art.

Conclusions

Through the examples of seven discourses of sustainability, we have examined arts-based methods as vehicles for creative and sustainable tourism (see Fig. 30.9). In doing so, we have argued that these concepts support the emergence of cultural vibrancy and sustainability for creative tourism, souvenir and service production, and new artistic expression in local communities. We have illustrated these in Fig. 30.9, where the middle circle presents the potential of creative tourism as a contribution to sustainability, the second circle presents the grouped version of the seven storylines of cultural sustainability, and the outer sectors describe the Arctic as a venue for creative tourism development.

Fig. 30.8. Fostering eco-cultural civilization in a dialogue with nature and art. Photos by Antti Stöckell, 2017, used with permission.

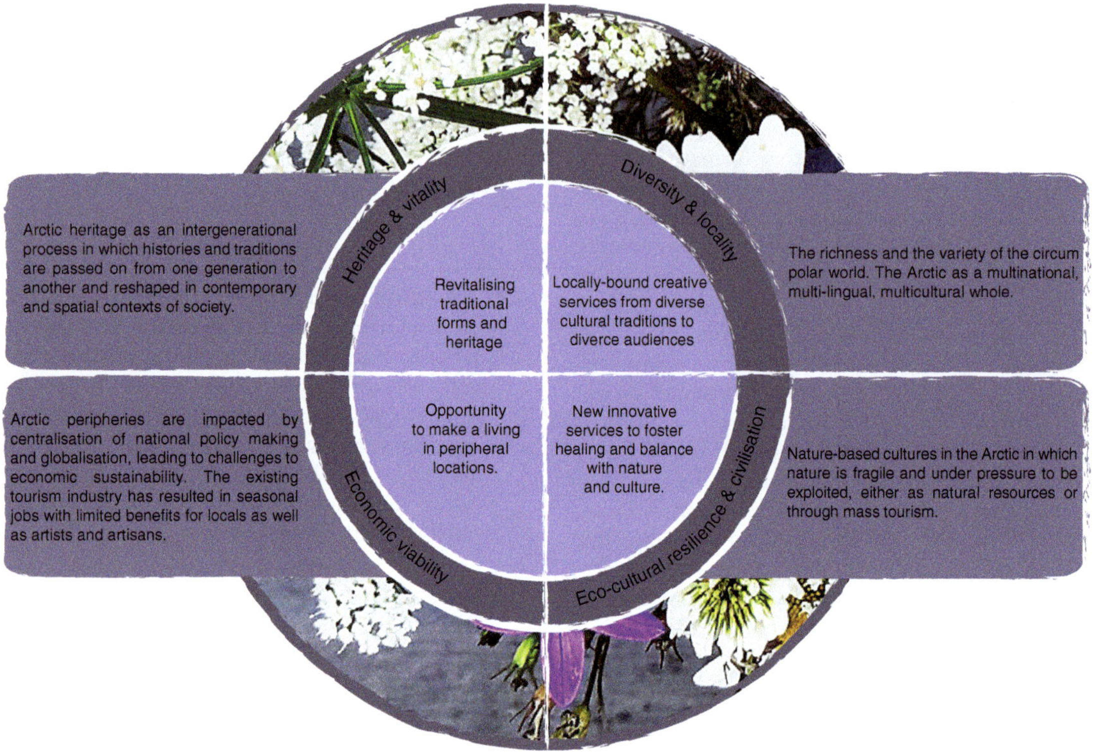

Fig. 30.9. The storylines of cultural sustainability: heritage, vitality, economic viability, ecocultural resilience, and eco-cultural civilization located into Arctic environment and culture.

We can argue that creative tourism is needed in Lapland for fostering knowledge exchange between host and guest. Moreover, it also has a potential to support the revitalization of local cultures and heritage. Producing new interpretations and expressions of place, culture, and human–land relationships can lead to celebrations of diversity as well as eco-cultural resilience and eco-cultural civilization. While designing creative experiences, social encounters should be attuned to local particularities and global concerns. Creative tourism activities must be driven by authentic cultures and map the place, community, heritage, and narratives. Nevertheless, the activities and services offered do not need to be same as those of the traditional culture: traditions can be transmitted respectfully into contemporary needs and practices. As the experiences are always subjective in creative tourism as well, tourists must be able to develop their own narratives, relations, and kinship into the hosts and places of encounter. This is a way for the traveller–learner to change their old ways of experiencing and find new meanings for life.

Heritage, culture, art, and craft offer diverse possibilities for creative tourism activities. Sustainable development, especially cultural sustainability, can be enhanced in collaborations of artists, crafters, and tourism businesses. The creative tourism activities need to be bound from local resources in order to be sustainable and offer steady income for crafters. The most important lesson here is that the creative tourism activities can be based on mundane activities, such as knitting mittens from local materials and patterns, or making wooden spoons. Many artists, crafters, designers, and art educators have the skills to renew traditions for contemporary expressions, needs, and circumstances and to design creative activities.

Furthermore, future initiatives in creative tourism should include aims to support empowerment, equality, agency, and participatory society for citizens. Artistic, creative, and critical interventions call for the generation of strategies and methods that make it possible to live in more humble ways, move towards carbon-neutral lifestyles, and decolonize nature and Arctic cultures. Respect for the fragile

Arctic nature and rich culture can increase via transformative experiences that have a positive impact on healing and promote personal and social balance between nature and heritage.

Note

1. The word *himmeli* has its roots in the Swedish and German languages; *himmel* means sky and heaven.

References

AHDR (2015) *Arctic Human Development Report II: Regional Processes and Global linkages.* TemaNord 2014, 567. Nordisk Ministerråd, Copenhagen, Denmark. Available at: http://norden.diva-portal.org/smash/record.jsf?pid=diva2%3A788965anddswid=-5478 (accessed 11 November 2019).

Auclair, E. and Fairclough, G. (2015) Living between past and future: An introduction to heritage and cultural sustainability. In: Auclair, E. and Fairclough, G. (eds.) *Theory and Practice in Heritage and Sustainability. Between Past and Future.* Routledge, Oxon, UK, pp. 1–22.

Barone, T. and Eisner, E. (2012) *Arts Based Research.* Sage, Los Angeles, California, USA.

Buckley, R. (2012) Sustainable tourism: Research and reality. *Annals of Tourism Research* 39(2), 528–546. doi: 10.1016/j.annals.2012.02.003.

Chartier, D. (2018) *What is the"Imagined North"?* Presses de l'Université du Québec, Québec City, Canada.

Cohen, E. (1988) Authenticity and commoditization in tourism. *Annals of Tourism Research* 15, 371–386.

Daniels, J., Baldacchino, G., and Vodden, R. (2015) Matters of place: The making of place and identity. In: Vodden, K, Gibson, R., and Baldacchino, G. (eds.) *Place Peripheral: Place-based Development in Rural, Island, and Remote Regions.* ISER Books, St. John's, Canada, pp. 23–40.

de Bernardi, C., Kugapi, O., and Lüthje, M. (2018) Sámi indigenous tourism empowerment in the Nordic countries through labelling systems: Strengthening ethnic enterprises and activities. In: Borges de Lima, I. and King, V.T. (eds.) *Tourism and Ethnodevelopment: Inclusion, Empowerment and Self-Determination.* Routledge, Oxon, UK, pp. 200–212.

Desmet, P.M. and Hekkert, P. (2007) Framework of product experience. *International Journal of Design* 1(1), 57–66.

Dewey, J. (1934) *Art as Experience.* The Berkeley Publishing Group. First Perigee printing 1980, USA.

Duxbury, N. and Richards, G. (2019) Towards a research agenda for creative tourism: A synthesis of suggested future research trajectories. In: Duxbury, N. and Richards, G. (eds.) *A Research Agenda for Creative Tourism.* Edward Elgar Publishing, Northampton, Massachusetts, USA, pp. 1–14.

Foster, R., Mäkelä, J., & Martusewicz, R. (eds.). (2019) *Art, ecojustice, and education: Intersecting theories and practices.* Routledge, Boca Raton, Florida.

Guttorm, G. (2015) Contemporary Duodji—a personal experience in understanding traditions. In: Jokela, T. and Coutts, G. (eds.) *Relate North: Art, Heritage and Identity.* Lapland University Press, Rovaniemi, Finland, pp. 60–77. Available at: http://urn.fi/URN:ISBN:978-952-310-971-1 (accessed 18 April 2020).

Haanpää, M., Hakkarainen, M., and Harju-Myllyaho, A. (2018) Katsaus yhteiskunnalliseen yrittäjyyteen matkailussa: osallisuuden mahdollisuudet pohjoisen urbaaneissa paikallisyhteisöissä. [Review of social entrepreneurship in tourism: Possibilities of participation in the Norther urban communities]. *The Finnish Journal of Tourism Research* 14(2), 44–58.

Hakkarainen, M. (2017) *Matkailutyön ehdot syrjäisessä kylässä* [*The Conditions of Tourism Work in a Remote Village*], Acta Universitatis Lapponiensis 357, Lapland University Press Rovaniemi, Finland.

Härkönen, E., Huhmarniemi, M., and Jokela, T. (2018) Crafting sustainability: Handcraft in contemporary art and cultural sustainability in Lapland. *Sustainability* 10(6), 1907. doi: 10.3390/su10061907.

Heyd, T. (2007) *Encountering Nature: Toward an Environmental Culture.* Ashgate, Aldershot, UK.

Hiltunen, M. and Rantala, P. (2015) Creating art-based approaches in working life development: The shift from success to significance. *International Journal of Education through Art* 11(2), 245–260.

Hitchcock, M. (2005) Introduction. In: Hitchcock, M., and Teague, K. (eds.) *Souvenirs: The Material Culture of Tourism.* Ashgate Publishing, Aldershot, UK, pp. 1–17.

Höckert, E. (2018) *Negotiating Hospitality: Ethics of Tourism Development in the Nicaraguan Highlands.* Routledge, London, UK.

Höckert, E., Kugapi, O., and Lüthje, M. (2021) Hosts and guests in participatory development. In: Harju-Myllyaho, A. and Jutila, S. (eds.) *The Future of Inclusive Tourism.* Channel View Publications, Bristol, UK.

House of Lapland (2018) 10 facts about tourism in Lapland. Available at: www.lapland.fi/uploads/2018/04/428a8f68-pdf_10-facts-about-tourism-in-lapland-2018.pdf (accessed 11 November 2019).

Huhmarniemi, M. and Jokela, T. (2019) Environmental art for tourism in the Arctic: From handicraft to integrated art and reform on artists' skills. *Synnyt/origins, Special issue on Arctic Arts Summit,* 63–80. Available at: https://wiki.aalto.fi/pages/viewpage.action?pageId=151504259 (accessed 11 November 2019).

Huhmarniemi, M. and Jokela, T. (2020) Arctic arts with pride: Discourses on Arctic arts, culture and sustainability. *Sustainability* 12(2), 604. doi: 10.3390/su12020604.

Jickling, R. and Sterling, S. (eds.) (2017) *Post-sustainability and Environmental Education: Remaking Education for the Future*. Palgrave MacMillan, London, UK.

Jokela, T., Härkönen, E., and Yliharju, A.-J. (eds.) (2014) *Snow Design in Lapland—Initiating Cooperation.* Lapland University Press, Rovaniemi, Finland.

Kramvig, B. (2017) Orientalism or cultural encounters? Tourism assemblages in cultures, capital and identities. In: Viken, A. and Müller, D.K. (eds.) *Tourism and Indigeneity in the Arctic*. Channel View Publications, Bristol, UK, pp. 50–71.

Kugapi, O., Huhmarniemi, M., and Laivamaa, L. (2020) A potential treasure for tourism—crafts as employment and a cultural experience service in the Nordic North. In: Walmsley, A., Åberg, K., Blinnikka, P., and Johannesson, G.T. (eds.) *Tourism Employment in Nordic Countries*. Palgrave, London, UK. pp. 77–99.

Leavy, P. (2009) *Method Meets Art: Arts-based Research Practice*. Guilford Press, New York, New York, USA.

Leavy, P. (ed.) (2018) *Handbook of Arts-based Research*. Guilford Press, New York, New York, USA.

Lehtola, V.-P. (1999) Aito lappalainen ei syö haarukalla ja veitsellä. Stereotypiat ja saamelainen kulttuurintutkimus [Real Sámi people do not eat with fork and knife: Stereotypes and cultural research on Sámi issues]. In: Tuominen, M., Tuulentie, S., Lehtola, V.-P., and Autti, M. (eds.) *Outamaalta tunturiin. Pohjoiset identiteetit ja mentaliteetit I.* [*From Field to Fells: Northern Identities and Mentalities*]. Kustannus-Puntsi, Inari, Finland, pp. 15–32.

Lempinen, H. (2019) Indigenous cultures, local lifestyles? 'Culture' in the northern strategies of the eight Arctic states. *Synnyt / Origins 1/2019. Special issue on Arctic Arts Summit*. Available at: https://wiki.aalto.fi/pages/viewpage.action?pageId=151504259 (accessed 15 November 2019).

Lin, J. and Oxford, R. (2011) *Transformative Eco-Education for Human and Planetary Survival.* Information AGE Publishing, Charlotte, North Carolina, USA.

Lüthje, M. (1998) The impacts of tourism on the Saami domicile area from the point of view of carrying capacity and Saami culture of the area. In: Aho, S., Ilola, H. and Järviluoma, J. (eds.) *Dynamic Aspects in Tourism Development. Proceedings of the 5th Nordic Symposium on Tourism Research*. Lapin yliopiston matkailun julkaisuja B 3, Rovaniemi, Finland, pp. 31–46.

MacCannell, D. (1973) Staged authenticity: Arrangements of social space in tourist settings. *American Journal of Sociology* 79(3), 589–603. doi: 10.1086/225585.

Miettinen, S. (2007) Designing the creative tourism experience. A service design process with Namibian crafts people. Doctoral dissertation. Publication series of University of Art and Design Helsinki A 81. Gummerus, Jyväskylä, Finland.

Miettinen, S., Erkkilä-Hill, J., Koistinen, S.-M., Jokela, T., and Hiltunen, M. (2019) Stories of design, snow and silence: Creative tourism landscape in Lapland. In: Duxbury, N., and Richards, G. (eds.) *A Research Agenda for Creative Tourism*. Edward Elgar Publishing, Northampton, Massachusetts, USA, pp. 69–83.

Müller, D.K. and Viken, A. (2017) Indigenous tourism in the Arctic. In: Viken, A. and Müller, D.K. (eds.) *Tourism and Indigeneity in the Arctic.* Channel View Publications, Bristol, UK, pp. 3–15.

Niemelä, N. (2019) Inarijärven risteilyihin ei enää sisälly rantautuminen Ukonsaarelle—päätös syntyi, kun keskustelu Ukonsaaren käytöstä alkoi. [Inarijärvi cruises do not stop in the Ukonsaari anymore]. Available at: www.lapinkansa.fi/inarijarven-risteilyihin-ei-enaa-sisally-rantautum/318985?fbclid=IwAR28ga7-MXRY-vDhjMiO6bA_Q8dWrY7cqwM1jYJ2BeH_yGrAGTRp6EMNJZhg (accessed 15 November 2019).

Niskala, M. and Ridanpää, J. (2016) Ethnic representations and social exclusion: Sáminess in Finnish Lapland tourism promotion. *Scandinavian Journal of Hospitality and Tourism* 16(4), 375–394. doi: 10.1080/15022250.2015.1108862.

Olsen, K., Avildgaard, M.S., Brattland, C., *et al.* (2019) *Looking at Arctic Tourism through the Lens of Cultural Sensitivity: ARCTISEN – a Transnational Baseline Report.* Publications of Multidimensional Tourism Institute, University of Lapland, Rovaniemi, Finland. Available at: http://urn.fi/URN:ISBN:978-952-6620-39-8 (accessed 18 April 2020).

Pihkala, P. (2017) *Päin helvettiä? Ympäristöahdistus ja toivo.* [*Eco-Anxiety and Hope*]. Kirjapaja, Helsinki, Finland.

Pine, B.J. and Gilmore, J.H. (1999) *The Experience Economy: Work is Theatre and Every Business a Stage*. Harvard Business School Press, Boston, Massachusetts, USA.

Rantala, O., Valtonen, A., and Markuksela, V. (2011) Materializing tourist weather: Ethnography on weather-wise wilderness guiding practices. *Journal of Material Culture* 16(3), 285–300.

Rantala, O., de la Barre, S., Granås, B, Þór Jóhannesson, G., Müller, D.K, Saarinen, J., Tervo-Kankare, K., Maher, P.T., and Niskala, M. (2019) *Arctic Tourism in Times of Change: Seasonality*. Tema Nord 2019, 528. Nordic Council of Ministers.

Richards, G. and Wilson, J. (2006) Developing creativity in tourist experiences: A solution to the serial reproduction of culture? *Tourism Management* 27(6), 1209–1223.

Sámi Parliament (2006) *Saamelaisten kestävän kehityksen ohjelma 2006.* [*Programme for sustainable development in Sámi Culture*]. Saamelaiskäräjät, Inari, Finland.

Sámi Parliament (2018) *Vastuullisen ja eettisesti kestävän saamelaismatkailun toimintaperiaatteet.* [*Principles of Responsible and Ethically Sustainable Sámi Tourism*]. [Inari]: Saamelaiskäräjät. Available at: www.samediggi.fi/wp-content/uploads/2018/11/Vastuullisen-ja-eettisesti-kest%C3%A4v%C3%A4n-saamelaismatkailun-toimintaperiaatteet_hyv%C3%A4ksytty_24092018-3.pdf (accessed 11 November 2019).

Schilar, H. and Keskitalo, E.C.H. (2018) Ethnic boundaries and boundary-making in handicrafts: Examples from northern Norway, Sweden and Finland. *Acta Borealia* 35(1), 29–48. doi: 10.1080/08003831.2018.1456073.

Searle, K. and Gow, K. (2010) Do concerns about climate change lead to distress? *International Journal of Climate Change Strategies and Management* 2, 362–379. doi: 10.1108/17568691011089891.

Similä, J. and Jokinen, M. (2018) Governing conflicts between mining and tourism in the Arctic. *Arctic Review on Law and Politics* 9, 148–173. doi: 10.23865/arctic.v9.1068.

Smith, V.L. (1996) Indigenous tourism: The four Hs. In: Butler, R. and Hinch, T. (eds.) *Tourism and Indigenous Peoples.* International Thomson Business Press, London, UK, pp. 283–307.

Soini, K. and Birkeland, I. (2014) Exploring the scientific discourse on cultural sustainability. *Geoforum* 51, 213–223. doi: 10.1016/j.geoforum.2013.12.001.

Stöckell, A. (2018) Making wooden spoons around the campfire: Dialogue, handcraft-based art and sustainability. In: Jokela, T. and Coutts, G. (eds.) *Relate North: Art and Design for Education and Sustainability*. Lapland University Press, Rovaniemi, Finland, pp. 80–97.

Swarbrooke, J. (2002) *Sustainable Tourism Management.* CABI Publishing, Wallingford, UK.

Tervo-Kankare, K. (2017) Kestävä matkailu (Sustainable tourism). In: Edelheim, J. and Ilola, H. (eds.) *Matkailututkimuksen avainkäsitteet.* [*Key Concepts in Tourism Research*]. Lapland University Press, Rovaniemi, Finland, pp. 235–240. Available at: http://urn.fi/URN:ISBN:978-952-310-952-0 (accessed 18 April 2020).

Timothy, D.J. (2012) Destination communities and responsible tourism. In: Leslie, D. (ed.) *Responsible Tourism: Concepts, Theory and Practice*. CABI, Wallingford, UK, pp. 72–81.

Veijola, S., Säynäjäkangas, J., Carlin, D., Honkasilta, J., Höckert, E., Light, A., and Vikman, N. (eds.) (2018) *Matkasanakirja hiljaisuuteen.* [*Travel Book to the Silence*]. ntamo, Helsinki, Finland.

Visit Finland (2019) Available at: www.visitfinland.com/destinations/ (accessed 17 May 2021).

31 Policy recommendations on creative tourism development in small cities and rural areas: Insights from the CREATOUR® project in Portugal

Alexandra R. Gonçalves,[a]* Maria do Rosário Borges,[b] Nancy Duxbury,[c] Cláudia Pato de Carvalho,[c] and Pedro Costa[d]

[a]Research Centre for Spatial and Organizational Dynamics, University of Algarve, Portugal; [b]Interdisciplinary Centre for History, Cultures and Societies, University of Évora, Portugal; [c]Centre for Social Studies, University of Coimbra, Portugal; [d]ISCTE - Instituto Universitário de Lisboa/DINÂMIA'CET-ISCTE - Research Center on Socioeconomic Change and Territory, Portugal

The CREATOUR® project in Portugal (see Box 1.1 in Chapter 1) focused on the ideation, design, and pilot implementation of small-scale creative tourism initiatives that aimed to bridge culture and tourism, contributing to the enhancement of the attractiveness of rural communities and the sustainability of local populations. We learned that creative tourism development outside large cities is feasible, but challenges of distributed geography and the diversity of organizations and initiatives require additional attention to capacity building, training, and mentorship; dedicated resources for collective marketing; and more time to deepen knowledge networking and the development of partnerships.

This chapter is a synthesis of proposed recommendations to authorities and agencies at all levels (national, regional, and local) relating to the development of creative tourism initiatives in small cities and rural areas.[1] It is centrally informed by the needs of the CREATOUR® pilot organizations in Portugal, which were articulated within discussions between researchers and practitioners. We also considered public policies related to rural tourism development and the types of public policies needed to foster and support cultural and creative organizations in small cities and rural areas. The recommendations recognize and reinforce the need to develop new structures to foster ongoing interconnections between the cultural and tourism sectors.

What is creative tourism?

The CREATOUR® project defined creative tourism as active creative activity encouraging personal self-expression and interaction between visitors and local residents, inspired by local endogenous resources (place and people), and designed and implemented by local residents. This perspective on creative tourism includes four dimensions: active participation, creative self-expression, learning, and community engagement.

Creative tourism is conceived as sustainable small-scale tourism that provides an active visitor experience by combining an immersion in local

*Corresponding author: marodrig@ualg.pt, creatour@ces.uc.pt

DOI: 10.1079/9781789243536.0031

culture with a learning and creative process. To distinguish 'creative' tourism from other types of 'experience' tourism, particular emphasis is placed on the creation process and capacity for the visitor to engage in the activity not only from the perspective of learning and skill development (or, alternatively, of entertainment and relaxation) but also one of potential for self-expression. These activities enable the generation of added value – economic, social, and cultural – at the local and regional level.

One of the greatest potentials of creative tourism is its transversality, complementing and extending the offers within the various types of existing tourism (such as cultural tourism, nature tourism, sun and beach tourism, gastronomic tourism, industrial tourism, military tourism, nautical tourism, etc.) to meet the diverse needs of travellers today, as well as those of cultural and territorial development. Some of the motivations of the 'new tourist' are linked with self-fulfilment, experimentation, emotional engagement, and the development of skills – all characteristics that are associated with creative tourism.

In Portugal, creative tourism activity is currently at a developmental phase, with creative tourism offers largely at embryonic/early stages. However, creative tourism has proven to be an inspiring perspective and direction for local culture-based development and for extending cultural tourism offers in smaller places. Both cultural/creative and tourism organizations, as well as territorial development agents, are still discovering ways to meaningfully engage with the new profile of visitors that search for authentic interactivity with locals and experiences that combine fun, learning about local cultures, and creative self-expression. Forging ongoing opportunities for inter-sectoral networking, co-learning, communication, and collaboration will be key in advancing this work.

What lessons did we learn working with creative tourism practitioners? Summarizing key development needs

Among many insights from the CREATOUR® project, we identified key areas of development needs that emerged in the course of the development of the pilot initiatives. We believe that addressing these needs is essential in order to develop successful creative tourism initiatives and that this could be incorporated in a future network supported by a creative tourism policy agenda.

Partnerships, networks, and clusters

Creative tourism practitioners expressed an urgent need to be involved in partnership approaches, networks, clusters, and territorially embedded creative milieus at both regional and national levels. Greater interconnections should be fostered between cultural/creative and tourism/hospitality professionals for different purposes (e.g. share knowledge, circulate ideas, stimulate innovation, co-develop activities, co-develop commercial strategies, attract new visitors, and link local and regional places and practitioners into national and global circuits of information and resources). At the regional level, practitioners expressed the need to develop closer relations with different types of partners/collaborators from various areas (e.g. artisans/artists, tourist animation companies, and public bodies and agencies) to co-develop creative tourism packages. Aligned with this work is the need to build distributed capacity for partnership management.

Inter-sectoral mediation

There is an expressed and urgent need to develop inter-sectoral mediation and communication skills to build bridges among the different sectors involved in this new niche of creative tourism: culture, creative industries, tourism, and local/regional development. The objective is to create mediators, knowledge brokers, mentors, and facilitators who can act as elements of articulation at two levels: at the intra-sectoral level (e.g. to identify needs, articulate priorities, and define roles) and at the inter-sectoral level, enabling and fostering cross-sectoral collaboration (e.g. to define a plan for a common vision for creative tourism in an area, explore resources and optimize uses, reach agreements on defined priorities and needs for financial and technical support, assign roles, etc.). This inter-sectoral mediation is required not only among agents working in the field, but also at the level of policy-making and regulatory agencies, where articulation and cross-fertilization between diverse sectoral policies and different levels of administration is often insufficient.

Qualified human resources

There is an expressed and urgent need to have qualified and motivated human resources involved in all cycles of the creative tourism value chain and to promote capacity building among the different players (e.g. entrepreneurs, employees, mediators, public cultural organizations, interpreters and guides, artisans, volunteer collaborators). The objective is to acquire or strengthen skills to be more competitive and more committed to the activities, and to better understand and deal with the specific contexts of creative tourism markets, while also caring for the development needs of their community and respecting sustainable tourism principles (e.g. develop skills in collaborative work solutions, and on the design, development, implementation, and promotion of creative tourism offers). Special attention was requested by CREATOUR® practitioners concerning the need to transfer knowledge on craft work practices and other traditional and rare/unique activities from elder craftspeople to the younger generations. Training new artisans and workshop providers is also a great concern, both in the crafts techniques as well as the development of workshops for creative tourism – without these qualified human resources, capacity building for planning and commercialization, in the medium- or long-term, may no longer be necessary.

Communication systems and processes

There is an expressed and urgent need to develop improved, targeted communication and marketing approaches, with particular attention to the areas of advertising (e.g. integrated presentation and promotion of activities by theme or territory), sales promotion (e.g. market segmentation), personal selling and public relations (e.g. building customer relationships, special events and festivals, integrating successful international networks of creative tourism, and attendance at international fairs where international tour operators and promoters can get to know the offer of creative tourism in Portugal), and direct and digital marketing (e.g. online and social media, catalogues, a specialized web platform for creative tourism offers). CREATOUR® practitioners also seek solutions to facilitate communications between tourism companies, institutions, and associations that usually do not combine efforts in order to promote synergies and the competitiveness of this sector. The latter challenge can be informed by communication solutions such as the CREATOUR® IdeaLabs and workshop initiatives in which pilots and researchers, together, regularly discuss, examine, and reflect on developments: unforeseen difficulties, challenges, and gaps as well as accomplishments and several points of improvement. The objective is to develop an integrated knowledge-building and -sharing 'community of practice' that can collaboratively seek solutions to pragmatic questions and reflexively discuss common concerns so that they can continuously add value to their activities, diversify markets, and attract both national and international travellers to their destination throughout the year.

Local community development

There is an expressed need for local involvement in creative tourism activities because the participation of local residents helps to create a sense of community, provides the connections with locals that visitors seek, and enables an open exchange of ideas and knowledge about the locale. At the same time, it can also contribute to preserving and renewing traditional knowledge, promote linkages with contemporary art expressions, and enhance cosmopolitism and openness in those communities, empowering the local population and 'refreshing' local identities. Since culture-based creative tourism initiatives are diverse by nature, they can be designed to promote social inclusion of particular groups of residents in designing and facilitating workshops and other activities, and to foster cross-cultural learning and dialogue. Overall, the objective of closely integrating creative tourism with local development is to encourage and share mutual direct and indirect benefits more widely among community residents – including those who might not 'traditionally' be involved in the tourism sector.

Creative tourism, given its nature, fosters an atmosphere that is attractive for a wide variety of local interests (e.g. creating employment and commercial opportunities for local businesses and associations, designing new types of occasions and platforms for sharing cultural traditions and skills, and expanding training opportunities for marginalized residents) while also offering unique and creative experiences to tourists, altogether contributing to improve the quality of life for both local residents and visitors and building social capital within the local community. However, this potential requires attentiveness to the needs of existing artisans and 'culture-keepers' in communities as well as the need to train new artisans and to provide potential workshop organizers with mentorship and training in creative tourism. Aligned with this work is the need to build distributed capacity for cross-sectoral collaboration and partnership management.

Policy framework for creative tourism development

The creation of policies and institutional frameworks can contribute to advancing the development of culture-based creative tourism in small cities and rural areas. We propose the development of an overarching national policy and programme framework with associated funding, which will provide a comprehensive structure for addressing and enabling different levels of intervention. The implementation of the proposed vision outlined in this document will require an integrated open-ended process engaging government at all levels, associations/organizations, businesses, non-governmental organizations, local communities, and consumers.

At a macro level, we recommend that tourism and cultural/creative sector policymakers:

- Define culture-based creative tourism as one of the strategic development priorities for tourism and culture.
- Develop an appropriate legal/regulatory framework to support the strategic development of culture-based creative tourism activities.
- Empower and engage practitioners from the tourism and cultural/creative sectors, local and regional development agents, communities, and government officials, both technical and political, to participate in an active and informed manner in the policy-making process concerning different levels of development of creative tourism.
- Create a governance structure to coordinate multi-level initiatives and inter-sectoral dialogue within the framework of public administration (e.g. between different ministries or different councillors). In addition, the development of national-level policies should directly involve practitioners already working in creative tourism at the regional level.
- Promote the ongoing involvement of all organizations identified in each region as 'creative tourism' practitioners in the development of creative tourism policies at the regional and local levels (e.g. in an autonomous consultant body, or as a creative tourism network), promoting participation in the design, operationalization, and monitoring of all relevant policies.

Within this policy framework, we identify five main interlinked domains that should determine short and medium-term strategic planning measures:

1. Product and enterprise development.
2. Inter-sectoral co-operation, networking, and soft infrastructure.
3. Capacity building and training.
4. Communications and marketing.
5. Research, monitoring, and industry intelligence.

1. Product and enterprise development

1.1. Develop a financing programme to support the conceptualization and operationalization of new creative tourism ideas and also to support current creative tourism enterprises to become more sustainable – that is, create benefits for micro-businesses and small- and medium-sized enterprises (SMEs) in the cultural and creative sectors, mainly in rural and interior areas, through specialized financing and stimulation programmes of regional and local development (i.e. through regional coordination and development commissions).

1.2. Stimulate the development of bottom-up creative initiatives by supporting frameworks and spaces for creative innovation, including social innovation, social design, Fab-Labs, etc., which can contribute to strengthening 'quadruple helix' structures (bringing together government, universities, private sector, and civil society actors), or facilitate the development of more 'organic' creative milieus based on territorially embedded dynamics and specific governance mechanisms.

1.3. Encourage creative tourism initiatives that co-develop creative tourism experiences with local residents in ways that build and strengthen social connections and networks across organizations and that aim to involve marginalized populations.

2. Inter-sectoral co-operation, networking, and soft infrastructures

2.1. Integrate policy approaches between the tourism and culture sectors, fostering and enabling closer linkages among actors and resources. Define a governance model that is able to strengthen inter-sectoral collaboration, at national, regional and local levels, in order to effectively address culture-based creative tourism challenges (i.e. regarding efficient and effective use of local resources and services for community benefit).

2.2. Improve governance mechanisms and interaction among cultural, tourism, and territorial development agencies and policies. Develop dialogue and articulation mechanisms between the actors/institutions responsible for the design of sectoral policies in the three fields (culture, tourism, and territorial development), at national/regional and municipal levels, for the purpose of encouraging and developing creative tourism.
2.3. Support a national network for creative tourism, involving creative producers, industry leaders, and communities, in order to build and strengthen creative clusters, promote and support entrepreneurship, and engage in capacity building, mentorship, and networking activities.
2.4. Create opportunities for making new connections, prioritizing personal and small-group exchanges, supported by appropriate technological approaches.
2.5. Nurture active collaboration and co-working between tourism and cultural sectors, supported by universities and research centres as well as local and regional tourism agencies and local/regional development agencies and organizations.
2.6. Regenerate rural areas based on enhancing cultural 'soft infrastructures' using local cultural spaces and rural creative assets.

3. Capacity building and training
3.1. Develop and promote tailor-made types of capacity-building activities spanning formal professional education, informal training and specific training programmes, as well as the fostering of a geographically distributed community of practice to ensure that practitioners continue to manage the development and evolution of their offers on a sustainable basis. At the regional level, the involvement of training institutions, schools, universities, and research centres is essential here.
3.2. Empower and develop the skills of practitioners identified as developers of creative tourism projects, at the local level, enhancing a common collective knowledge and know-how base, mediated/facilitated by regional authorities and/or universities and research centres. These entities are also essential in the promotion and sharing of that collective knowledge at an inter-regional and international level, assuring bridging between agents with similar interests in different regions.
3.3. Develop a support structure at the regional level to help SMEs and other kinds of organizations in creative tourism access and apply to financing and training programmes available (for example, Local Development Associations can be focal points).
3.4. Provide knowledge and skill-development adapted to the different needs of the actors involved (e.g. prototyping and renewal processes, creative tourism project management, community engagement, the integration of new technologies that provide enhanced links between producers, consumers, and places).

4. Communications and marketing
4.1. Reinforce the use of new technologies and digital media solutions for B2B (business-to-business) and B2C (business-to-consumer) systems, for example, in shaping and delivering creative contents and tourism experiences, and improving connections between the different actors of the creative tourism value chain.
4.2. Define a national/regional communication plan and strategy (i.e. 'branding places and experiences' to connect creative activities with their destinations).
4.3. Integrate marketing and communications about creative tourism offers into local cultural and tourism promotional systems, information centres, and other communication products.
4.4. Enhance marketing of creative tourism (and specific activity branches within creative tourism) to generate an image at national/regional/local levels and create an atmosphere within which individual initiatives are developed to capture more producers and consumers and create a more competitive sector.
4.5. Develop an informative and awareness campaign and materials on the potential of creative tourism at the local level, communicated to local actors in tourism, culture, governance and local development sectors.
4.6. Raise consumer awareness of the nature of creative tourism experiences (e.g. through educational/promotional campaigns, setting guidelines to identify suitable creative tourism initiatives, and to choose creative experiences that truly fit creative tourism).
4.7. Encourage consumers to use local products and support the development of the local economy through consuming creative tourism products and services (e.g. specific mediation on this can be developed through local development associations).
4.8. Improve access to creative tourism experiences within traditional tourism marketing channels and also develop a tailor-made distribution system based on ICTs and social media.

5. Research, monitoring, and industry intelligence

5.1. Develop incentives in order to build up knowledge about creative tourism within the tourism and cultural sectors, both at regional and national levels, involving universities, research centres, and culture and tourism regional institutional structures. Continue to use research-and-application approaches to foster ongoing knowledge-sharing and knowledge-building between academia and practice.

5.2. Develop and promote a system of information tools (i.e. mapping creative tourism organizations, initiatives, resources, and experiences by geographical region, scale, and thematic areas) to support business and policy decision-making, involving both culture and tourism regional institutional structures.

5.3. Establish a series of core baseline indicators to collect data and information, and regularly review progress and report on changes and trends at the national level.

5.4. Develop and promote a system for assessing the territorial development impacts of creative tourism activities, collecting and monitoring information on a continuous basis in terms of the social, cultural, economic, and environmental impacts of these activities, and providing feedback to the creative tourism operators, raising awareness of its importance for the local communities and for those operators.

5.5. Adapt monitoring and evaluation approaches to different situations, considering the geographic region and thematic specificities of different creative tourism experiences.

5.6. Organize collective discussions and networking (locally/sub-regionally, regionally, and nationally) so that it is possible to survey the needs and expectations of creative tourism practitioners on an ongoing basis, at a local and sub-regional level, as well as to better understand local dynamics.

5.7. Create an observatory of creative tourism to conduct and manage this work.

Conclusion

This integrated set of recommendations is proposed to authorities and agencies at all levels (national, regional, and local) relating to the development of creative tourism initiatives in small cities and rural areas. It is centrally informed by the needs of the 40 CREATOUR® pilot organizations, articulated within the first national multidisciplinary research-and-application project on creative tourism, active between November 2016 and June 2020. Through CREATOUR®, we learned that creative tourism development outside large cities is feasible in Portugal, but challenges of distributed geography and the diversity of organizations and initiatives involved require additional attention to capacity building, training, and mentorship; dedicated collective marketing resources; and more time to deepen networking and develop partnerships.

Creative tourism has the potential to develop new forms of collaboration between a wide range of partners at a destination level. Rather than proposing ready-made models, it advocates a locally sensitized approach to development that is bottom-up and collaborative. Achieving the necessary co-operation between public stakeholders, businesses, local authorities, cultural associations, and tourism agents is one of the biggest challenges.

An adequate framework that enables the effective involvement and participation of local actors is essential. To build links and synergies between culture-based creativity and tourism, we propose a creative tourism policy, programmes, and funding framework with a national approach, but attentive to regional specificities/needs. The core of this framework is a content-generation-and-development model that enables and supports the five key dimensions previously outlined in this chapter.

As a transversally applicable approach to community-sensitive tourism development, creative tourism can enable the diversification and differentiation of tourism offers in small towns and rural areas, strengthening them as tourist destinations that promote and value cultural and natural heritage and sustainable tourism practices.

In view of the challenges currently facing both the cultural/creative and tourism sectors it is vital to develop tourism in ways that will increase the potential for local economic development, social cohesion, and creative development. There is a need to view the tourism and the cultural/creative sectors as mutually dependent parts of the local ecosystem, rather than separate economic spheres. A broader placemaking approach can be developed to make creative use of local resources, to charge them with meaning for local and other stakeholders and provide the basis for innovation.

Particular attention to addressing any negative externalities of these activities is also essential, however small they may be in creative tourism compared to mass tourism (e.g. noise, environmental impacts, social impacts, etc.). Public agencies and authorities should assume political responsibility to deal with situations that arise and manage the different public interests within the community to assure the sustainability of territorial development in all its dimensions (social, economic, environmental, and cultural).

Our ambition is not to promote creative tourism as an alternative and autonomous market to mass tourism and sun-and-beach-based offers, but to develop more sustainable forms of tourism that can generate new income for small towns and rural areas, promote the (re)valorization of cultural traditions through creative experiences, and provide networks and co-operation between different levels of actors, institutions, and businesses, ensuring the functioning of a durable creative tourism ecosystem.

Creative tourism is an option that does not require the physical development of new facilities, but more effective use of existing resources to meet current and future needs of residents and visitors. The public sector has a central role in setting an enabling policy environment to foster the networks and cross-sector linkages. Creative tourism can add value – economic, social, and cultural – at local and regional levels, and foster inclusive, integrated, and sustainable development for greater prosperity, cohesion, and equity across the regions.

Acknowledgements

This chapter was developed within the CREATOUR® project (no. 16437), which is funded by the Portuguese Foundation for Science and Technology (FCT/MEC) through national funds and co-funded by FEDER through the Joint Activities Programme of COMPETE 2020 and the Regional Operational Programmes of Lisbon and Algarve.

Note

1. This chapter is based on a policy recommendations document created by the CREATOUR® project in 2020 for local authorities, regional agencies, and national ministries in Portugal.

Part VI: Concluding remarks

32 Pursuing creative tourism for positive transformation and sustainable development

Nancy Duxbury,[a*] Sara Albino[b] and Cláudia Pato de Carvalho[c]

[a]Centre for Social Studies, University of Coimbra, Portugal; [b]CIDEHUS, University of Évora, Portugal; [c]Center for Social Studies, University of Coimbra, Portugal

We believe that the potential of creative tourism goes far beyond a marketing approach or a response to tourism trends. Internationally, there is growing attention to how creative tourism can be strategically designed and leveraged for both visitor and local benefit. And it aligns with community-based tourism, regenerative tourism, community development thinking, and creative placemaking – all areas that are receiving a resurgence of interest.[1] All this provides a robust platform from which to envision, design, and manage creative tourism initiatives.

Creative tourism is a place-embedded, relational, and potentially transformative approach – for both travellers and resident communities. In moving forward, creative tourism's multiple threads and trajectories must explicitly recognize and meaningfully engage with the diverse contexts in which it is being developed, while being more connected internationally. Creative tourism must incorporate a high degree of care for those involved and the broader communities, continually improve how projects are planned and managed, attend to how initiatives can be made more sustainable, and deeply consider the implications of the visitor–local relations that are enabled. Creative tourism must also be guided by the generative possibilities brought forth through processes of creation, sharing, and conversation.

This chapter is divided into three parts. It begins by discussing some strategic areas to consider to advance creative tourism development locally and internationally. Then, we consider creative tourism in light of the United Nations Sustainable Development Goals. Filling out the chapter, we close by exploring some of the potential trajectories of creative tourism in the future, in a section entitled 'Going forward with transformative aspirations'.

Advancing creative tourism development

This book contains an array of insights, examples, and advice on creative tourism development and practice in many geographic and cultural contexts. While it is not possible to synthesize all this here, we have included a list of general points that resonate with us on the practice of creative tourism development (see Box 32.1). A few words on advancing creative tourism as a field of international practice are also in order. We propose three themes as key areas for collective attention: developing innovative milieu, public policy and support frameworks, and knowledge networking and collaboration.

Developing an innovative milieu

In order to continually improve and refresh creative tourism initiatives as well as to imagine and formulate new ones, processes of continuous innovation must be supported and sustained. These processes help to propel new ideas and 'expand the

*Corresponding author: duxbury@ces.uc.pt

 Creative Tourism: Activating Cultural Resources and Engaging Creative Travellers (eds. N. Duxbury, S. Albino, and C. Pato de Carvalho)
DOI: 10.1079/9781789243536.0032

Box 32.1. Key points for creative tourism development

The book's authors provide a wide range of advice and guidance, and it is impossible to summarize the depth and breadth of these insights in a concluding chapter. We have compiled a few points that resonate with us in moving forward:

General

- The potential of creative tourism goes far beyond a marketing approach or a response to tourism trends.
- Use creative tourism activities to contribute to local socio-economic and cultural development.
- Use creative tourism activities to engage with broader societal and global issues.

Cultural resources

- Creative tourism values local resources and enhances place identity.
- Continuously identify interesting aspects and embed ideas and actions in local resources.
- Pay attention to the invisible, the everyday, and less prominent aspects of communities and places.
- Creative tourism can help keep traditions alive and revitalize local cultural practices.

Entrepreneurship

- Continuously innovate new approaches to keep creative tourism vital, interesting, and relevant.
- Consider how creative tourism can support and extend other economic activities, traditions, and community-based initiatives.
- Involve a range of individuals from local communities to make local tourism meaningful to both residents and visitors.
- Diverse types of cross-sectoral collaborations are challenging but necessary. Continuously build capacities, knowledge, and inter-sectoral mediation skills.
- Regularly engage in knowledge-gathering about local needs and opportunities.
- Articulate motivations and strategic objectives, and integrate moments for systematic self-assessment of project developments and monitoring impacts.

Engaging creative travellers

- Creative travellers are diverse in nature. Market segmentation exercises and building profiles can guide the development of creative tourism initiatives for different types of travellers.
- The motivations and interests of creative travellers must be linked to a close understanding of what may be interesting to them in your community.
- Individuals 'travel to be creative' in both touristic and professional contexts. Travelling artists and digital nomads are complexifying the profile of creative travel.
- Destination attractiveness is linked to the cultural, creative, and social scenes as well as the spaces and infrastructures available.

Moving forward

- The value of a creative tourism offer is linked to its relevance to the local community and its specific context. This vital connection to place must be active and dynamically cultivated.
- Bridge research and practice to co-build knowledge and foster knowledge networking. Together is stronger – learn from one another.

potential' of a culture, with exchanges increasing 'collective knowledge, skills, and understanding' (Wisansing, Chapter 12, p. 86). This speaks to organized moments of co-learning and co-creation, knowledge and skill exchanges, opportunities for experimental explorations, and processes of creative tourism ideation, prototyping, and testing.

As Lénia Marques notes, the future of creative tourism requires 'more structural, managerial and governance innovations' that can facilitate capacity-building processes and provide 'the space for continuous innovation to occur' (Chapter 23, p. 82). Marília Lúcio of the Municipality of Loulé also stresses the importance of encouraging continuous innovation in a structured way. Beyond networking and capacity-building sessions that the municipality organizes, Loulé Criativo creates 'enabling conditions for research in arts and crafts and related topics, to foster new knowledge and flows of information that can innovate working practices and perspectives' and supports a 'creative laboratory' that includes research, experimentation, and design, incubates companies, and fosters connections within the creative community and more widely (Chapter 21, p. 164). These types of structures are still rarely linked to creative tourism.

Public policy and support frameworks

Public policy frameworks to nurture, support, and foster a generative milieu for creative tourism are integral to advancing creative tourism development. As Larissa Almeida observes in her work in Brazil,

> If public policies offer structure, people become more confident to invest in the process. Public policies are necessary for the creation of infrastructure. When people see the ongoing construction of structures, they feel more confident about investing in the development process because they realize that they are not alone and there is a greater chance that their efforts will be rewarded. (Chapter 24, p. 186)

Based on the experiences of 40 pilot creative tourism initiatives geographically dispersed in small cities and rural areas of Portugal, the CREATOUR® project developed a series of public policy recommendations, which are organized into five interlinked domains:

1. Product and enterprise development.
2. Inter-sectoral co-operation, networking, and soft infrastructure.
3. Capacity building and training.
4. Communications and marketing.
5. Research, monitoring, and industry intelligence (Gonçalves *et al.*, Chapter 31).

We note that achieving the necessary co-operation between agents as diverse as public stakeholders, businesses, local authorities, cultural associations, and tourism agents can be a significant challenge but is also a necessary one (Gonçalves *et al.*, Chapter 31).

At a local level, with community-based creative tourism integrally connected with the development of multi-sector networks and partnerships, public authorities can play a key facilitating and supporting role. For instance, Rita Salvado and Guida Rolo see that the continued development of the COOLWOOL creative tourism initiative would greatly benefit from the municipality's involvement:

> As the project is focused on networking and on the city and its industrial heritage, the involvement of the city council appears to be a determining factor to its further evolution and development, to increase the visibility of the project, to articulate it with local policies and strategies for tourism, and to enhance its impacts in other local growth and development strategies. (Chapter 17, p. 128)

With regional development objectives in mind, building an extended and ongoing network is imperative. As Larissa Almeida writes,

> In order to develop creative tourism initiatives capable of developing a territory, it is essential to establish a trust relationship not only within the community but with partners who can help to support initiatives in that space. Tourism needs to be integrated within many areas and its development requires joint efforts from both public and private sectors. Development agents such as governments, promotion agents, and big enterprises need to play the role of supporting creative entrepreneurs and connecting stakeholders who can catalyse the activities. (Chapter 24, p. 190)

From a broader perspective, programmes available from other levels of government can be an essential catalyst. For example, Elena Paschinger notes how European and regional programmes have enabled networking and knowledge exchange meetings that have helped 'stimulate the economic process of getting small-scale artisans off the ground in order to become ready to embrace creative tourism as a new market opportunity' (Chapter 3, p. 27). The international Creative Tourism Network® originated as a European initiative in 2011 with the support of the European Union (Couret, Chapter 8). The CREATOUR® project in Portugal was made possible by support from FEDER (European Fund for Regional Development) through the Joint Activities Programme of COMPETE 2020 and the Regional Operational Programmes of Lisbon and Algarve, as well as the Portuguese Foundation for Science and Technology.

Knowledge networking and collaboration

Forging ongoing opportunities for inter-sectoral networking, co-learning, and collaboration is central to advancing creative tourism development in a socially embedded and cross-sectoral way. To avoid the risk of traditionalism, it is important to secure local diversity within creative tourism initiatives, attending to the inclusion of different groups with stakes in the project. A solid platform for collaboration can facilitate entrepreneurs to innovate together, among themselves, as well as with the tourists, in processes of 'value (co-)creation' (Marques, Chapter 23, p. 175).

Creative travellers can become an extended community, connected to the local agent for some time after the creative tourism experience. As Diana Zuluaga and Diana Guerra note, 'creative tourism strives to shape a community ... the success of creative tourism is people dealing with people, and that means customers will belong to our community and they will want to keep in touch after their travel experiences' (Chapter 14, p. 110). Jaime Serra *et al.* found that over 85% of tourists participating in creative tourism activities indicated they would be interested in participating in creative tourism activities elsewhere in Portugal. These findings suggest 'the value of operating in a network where travellers' participation in a creative tourism activity at one location can lead to participation in other activities elsewhere' (Chapter 4, p. 35). Looking beyond individual, localized projects, 'coordinated creative tourism initiatives [can] generate international networks that encourage extended sharing and discussion of the ideas that are important to people' (Feeney, Chapter 5, p. 38).

Networking and knowledge sharing is also important for the creative tourism developers and innovators. We agree with Larissa Almeida and the members of Recria, the Creative Tourism and Experiences Network of Brazil, who 'strongly believe the full potential of local knowledge and experiences will only be harnessed when there is an exchange of knowledge between territories' (Chapter 24, p. 187). We view this collection as an important steppingstone in the development of creative tourism internationally, and a bridge to connect the many proponents and emerging initiatives within this dynamic field. We invite you to participate in a 'companion' international network aiming to foster ongoing knowledge sharing and capacity building among creative tourism practitioners and researchers – CREATOUR® International (see Box 32.2).

Box 32.2. Towards ongoing connections: CREATOUR® International

As a beginning to inform and inspire international knowledge sharing, we are teaming the development of this book with the launch of CREATOUR® International (www.creatourinternational.net; accessed 17 May 2021), an informal network to further knowledge exchange and capacity building among creative tourism researchers and small-scale practitioners internationally. It aims to encourage both researchers and practitioners to discuss experiences and emerging issues in the area of creative tourism, toward the development of a diverse, geographically distributed community of practice.

It aspires to build closer research–practice relations through providing a platform for sharing experiences, practice-based knowledge, and insights from research; articulating and discussing issues and solutions; and learning from one another to collectively build capacity in creative tourism internationally. Ultimately, it aspires to advance both research and practice in the field of creative tourism, with special attention to initiatives in smaller places, and to foster and inform progressive connections between culture, tourism, and local sustainable development.

Creative tourism and the Sustainable Development Goals

As discussed throughout this volume and embedded in many of the examples presented, creative tourism can have a practical relation to local and regional sustainable development.[2] At the international level, the United Nation's 2030 Agenda for Sustainable Development provides 17 Sustainable Development Goals (SDGs) to guide the world's collective development over the next decade (UN, 2015). As a 'plan of action', the Agenda and the SDGs provide a comprehensive framework to inform and shape public policy, development programmes, research agendas, and public discourse.

Thus, as part of reflecting on the chapters in this book, we reviewed and analysed the authors' contributions against the SDGs. Four SDGs emerged as most relevant to the current collection:

- No. 8 – Promote sustained, inclusive and sustainable economic growth, full and productive employment, and decent work for all.
- No. 10 – Reduce inequality within and among countries.
- No. 11 – Make cities and human settlements inclusive, safe, resilient and sustainable.
- No. 12 – Ensure sustainable consumption and production patterns.

We developed a table that aggregates some of the topics discussed in the chapters that relate to these four goals (see Table 32.1). A synthesis of the main narratives relating creative tourism to the SDGs 8, 10, 11, and 12 is presented in Box 32.3.

While we believe that creative tourism presents an opportunity to contribute to the SDGs and to alleviate poverty, we also recognize 'it is not a panacea to all problems – not on paper, and not in the field' (Marques, Chapter 23, p. 175). This analysis is put forward for two reasons. Firstly, to help entrepreneurs and public agencies be more aware of connections between local creative tourism development and the SDGs. Secondly, to inspire new creative tourism initiatives that may be explicitly designed to tackle issues that have been identified as integral to building a more sustainable world.

Going forward with transformative aspirations

Thinking about the future of creative tourism can be facilitated through a process of placing intentional and careful attention to the dimensions of creative tourism that provide avenues for generative and potentially transformative possibilities. In this light, in closing, we highlight four topics emerging from contemporary creative tourism practices that are discussed in this collection: fostering cultural revitalization and sustainability, opening a space for intercultural dialogue and understanding, expanding narratives and innovating new practices, and engaging with the environmental crises.

Fostering cultural revitalization and sustainability

While cultural revitalization is 'a complex aim', as Maria Huhmarniemi *et al.* observe, 'creative tourism can breathe new life into active practices and foster appreciation for traditional skills and designs' (Chapter 30, p. 243). Lénia Marques

Table 32.1. Connecting creative tourism to the SDGs – key points emerging from the chapters

Decent work and economic growth (SDG 8)	Reduce inequalities (SDG 10)	Sustainable cities and communities (SDG 11)	Responsible production and consumption (SDG 12)
• Economic development of traditional communities • Tourism start-ups and development • Capacitation of cultural and artisan entrepreneurs • Creative tourism product development • Increasing the value of community-based tourism • Building infrastructures for creative tourism and cultural activities • Valuing stakeholder relations (e.g. artisan entrepreneur-mediator activities) • Co-operation among 'competitors' • Co-creation processes in tourism • Enabling public policies for cultural development and entrepreneurship • Nurturing a creative experimentation setting (public policy)	• Creative tourism leveraged from local initiative • Creative tourism as basis for local collaboration • Capacitation for social and economic cohesion • Valuing peripheral places • Recognizing peripheral cultures and informal knowledge • Support for cultural workers through grants and public policies	• Sustainability of creative tourism linked to economic competitiveness • Sustainability linked to uniqueness of place identity and values • Survival of traditions through creative tourism entrepreneurship • Creative tourism adding value to local resources	• Tourism in a market of proximity • Participative models of production • Responsible production through youth engagement • Intercultural dialogue between locals and visitors • Fostering a creative ecosystem • Responsible consumption through meaningful experiences • Local development through equitable accommodation

Box 32.3. A synthesis of key topics of relating creative tourism to SDGs 8, 10, 11, and 12

SDG 8 – Decent work and economic growth. Creative tourism can be a source of meaningful work opportunities for artisans and other creators, enable the emergence of artisan entrepreneur-mediators, and help to enhance co-creation processes in tourism development. The importance of building capacities of entrepreneurs, as essential to attain economic growth and decent work and income for local communities, arose from several authors and we see this as a transversal theme. For instance, Jutamas (Jan) Wisansing (Chapter 12) notes that in Thailand creative tourism has been specifically linked to the notion of adding value to community-based tourism, which is a mainstay of the Thai economy.

SDG 10 – Reduce inequality within and among countries. At a local/organizational level, a number of authors discuss the importance of creative tourism in building the capacities of entrepreneurs, artisans and artists, and local communities dependent on traditional activities (for example, the Masaai tribe in Kenya, discussed by Lénia Marques in Chapter 23). Broader in scope geographically, the valorization of peripheral places, referred by Larissa Almeida (Chapter 24) as well as the CREATOUR® project, is important to bring recognition to rural and remote locations as well as subcultures and practices that are outside the mainstream. Creative tourism can activate informal knowledge resources to generate value in these places.

The nature of creative tourism itself also aims to address traditional inequities between tourism and 'receiving' residents. Creative tourism developed and leveraged from local initiatives (Wisansing, Chapter 12) can be important to combat the impacts that traditional mass tourism and transnational tourism corporations have on developing countries. At a local scale, creative tourism initiatives are also discussed as a possible solution for better local collaborations between visitors and residents (Scherf, Chapter 25).

SDG 11 – Sustainable cities and communities. Tourism development in sustainable cities and communities cannot happen with unequal social systems and deteriorated landscapes. For this reason, three key topics emerged related to this topic: sustainability of creative tourism; economic competitiveness; and uniqueness of place identity and values. The survival of traditions through entrepreneurship by adding value to local resources is also transversal to several chapters. Some are more practice-based chapters, such as the entrepreneurial guide of Marie-Andrée Deslisle (Chapter 19), with a marketing focus, and the entrepreneurial story of the company 5Bogota in Colombia by Diana Zuluaga and Diana Guerra (Chapter 14). Kathleen Scherf's more research-based chapter (Chapter 25) provided insights on managing encounters between locals and visitors Barcelona and Copenhagen in ways that value local practices without 'touristifying' them.

SDG 12 – Responsible consumption and production. Creative tourism can promote a dialogue between local communities and visitors, building an ecosystem of creativity and participation. Such an ecosystem, if composed of a variety of stakeholders and characterized by cooperative relations, can propose new interactions and collaborative modes of production. In terms of consumption, creative tourism designs and generates opportunities for active, learning, creative, immersive experiences that are responsibly and sustainably connected to the place they occur. For example, Sabrina Pratt (Chapter 22) and Britta Timm Knudsen and Jan Ifversen (Chapter 18) refer to how responsible consumption can reframe tourism through meaningful experiences. While Pratt refers to creative local arts and crafts workshops as an example of visitor engagement and interest in local cultures, Knudsen and Ifversen refer to co-creation practices that rethink tourism in coastal spaces.

Moreover, it is important to recognize that creative travellers and creative tourism initiatives are, in practice, dynamically inventing new responsible and sustainable modes of consumption and production. For example, as young people today question their roles in the production and political systems, they are inventing new modes of expression, connecting, and taking action – all while envisioning ways to mediate their digital connections with 'grounded' local communities. As an example, in Chapter 26, Fernando Almeida considers how contemporary youth co-develop spaces of imagination and action through creative practices that play out in the experimental spaces of the participatory artistic festival Encontrarte Amares, which takes place in a small community in the north of Portugal.

explains how creative tourism can provide 'a significant local economic impact, enabling certain community values and practices to endure in time' (Chapter 23, p. 175). It can also allow the same community to evolve, pushing the tourism activities it organizes 'beyond framed ritualistic enactment' (p. 175). As Jutamas (Jan) Wisansing remarks,

> Creative tourism, if designed and implemented well, could be a mechanism to create local pride and stimulate cultural preservation, to reinforce a unique local sense of place based on visual quality and functional efficiency and, ultimately, to provide a better platform to engage different generations – 'the old meets new'. (Chapter 12, p. 92; see also L. Almeida, Chapter 24)

Creative tourism apprenticeships with master craftspeople can have an impact on 'the preservation of global crafts' (Agrawal, Chapter 10, p. 69) through contributing to knowledge transfer between generations and internationally.

Opening a space for intercultural dialogue and understanding

Embedding creative tourism activities into specific aspects of the culture of a place enables visitors to learn about a culture and locale in memorable ways, sharing perspectives, meaningful experiences, and creative expressions. As Larissa Almeida observes, the encounters and learning experiences provided through creative tourism tend to favour connections 'in gentle contacts with different realities' and thus 'is a valuable tool for human development that brings people together and reduces barriers' (Chapter 24, p. 190). Whether viewed from a domestic tourism context or a global one, creative tourism can deepen the experience of cultural diversity and richness (Huhmarniemi *et al.*, Chapter 30).

Sensitivity to fostering cultural plurality among creative tourism offerings can counter monovisions and tourism stereotypes. Creative tourism activities can be intentionally designed to enhance understanding of the cultural diversities in a place and to foster intercultural dialogue. For instance, craft-based creative tourism activities featuring 'place-specific intercultural approaches' could 'create an open space for dialogue in which the values and the perceptions of cultural heritage can be negotiated' (Huhmarniemi *et al.*, Chapter 30, p. 246, referencing Härkönen *et al.*, 2018).

Expanding narratives and innovating new practices

As Melanie Sarantou and Satu Miettinen elaborate in Chapter 27, creative tourism activities can be designed that enable 'power shifts' through proposing 'new roles and ways to experience and comprehend the narratives by local communities' (p. 209). Narrative- and arts-based practices, sensitivity to embodied learning, and attention to local knowledges can provide room for 'creating and expressing personal stories, transformative social practice, and counter-narratives for past oppression and/or animosity' (p. 206, referencing Sonn *et al.*, 2013). Creative tourism can also serve as a ground for experimentation, a space in which, for example, youth 'try out alternate models of civic and political participation', blending arts practices, digital connectivity, intergenerational community encounters, and reciprocal dialogues (F. Almeida, Chapter 26, p. 203). Such dynamics can inform and impact the creative-traveller participants as well as contribute to host-community reconfigurations.

Engaging with the environmental crisis

An intentional engagement with the need to address the environmental crisis is another of the possible innovation trajectories for creative tourism. As Maria Huhmarniemi *et al.* elaborate, 'there is a need for transformative experiences that change our values to being holistic and respectful towards nonhuman nature or, in other words, to encourage our lifestyles to be humbler' (Chapter 30, p. 240, referencing Lin and Oxford, 2011). Similarly, Britta Timm Knudsen and Jan Ifversen suggest that 'creative activities in nature could enhance and confirm nature connectedness on a whole range of levels for both residents and tourists, connecting creative tourism to criticism of instrumental uses of natural resources in general' (Chapter 18, p. 134). Carefully envisioned and designed creative tourism initiatives can enhance understanding of 'the interdependence of human and non-human nature as well as support participants' capacities to face environmental crises, stay resilient, and act for environmentalism' (Huhmarniemi *et al.*, Chapter 30, p. 248). This could build upon existing community-engaged arts practices that address environmental and sustainability issues (see Feeney, Chapter 5).

This book is a beginning

We invite readers to explore the possibilities in your local communities for creative tourism development, learning from the many contributors to this volume and innovating new approaches, activities, and development processes. We encourage you to consider the greater impacts and implications of these activities and to search for ways to pursue them, embedded in the design, planning, operationalization, and ongoing management of your creative tourism initiatives. Furthermore, we urge you to develop inclusive, participative, and effective structures that can inspire, support, and coordinate efforts – and make your creative tourism programme greater than the sum of its individual parts. We hope this book can be a useful companion on this journey, and we look forward to hearing about your experiences as we move forward together.

Acknowledgements

This chapter was developed within the CREATOUR® project (no. 16437), which is funded by the Portuguese Foundation for Science and Technology (FCT/MEC) through national funds and co-funded by FEDER through the Joint Activities Programme of COMPETE 2020 and the Regional Operational Programmes of Lisbon and Algarve.

Notes

1. For information on how different operational models of creative tourism relate with regenerative tourism, see Duxbury *et al.* (2020). For insights on connecting creative tourism with creative placemaking, see Richards (2020).

2. As Maria Assunção Gato *et al.* elaborate in Chapter 29, a multidimensional consideration of sustainable development, holistically conceived, can take into account 'economic, social, cultural, and environmental factors as well as the quality of life of local communities, their cultural involvement, civic participation, and the conservation and preservation of both tangible and intangible cultural forms' (p. 225). This scope sets a wide foundation for considering and identifying multiple effects that can arise from creative tourism activities embedded in local communities. See also Maria Huhmarniemi *et al.*, Chapter 30, for an elaboration of how creative tourism can address the cultural dimension of sustainable development in different ways.

References

Duxbury, N., Bakas, F.E., Vinagre de Castro, T., and Silva, S. (2020) Creative tourism development models towards sustainable and regenerative tourism. *Sustainability* 13(2), 1–17. doi: 10.3390/su13010002.

Härkönen, E., Huhmarniemi, M., and Jokela, T. (2018) Crafting sustainability: Handcraft in contemporary art and cultural sustainability in Lapland. *Sustainability* 10(6), 1907. doi: 10.3390/su10061907.

Lin, J. and Oxford, R. (2011) *Transformative Eco-Education for Human and Planetary Survival.* Information AGE Publishing, Charlotte, North Carolina, USA.

Richards, G. (2020) Designing creative places: The role of creative tourism. *Annals of Tourism Research* 85, 102922. doi: 10.1016/j.annals.2020.102922.

Sonn, C.C., Stevens, G., and Duncan, N. (2013) Decolonisation, critical methodologies and why stories matter. In Stevens, G., Duncan, N., and Hook, D. (eds.) *Race, Memory and the Apartheid Archive.* Palgrave Macmillan, London, UK, pp. 295–314.

United Nations (UN). (2015) *Transforming our World: The 2030 Agenda for Sustainable Development.* UN, New York, New York, USA. Available at: https://sdgs.un.org/2030agenda (accessed 17 May 2021).

Index

Note: The page numbers in bold and italics and with letter n represents tables, figures and note numbers respectively.

This book is published by **CABI**, an international not-for-profit organisation that improves people's lives worldwide by providing information and applying scientific expertise to solve problems in agriculture and the environment.

CABI is also a global publisher producing key scientific publications, including world renowned databases, as well as compendia, books, ebooks and full text electronic resources. We publish content in a wide range of subject areas including: agriculture and crop science / animal and veterinary sciences / ecology and conservation / environmental science / horticulture and plant sciences / human health, food science and nutrition / international development / leisure and tourism.

The profits from CABI's publishing activities enable us to work with farming communities around the world, supporting them as they battle with poor soil, invasive species and pests and diseases, to improve their livelihoods and help provide food for an ever growing population.

CABI is an international intergovernmental organisation, and we gratefully acknowledge the core financial support from our member countries (and lead agencies) including:

UKaid from the British people

Ministry of Agriculture People's Republic of China

Australian Government Australian Centre for International Agricultural Research

Agriculture and Agri-Food Canada

Ministry of Foreign Affairs of the Netherlands

Schweizerische Eidgenossenschaft Confédération suisse Confederazione Svizzera Confederaziun svizra

Swiss Agency for Development and Cooperation SDC

Discover more

To read more about CABI's work, please visit: **www.cabi.org**

Browse our books at: **www.cabi.org/bookshop**, or explore our online products at: **www.cabi.org/publishing-products**

Interested in writing for CABI? Find our author guidelines here: **www.cabi.org/publishing-products/information-for-authors/**